UNTHINKABLE

UNTHINKABLE

TRAUMA, TRUTH, AND THE TRIALS OF AMERICAN DEMOCRACY

JAMIE RASKIN

HARPER

An Imprint of HarperCollins*Publishers*

HarperCollins books may be purchased for educational, business, or sales promotional use. For information, please email the Special Markets Department at SPsales@harpercollins.com.

FIRST EDITION

Library of Congress Cataloging-in-Publication Data has been applied for.

ISBN 978-0-06-320978-7

22 23 24 25 26 FB 10 9 8 7 6 5 4 3 2 1

To Thomas Bloom Raskin, Hannah Grace
Raskin, and Tabitha Claire Raskin

One word frees us of all the weight and
pain of life. The word is love.
—Sophocles

I realized, through it all, that in the midst of winter, there was, within me, an invincible summer.

—ALBERT CAMUS

CONTENTS

Preface xi
Prologue: Democracy Winter 1

Part I | 51

Part II | 107

Part III | 289

Acknowledgments 425

PREFACE

I n the week between December 31, 2020, and January 6, 2021, my family suffered two impossible traumas: the shattering death by suicide of my beloved twenty-five-year-old son, Tommy, and the violent mob insurrection at the U.S. Capitol that left several people dead, more than 140 Capitol and Metropolitan Police officers wounded and injured, hundreds of people (including several in our family) fleeing for their lives, and the nation shaken to its core. Although Tommy's death and the January 6 insurrection were cosmically distinct and independent events, they were thoroughly intertwined in my experience and my psyche. I will probably spend the rest of my life trying to disentangle and understand them to restore coherence to the world they ravaged.

Each of these traumas was itself the product of an underlying crisis. Tommy's death by suicide followed a merciless advance of mental illness that seized and ultimately controlled the dazzling mind and pure heart of this brilliant and empathetic young man. Like millions of other young Americans, he grew despondent during the COVID-19 pandemic, which left him vulnerable to the darkest impulses created by his illness. Similarly, before the attempted coup of January 6 destroyed our fundamental expectations about the peaceful transfer of power in America, the norms of our constitutional democracy had already been overrun by years of political propaganda, social media disinformation, racist violence, conspiracy theorizing, and authoritarian demagoguery.

When these underlying crises turned into the private and public traumas of suicide and violent insurrection, they demolished all the core assumptions I carried around with me each day—that my children would be healthy and alive, that they would let Sarah and me know if they needed anything, that no political party or power elite would try to overthrow our constitutional democracy, that the country would continue to successfully grow beyond its historic baseline of violent white supremacy and a racial caste system.

I was devastated and crushed by these traumatic events.

And yet, you will find that this is *not* a story of unyielding despair and destruction.

On the contrary, at a moment of impenetrable darkness, the lowest point I have ever experienced—a time when I went for days without sleeping or eating a real meal—the Speaker of the House of Representatives, Nancy Pelosi, threw me a lifeline: acting on astounding faith and something like political clairvoyance, she offered me an invitation that was akin to a challenge, a dare to rise up from my despondency and to bring others along with me. On the day before we voted in the House of Representatives to impeach Donald Trump, when it was clear the votes were there, she asked me to be the lead impeachment manager, to organize and lead the team of House members we would send over to the U.S. Senate to prosecute Trump for inciting violent insurrection against Congress on January 6. It was the hardest thing I have ever been asked to do professionally, at the most difficult time I have ever experienced personally, but the assignment became, paradoxically, a salvation and sustenance for me, a pathway back to the land of the living and a fountain of hope that renewed and strengthened my radical faith in democracy, the system of beliefs and practices that upholds the equal rights of the individual and demands that we all work together to take care of our common inheritance.

Speaker Pelosi's invitation forced me to draw upon the deepest springs of meaning and clarity I have in life: the love of my three children and my wife, Sarah; the wisdom of my late parents; Tommy's remarkable political values; the strength of my siblings; the dreams and fears of my childhood; the insight of my teachers and creativity of my students; the inextinguishable resiliency and solidarity of my Maryland constituents; the genius of my political and academic colleagues; the vision of my mentors and boundless generosity of my friends; the amazing constancy of my staff; and the moral courage of my fellow citizens past and present. These things inspired me to make a case not just *against* a savagely corrupt president, but *for* American constitutional democracy itself.

In American history there have been just four presidential impeachment trials in the U.S. Senate—of President Andrew Johnson, of President Bill Clinton, and of President Donald Trump in 2020 (for pressuring a foreign leader to interfere in our presidential election) and then again in 2021 (for inciting a violent insurrection against Congress to block and overthrow the counting of Electoral College votes). The bipartisan votes to impeach and convict Donald Trump in 2021 (a vote of 232–197 to impeach in the House and of 57–43 to convict in the Senate) marked the most sweeping bipartisan votes against a president for committing high crimes and misdemeanors in the history of the United States. The House impeachment vote on January 13, 2021, and the Senate trial, which ended a month later, on February 13, 2021, together constituted an episode of passionate democratic lucidity and a sweeping rejection of violent authoritarianism.

But, ultimately, this is *not* a book about Donald Trump. Quite the opposite. It is about the kind of people whose dreams and actions have allowed us to *survive* Donald Trump and his sinister incitement of racism and hatred among Americans who feel displaced and threatened by the uprooting of America's racial caste system. Within these pages, you will find citizen activists, public servants, and seriously engaged political leaders, people who are democratic heroes and constitutional patriots of different political parties. You will find young moral visionaries like Tommy Raskin and change agents like his remarkable sisters, Tabitha and Hannah, and their mom, Sarah, whom I met long ago in a Constitutional Law class taught by Laurence Tribe, another luminous hero of my book. You will find great democratic freedom fighters like Bob Moses, Frederick Douglass, Abraham Lincoln, Marcus Raskin, Judith Shklar, Michael Anderson, Timothy Snyder, Jim Clyburn, and Julian Bond; all the fine impeachment managers; and so many other colleagues. You'll find brave officers Harry Dunn, Michael Fanone, Eugene Goodman, Daniel Hodges, Aquilino Gonell, and all the other men and women in blue who risked and sometimes gave life and limb in the bitter, protracted, hand-to-hand "trial by combat" (as Rudy Giuliani eagerly described what was to come a few hours before it happened) at the Capitol; the fearless at-large representative from Wyoming, Liz Cheney, an uncompromising partisan of the Constitution; our stunningly compassionate president, Joe Biden; and of course, the indomitable Nancy Pelosi, whose patriotic love for our country and our people is a force of nature that future generations of Americans will marvel at and celebrate long after we are all gone.

These democratic heroes, people who rise up to resist and oppose tyrants and fortify their communities against them, are not "larger-than-life figures," as Donald Trump is always described—an expression that makes me think of the gargantuan, distorted cartoon characters bouncing around Disney World. Tyrants tell stories only about themselves because history for them begins and ends with their own insatiable appetites. But my own story of despair and survival depends entirely on *other* people, above all the good and compassionate people, the ones like my son Tommy, the non-narcissists, the feisty, life-size human beings who hate bullying and fascism naturally—people just the right size for a democracy in which each person has one vote and one voice, where we are all "created equal" and thus given an equal chance to lead a life of decency and integrity.

I have learned that trauma can steal everything from you that is most precious and rip joy right out of your life. But, paradoxically, it can also make you stronger and wiser, and connect you more deeply to other people than you ever imagined by enabling you to touch their misfortunes and integrate their losses and pain with your own.

If a person can grow through unthinkable trauma and loss, perhaps a nation may, too.

If you are one of millions of Americans who have suffered, in these hard days of plague, violence, and climate emergency, a trauma and rupture like the ones we have experienced in our family, I bid you and your family deep healing and recovery for the battles ahead. I hope you will find in these pages some grounds for personal solace and revival.

But the truth is that all Americans have been shaken by the inter-locking dysfunctions and disorders of our times. If we can be honest about the causes and culprits, if we can recapture the spirit of American pragmatism, then we will find ways to redeem the vast promise of our country against the dangerous lies and fantasies holding us back. We can renew. We can rebuild.

PROLOGUE

Democracy Winter

There are years that ask questions and years that answer.

—ZORA NEALE HURSTON

Wednesday, January 6, 2021
Takoma Park, Maryland

Out on the front porch, the winter morning air cuts through my jacket, my suit, and my gloves, chilling me to the bone. From the corner of my eye, I see people who have just walked past our house turning back and looking at me. From across the street a neighbor blows a forlorn gloved kiss.

I call to the dogs, who are curled up together on their lumpy porch bed. Toby, who is Tabitha's Siberian Husky, barely moves but growls something plaintive. Potter, our lovable mutt, gets up and walks over to wag his tail and give me a nuzzle. I rub his ears with my frozen fingers.

Sarah says the dogs know.

Our porch is now like one of those makeshift memorials on a highway to someone lost in a car accident. There are white roses and red roses everywhere, stacks of books, wrapped packages, long handwritten letters touched by rain. I empty our wicker mail basket and sort through piles of notes with hearts drawn on them from neighborhood kids, rose-scented candles, more of the high school prom photos of Tommy in tuxedo that his friends have been leaving us, accompanied by long, reflective notes. Our friends have left on our milk crate lasagna for dinner, chocolate chip cookies, books on grieving. I peel back the tin foil wrapping on a loaf of banana bread to find it still warm.

These sunken treasures of mourning tug distantly for my attention, but I am in a weeklong daze, asking: *Tommy, Tommy, Tommy, where have you gone, my dear boy? Where have you gone?*

Scenes from his life replay in my head like shorts at an all-night film festival.

He is on a swing, laughing.
He is playing on the seesaw with his sisters.
He is kicking a soccer ball near the Eiffel Tower.
He is performing a poem.
He is doing improv using a wise guy accent.
He is introducing me when I announce my campaign for the state senate.
We are marching together against the Iraq War.
Sarah is holding him as a baby at the beach.
We are at his girlfriend's house in Bethesda, making dinner with her family.
I am cutting his hair on the back porch during COVID-19.
He is doing stand-up comedy at a family reunion beach vacation.
We're walking on the beach, and I accept his challenge to debate whether violence is ever justified.
We are hiking in Rock Creek Park in the dark woods together.

He is making fun of his cousin Boman for ordering chicken wings at
   the hotel shortly after Thanksgiving dinner.
He is smiling broadly and standing with his arms around his sisters.
He is laughing with his cousins Phoebe and Lily.
He is on a sofa laughing hysterically with his cousins Emily, Jason,
   Zacky, Maggie, Asa, Daisy, Mariah, and Boman, his aunt Eden,
   and Brandon.
He is giving Sarah and me a lecture on the First Amendment and
   viewpoint discrimination and drawing a giant diagram.
He is playing with his baby cousins, Gray and Tess and Emmet.

I continue to arraign and prosecute myself for every sign, every
clue, I missed.

Yesterday was darker, the coldest day yet. At the grave site at the
Gardens of Remembrance Jewish cemetery in upper Montgomery
County, tiny freezing raindrops nipped at our cheeks, and the winter
crawled deep into my gloves and my shoes.

Shivering and taking turns with the shovels, we dropped mounds
of orange-brown dirt down on to Tommy's coffin, each large, frozen
clump landing hard and loud on the polished wood and scattering
down the sides and into the earth below, bringing forth wails and
sobs from Sarah and our daughters, Hannah and Tabitha.

Since I was a boy, I had vaguely assumed that I would be cremated
one day. Erika, my eldest sibling and the most convincing, always
warned us that traditional burial meant eternal claustrophobia. Yet
now, when we talked over the weekend about Tommy's burial, I insisted that Sarah purchase the plot alongside Tommy's, so I could
be buried next to my boy for eternity and we could talk philosophy
and politics and make jokes forever, starting as soon as I got there
to be with him—*and sooner rather than later, I hope*, I remember
adding darkly in my mind. Never before had I felt so equidistant, so
vacillating, between the increasingly unrecognizable place called life
and the suddenly intimate and expanding jurisdiction called death.

I hear a light beep-beep from our driveway.

My chief of staff, Julie Tagen, who lives ten minutes away, in Silver Spring, has arrived to collect me for our ride to work.

Time to go to the United States Capitol.

I am a member of the U.S. House of Representatives, and I have to go represent the 776,393 people of Maryland's Eighth Congressional District.

I throw my briefcase in the rear of Julie's car and open the front door to get in.

"Hey, how are you?" Julie says.

"I'm all right," I say listlessly.

This day, January 6, is my first day back at real work in the New Year after taking a few disoriented minutes to be sworn in three days ago.

Tabitha, our youngest child at twenty-three, last night tried to convince me not to go in today, but I told her the Constitution compelled it. Today is the day we count the 2020 Electoral College votes in a joint session of Congress, as required under the Twelfth Amendment and the Electoral Count Act. Joe Biden has won 306 electoral votes in the states to 232 for Donald Trump in an election that was conducted under the grisly pandemic conditions of COVID-19 but was nonetheless the "most secure" election in American history, according to Trump's own Department of Homeland Security.

After four years of rampant official lawlessness, spectacular cruelty, and narcissistic drama, and after a year of this galloping plague of mass death, today will be a significant day for the American people, a day we hope of redemption for our democratic institutions. Speaker Nancy Pelosi has asked me, along with my colleagues Rep. Zoe Lofgren, Rep. Joe Neguse, and Rep. Adam Schiff, to lead our floor response to anticipated GOP objections to electoral votes' being cast by certain swing states where Biden won. I have longed for this day of a peaceful transfer of power out of the Trump darkness, and I am resolved to see it through, to make the electoral process complete.

At our dining room table last night Tabitha said she would come with me to watch. If we couldn't be together at home, we would be together on the Hill.

This made me what passes for happy. I thought Tabitha did not want to be alone, but will later learn she decided to come to look after me.

Hank, my brand-new son-in-law, a high-tech entrepreneur who eloped (COVID-19 style) in Nevada with our older daughter, Hannah, and thus quietly joined our family in the summer of 2020, says he will come with us too. This also makes me happy.

Everyone else will be home. Sarah said she will stay with her mom Arlene and brother Kenneth, who have been stalwart with us for a week. Hannah, who is twenty-eight and finishing business school at Berkeley, will stay home in Takoma Park to conduct a Zoom meeting with the people at RFK Stadium, to explore whether we can conduct Tommy's public memorial ceremony as a drive-in service in its parking lot. Ryan, Tabitha's boyfriend, from Pennsylvania whom she met at the University of Maryland, in College Park, has volunteered to help Hannah get us the RFK site; he was close to Tommy and shares his quirky sense of humor and gentleness. He has volunteered to help Hannah organize the service, which the girls are insisting we find somewhere to do in person, not on Zoom.

"How are you?" I remember to ask Julie as she pulls out of the driveway.

Julie was the first person to arrive at my house last Wednesday after the police came. When I was all alone, hours in and swept away by anguish, drowning in agony, the police got me to say someone's name and number; they called Julie, and she came over to try to help me.

"I'm okay," she says. Julie and her wife, Di, have two teenage daughters, Natasha and Carly. "How was the service?" she says.

"Hard," I say.

I could hear that frozen dirt falling hard on the wood, crunching and sliding.

Riding down North Capitol Street, I try to focus on the coming showdown in joint session, but I am "traumatized" for the first time in my life. Everyone sees it.

I stare out the window.

Traumatized. The word, I have learned, comes from the ancient Greek *trauma*, meaning "wound," like a warrior might take on the battlefield.

In a clinical sense, *traumatized* means I have experienced a violent and comprehensive shock to the foundations of my existence, a rupture in my most basic assumptions and beliefs about life—like the assumption that I would have my son with me forever.

Exactly one week ago, on the evening of Wednesday, December 30, I saw Thomas Bloom Raskin for the last time.

Our dazzling, precious, brilliant twenty-five-year-old Tommy—pure magic and pure love, our middle child in his second year at Harvard Law School, a moral visionary, a slam poet, an intellectual giant slayer, the king of Boggle, a natural-born comedian, a friend to all human beings but tyrants and bullies, a freedom fighter, a political essayist, a playwright, a jazz pianist, and a handsome, radical visitor from a distant future where war, mass hunger, and the eating of animals are considered barbaric, intolerable, and absurd.

With Sarah visiting her mom in Connecticut the last week of December, Tabitha still in Philadelphia where she was teaching Algebra to ninth graders with Teach for America and now spending the holiday with Ryan's family, and Hannah out in Nevada with Hank, Tommy and I were home alone. This was ordinarily a recipe for fun and games for the boys; I would generally regress several decades in Tommy's company.

Wednesday night, we had dinner in the dining room. Tommy ordered Impossible Burgers from HipCityVeg, one of his favorites. We played his game: Boggle. He beat me solidly but showed little energy for litigating over words, a small telltale sign of mental disengagement that I notice now, only in hindsight. Looking back, I find it hard to remember any trademark ridiculous joy or uproarious laughter in his actions of that evening. We laughed together at *Family Guy*, but now when I remember it, his laughter strikes me as listless and forced, a ritual performance. I entreated him to watch one more episode, a movie, or at least an old Iraq War debate between Christopher Hitchens and George Galloway, the "Grapple in the Apple," which he knew by heart. But no, not tonight, he said; he was too tired and going to go to bed. He said he liked the idea of going on a hike the next day with the dogs, before New Year's Eve.

Today, I can see that his affect that evening was flat, just as I can still see his crystalline-blue eyes. At the time, I did not think anything was wrong. Usually, we might hang out a bit longer, and I thought maybe he had just been humoring me by hanging out at all. Maybe his new girlfriend was coming over; or maybe he was going to slip out and go to her apartment near Catholic University.

I could have stopped him. I could have walked him downstairs to his section of the house. I could have pressed him with a dozen questions about his state of mind, how he was feeling. Maybe I could have blown the whole plan up if I had said, "You're not thinking about suicide, are you?" Instead, I gave him a tight hug—this was his last hug and my last hug with him—and a kiss good night and told him I loved him, and he told me he loved me. I watched a little MSNBC, dozed off, and then went upstairs, where I studied the dense, opaque Electoral Count Act of 1887. I reread chapters in my favorite book, *Ordinary Vices*, by my college thesis adviser, the late political and moral philosopher Judith Shklar. I started to read the powerful new book that my friend Alexander Keyssar, its author, had just sent me, *Why Do We Still Have the Electoral College?* I fell asleep at around 1 a.m.

When I woke up on Thursday, December 31, 2020, the last day of the year, I went to the kitchen and called downstairs for Tommy to join me for a banana-and-peanut-butter breakfast smoothie. When he didn't respond, I descended to the basement apartment where he had been living for his second year of law school since the COVID-19 quarantine sent him home from Cambridge in March.

To wake him up, I yelled, "Tombo, Tombo!" When he did not respond, I knocked and went in.

And there, in a dreadful, indelible, irreversible instant of ghastly horror and disbelief, I found my boy, my extraordinary only son, lying motionless on his bed, and my life—Sarah's life, Hannah's life, Tabitha's life, our family's life, all of life—moved into a frightful and unrecognizable horror state.

After searching frantically for my phone—which I had thrown high in the air when I came upon the scene—after dialing 911 and screaming; after I tried to resuscitate him and get him to breathe by pressing repeatedly on his hard, beautiful chest; after speaking to the Takoma Park Police who arrived; after the first few hours of time freeze and shock; after speaking words from my mouth on the phone to my wife and my two daughters, words that I listened to myself speak but could barely hear and did not believe, words I had to convince them were true ("We lost Tommy last night; he took his life"), provoking shrieks and screams and sobs and "What are you saying to me? What are you saying?" After all this, for several hours, without family present—Sarah now struggling desperately to get home on the I-95 with her mother; Hannah now dissolved in grief with Hank in Lake Tahoe, looking for plane tickets and terrified to fly unvaccinated in the COVID nightmare; Tabitha in Malvern, Pennsylvania, heartbroken and collapsing with Ryan but preparing to drive back—I floated through the house and under the grey winter sky, thinking perhaps I was gone forever, too, and then I sat concretely traumatized and catatonic in my T-shirt and sweatpants and my "Democracy Summer" baseball cap, rocking back and forth like a baby, like my own baby son I held twenty-five years ago. Surrounded by patient Takoma Park police officers in Tommy's apartment downstairs, I did not even try to assimilate this dumbfounding, cognitively inadmissible, unthinkable fact, but just spoke words repetitively, compulsively, pathetically, robotically.

Rocking back and forth sobbing, all I could say was "My boy, my dear Tommy. My boy, my dear boy. I have lost my boy. My life is over. My life is over. I have lost my Tommy, I have lost my son. My life is over. My boy, my dear boy."

Julie later told me she spoke to the officers about the possibility of calling an ambulance because she thought I was in shock and not breathing right.

Later in the day, the police would find the farewell note Tommy left for us. We might have waited for several days to actually receive it because of the New Year holiday, but my friend Brian Frosh, Attorney General of Maryland, and Takoma Park Mayor Kate Stewart made sure we got it. It was written on the back of a Boggle word sheet in Tommy's brilliantly clear and eternally boyish print:

Please forgive me.
My illness won today.
Look after each other, the animals, and the global poor for me.

All my love,
Tommy

Family and friends came.
Flowers and food arrived.
The weekend passed like a dark winter storm.
And we buried our son yesterday, on Tuesday, January 5, 2021.

Today is Wednesday, January 6, 2021. Joe Biden and Kamala Harris won the 2020 presidential election more than two months ago, beating Trump and Vice President Mike Pence by more than 7 million votes, a popular vote romp that translated into an Electoral College victory of 306–232, a margin that Trump had declared a "landslide" when he beat Hillary Clinton by the same numbers in 2016.

But Trump, a profoundly transgressive figure in American politics, still refuses to concede. On the contrary, he has intensified his indignant, baseless claims that *he* scored a landslide victory on November 3 that is being mysteriously reversed and "stolen" from him and his long-suffering supporters. After propagandizing tens of millions of people with conspiracy theories and fantastical claims that even his slavishly loyal attorney general, William Barr, would come to call "bullshit"; after raising a fortune on-line off this "Big Lie" from his true believers; and after browbeating election officials and bullying Republican officeholders, Trump has now summoned his most avid followers to Washington for a "wild" protest against the counting of electors, with far-right groups in flank to confront Congress directly and "stop the steal."

Most politicians think Trump is playacting, raising money on paranoia and playing the polarizing carnival barker until the bitter end. In hindsight, it will seem painfully obvious that Trump was not playing. He had a specific plan to get Vice President Pence to declare and exercise wholly unprecedented and lawless powers to nullify electoral votes from Biden states and then force the contest into a contingent election in the House, where Trump was favored by a majority of state delegations. But Trump wrapped this political coup in the House in the chaos of insurrection by inviting his most extreme followers to Washington for a "wild" protest to "stop the steal" at the exact time we would be counting Electoral College votes. The Proud Boys, the Three Percenters, the Oath Keepers, the Aryan Nations, QAnon followers, different Christian white nationalists and militia forces would all show up in force in response to President Trump's invitation and engage in a mass violent assault against the Congress and the Capitol the likes of which had not been seen since the War of 1812. And just as I will condemn myself for missing multiple glaring clues that Tommy was on the path to taking his own life, I will condemn myself for missing multiple glaring clues that Trump and his forces were on a path to overthrow the 2020 election and would come dangerously close to doing so.

For pretty much my whole life, my friends and family have teased me about my insanely ridiculous optimism, about being "Mr. Rose-colored Glasses," as both Sarah and my sister, Erika, call me sometimes, for seeing only good news and perceiving only the best in people (including misanthropic right-wing Republicans who don't deserve it), for assessing only the positive dynamics in political situations and always elevating hope over realism—even now, I must resist the strong impulse to enclose that last word in quotation marks, because it is such an ingrained part of my nature to believe that realism is often just cynicism and pessimism dressed up in business clothing. I have been my whole life a *constitutional optimist*, in a double sense: it is in the irreducible constitution of my personality to be an optimist, and I am radically optimistic about how the Constitution of the nation itself can uplift our social, political, and intellectual condition.

But suddenly, this constitutional optimism shames and embarrasses me. I am tormented by the dreadful thought that this overflowing cheerfulness, a quality that has saved me through so many episodes of adversity and hardship in normal times, has turned into a massive and dangerous character flaw in the darkest of times, when all of us need to be soberly attuned to everything going on. Blocking out negative information, giving everything a positive spin, looking on the bright side—these habits of the heart have become a deadly peril to my family and to the people and nation I love, blinding me to danger, leaving me and, more important, anyone who depends on me, exposed to violent authoritarianism and fascism, the frightful biological and ideological plagues unleashed by autocrats and demagogues, and the mortal threats of depression, despondency, and despair. I fear that my sunny political optimism, what many of my friends have treasured in me most, has become a trap for massive self-delusion, a weakness to be exploited by our enemies. Yet I am also terrified to think about what it would mean to live without this buoyancy—and also without my beloved, irreplaceable son. The two always went hand-in-hand, and now I may be alive on earth without either of them.

I am, of course, not alone in my political myopia on the morning of January 6. Even after four years of continuous norm destruction and daily shredding of the Constitution by Trump and his henchmen, most American citizens have a defiant faith in the rule of law, what we in Congress call "regular order," and in the ideal of American "exceptionalism" from the rule of tyrants and bloodthirsty mobs. The counting of electoral votes has been accomplished as a formality for centuries. These low-key proceedings have traditionally set off celebration in the federal city, as my Rules Committee buddy Rep. Ed Perlmutter of Colorado tells me, a day of exuberant and proud bipartisan drinking in Capitol Hill bars.

Everyone I know is slightly on edge but determined to get through the proceedings safely and acting on the assumption that we will. America has gone more than two centuries without experiencing a violent interruption of the quadrennial counting of the electors. We have never seen a sweeping attack on the constitutional order at this key moment in the peaceful transfer of power from one president to another, one party to another, or one term to another.

The nearest we came to danger was on the eve of the Civil War, in February 1861, but even then, Abraham Lincoln's electoral count was never violently interrupted or deviously diverted by enemies of the republic. Back then, the *Baltimore Sun* reported on the chances of the Capitol's being blown up, and there were large and unruly pro-secessionist crowds trying to force their way into the building. But Gen. Winfield Scott kept the turmoil at bay with armed guards and his vivid warning that any disrupters "should be lashed to the muzzle of a twelve-pounder and fired out of the window of the Capitol," adding that "I would manure the hills of Arlington with the fragments of his body." According to Ted Widmer's authoritative *Lincoln on the Verge*, the vehemently pro-slavery vice president of the United States, John Breckenridge, personally carried the famous mahogany boxes containing the electoral votes from the Senate over to the House and proceeded to execute his duties faithfully, despite the fact that he was fiercely anti-Lincoln and would soon come not only to betray the Union but eventually to serve as the Confederacy's secretary of war.

Lincoln, who was traveling cross-country toward Washington to assume the presidency, received, at around 4:30 p.m., a telegram in Ohio that read, "The votes were counted peaceably. You were elected." He would still have to watch out for secessionist sharp-shooters bent on his assassination and avoid mobs filled with brawling street criminals like the Plug Uglies and the Bloody Tubs when his entourage passed through rowdy Baltimore. But there had been a fairly serene transfer of power in Washington when the electoral votes were counted. No violent mobs smashed the windows, attacked the police, or tried to change the results of the election.

Why should anything be different today?

We keep driving down North Capitol Street, passing the acres of grey tombstones in the hilly Rock Creek Cemetery and President Lincoln's summer cottage, on the grounds of the Soldiers' Home. It makes me think of Lincoln's son, Willie, who at age eleven contracted typhoid fever and suffered and died in the winter of 1862, plunging both Abe and Mary Todd Lincoln into profound depression, or what they called in those days "melancholia." Last summer, I read a work of fiction that was sent to me by our friend Diana Clark, *Lincoln in the Bardo*, by George Saunders, which reimagines President Lincoln's actual nightly visits to the Oak Hill Cemetery in Georgetown to visit the body of his precious, perfect son and imagines the response to his arrival among restless spirits in the "Bardo," the nether place where, according to some traditions, souls gather between death and rebirth. That book just about wrecked me when I read it, and even thinking of the title now is nearly unbearable as I feel myself hovering more and more in that indistinct passageway between life and death which Lincoln always seemed to frequent and traverse.

Now, we are mere blocks away from Ford's theater where Lincoln was assassinated. Like his assassin John Wilkes Booth, Lincoln was obsessed with *Macbeth*, a play about violent ambition and bewitched prophecy so dark and bloody that many people in the theater refuse even to speak its name. The president and Mrs. Lincoln saw Booth, a Marylander and successful Shakespearean actor from a distinguished family of actors, perform numerous times. When Booth saw Lincoln call for voting rights for some of the freed slaves two days after Robert E. Lee surrendered at Appomattox, Booth resolved that this would be Lincoln's last speech and got up the scheme to shoot him in the back of the head at Ford's Theater on April 14, 1865. A vehement racist and Confederate sympathizer, the actor became the first person ever to assassinate an American president. After shooting Lincoln in his box at the theater, Booth jumped to the stage and shouted, "Sic semper tyrannus," the Virginia state motto, which means "thus always to tyrants." He fled first to southern Maryland but was apprehended a few weeks later in a giant manhunt and killed in a barn in Port Royal, Virginia. When I was a boy and learned of Lincoln's assassination, I was spellbound and made my parents take me to Ford's Theater so I could see the scene of the crime with my own eyes.

We pass by New York Avenue, which makes me even more melancholy because it leads straight out to Route 50, the Bay Bridge, and the Eastern Shore. I spent a decade driving the first ninety days of each year from the Beltway to Route 50 to Annapolis, where I represented the people of Silver Spring and Takoma Park in the Maryland State Senate, serving several years as senate majority whip, leading floor fights and backroom brawls to abolish the death penalty, pass marriage equality, repeal mandatory-minimum drug sentences, increase the minimum wage, and enact the nation's first National Popular Vote Interstate Compact. Hannah, Tommy, and Tabitha used to love to come and spend the weekend with me and Sarah there, staying at one of the haunted local inns like the Governor Calvert House or the Robert Johnson House. Those were the days that convinced me that politics can be a sensational instrument of public education and government an extraordinary instrument of the common good. I also found I had a specific role to play because of my passion for the Constitution, my middle child's penchant for bringing people together for change (Dr. Martin Luther King Jr. is my middle-child's hero), and my inexplicable tolerance and even love for almost everybody I meet.

Now something new presents itself on our route. We see red MAGA hat–wearing protesters flowing in from all directions toward the Capitol and lots of signs hung along the road and over an underpass: "Stop the steal!" "Trump won!" This is the rallying cry Trump has been repeating like a manic cult leader from the White House and from Mar-a-Lago. The dogma has been decreed to his followers: *He won by a landslide* and *This is 1776!*

A car with Delaware plates flies by with an attitude, its bumper sticker reading, "If Guns Are Outlawed, How Am I Going to Shoot Liberals?" When we get to Third and Independence, Julie and I look over to see a rowdy assemblage of people in military-style camo gear, MAGA caps, and Trump T-shirts on the corner. They are screaming at the motorist in the car right next to us, a young African American guy in a suit. One of the people in MAGA hats takes our picture.

A woman carrying a sign that says "Fuck Your Feelings" yells at our motorist friend: "Stop the steal! Trump won it! We're gonna take it back, y'all!" There is a sarcastic lilt to her voice when she says "y'all." The people with her hoot and holler and explode into laughter; some drink from a flask.

Our next-lane driver rolls up his window and locks his doors, then looks over at me and Julie and shakes his head, smiling gently and shrugging. The light changes, and we roll forward. It's a moment. I take some comfort in the motorist's relaxed dismissal of the scene, as if he has seen this movie several times before but is determined to go on with his day. When I look back at the MAGA grouping, two of them are grimacing and giving our cars the finger.

We pass more protesters streaming in, many carrying flags: "Don't Tread on Me" flags adorned with snakes; huge Trump flags; flags with the large, smiling face of Donald Trump on them; and then, of course, Confederate battle flags and American flags, often carried in tandem, in a vulgar desecration of what Union soldiers fought and died for.

I tell Julie what our daughters told Sarah and me about the Churchill Hotel, where they are staying, on Connecticut Avenue in DC. When they arrived, the place was crawling with Proud Boys and right-wing MAGA-capped guests who brought a leering, menacing atmosphere to the lobby. Hannah and Hank said they had been told by the front desk clerk, "No smoking," but when some of the Proud Boys checked in, police officers hovered in the background and the clerk told the guests, "No fighting, no loud partying, no breaking furniture, and no smoking, which is a two-hundred-fifty-dollar fine and grounds for immediate removal. We're not putting up with any of it this time." The hotel clearly had some history and a beef with these people, who seemed to be settling in for the week.

"We're going to fuck you up on Yelp," one of them answered over his shoulder as the group made its way, laughing, to the elevators.

"See you on January seventh," another one yelled, "and welcome to the New World Order, asshole."

These little omens, dropped like fascist bread crumbs throughout the capital city, should be activating some kind of cultural emergency alarm. But working in Washington and living next door, we as liberal people have grown accustomed to protesters of all types. The Trump period has conditioned us to take in stride the outlandish and increasingly threatening language of right-wing zealots. None of these things has yet crossed the threshold in my consciousness as an imminent violent danger to the January 6 proceedings, much less to the republic itself. The threats on my mind today are internal to Congress in this unusual quadrennial proceeding: the aggressive but, I think, readily answerable and beatable challenges to the counting of electors from particular states.

I have not put the strategic clues together, but I do hear a voice in my head: *Attention must be paid to the mundane details. This is how authoritarianism will infiltrate your society and control your life, one little aggression at a time.* It is the voice and accent of my friend Claudio Grossman, who was dean at American University Washington

College of Law for most of the time I taught Constitutional Law there (1990–2016). Claudio had worked in Salvador Allende's democratic government in Chile before Gen. Augusto Pinochet overthrew it in a military coup in 1973 and instituted a reign of terror and torture against democracy's defenders. After Chile's return to democracy, I spent part of the summer of 2005 teaching there, with Tommy in tow, along with my niece Maggie Littlewood and quasi-niece Julia King. When Claudio visited us on that trip and we had lunch, he told me that when it comes to your house, fascism invades every nook and cranny, every fine detail of your existence. For that reason, it will be visible early on if you just open your eyes to it. That message stuck in my mind like Madeleine Albright's point in her 2018 book *Fascism: A Warning*, where she argued that fascism is not a fixed ideological system but rather a strategy for taking and holding power.

When Julie and I arrive at the Rayburn Building, the two seated guards at the garage entrance wave us in, and we park. We greet the next two guards at the metal detector entrance in the building, and I show them my congressional pin; they motion us through. I take note that there are only two guards at each checkpoint, but I assume that the real muscle will amass outside.

We make it to Room 2242, our nearly unfindable but spacious new office in the maze of the Rayburn House Office Building. Everything is in shambles here like a college dorm during move-in week. After two terms in the Cannon Building, where the most junior members are clustered, I have graduated in the 117th Congress to Rayburn, where two of my main committees, Judiciary and Oversight, have hearing rooms. The congressional movers have brought our stuff over, but it is in boxes stacked on furniture, with framed posters, photos, and campaign signs sitting on the sofas and leaning against the walls. There are hundreds of condolence letters I need to open.

Seated in the chaos of my new office but at my clean new desk, I work on two five-minute speeches. The purpose is to refute expected

GOP objections under the Electoral Count Act of 1887 to the counting of certified Electoral College votes that have been sent in by the
governors of Arizona, Georgia, and Pennsylvania. If the objections
are made in writing by at least one House member and one senator,
they will trigger a recess of our joint session and send the members of
the House and Senate back to our respective chambers to debate them
for no more than two hours per state (one hour per side) and then to
vote on them. The objections are sustained (and the electors rejected)
only if concurrent majorities in both chambers vote to uphold them.

My central point is that the election is over. The people have
spoken in their states, and the state governments have carefully
certified their electors. No fewer than sixty-one federal and state
courts throughout the land have rejected claims of fraud and corruption, and no legislature, including the GOP-run legislatures of
Arizona, Georgia, and Pennsylvania, is asking for its own electors
to be nullified by Congress.

The point of this proceeding is neither to question nor endorse the
popular will, much less to overturn it, but simply to formally *receive
and count* the electoral votes certified by the states and effectuate the
peaceful transfer of power. We are acting as election officials, not
voters. I waver a bit between ridiculing the absurd things House GOP
leaders are saying about electoral results in the states, like the claim
the election was stolen by a conspiracy run by an Italian security firm,
and simply trying to recapture the solemnity and gravity of the occasion. Before Tommy's death, I might have been tempted to mock
Trump and the Big Lie he has been promoting every day, but now,
filled with grief and a kind of terrible emotional weight I have never
felt before, I ditch the potentially crowd-pleasing digs about outlandish fantasies and simply state in a straightforward way that the courts
have overwhelmingly repudiated every baseless claim being pressed
in Congress about election fraud. My argument is solid, but it leaves
me a bit cold. I know it will not fully dispel the ambient ideological
noise and combustible mood of my GOP colleagues.

At around 11:45 a.m., we leave Rayburn to go over to the Capitol Building, where we will be joined by Tabitha and Hank.

When Julie and I arrive at the Capitol, the air is charged and electric, disorienting. I see GOP colleagues in the hallway staring at me, whispering. My friend Rep. Anna Eshoo gives me a great big hug. So does Massachusetts Rep. Katherine Clark. Julie and I enter Steny Hoyer's hideaway office, just off the Capitol floor, where Steny welcomes me on the way in. Julie forages for food and drink, finding me some orange juice in Steny's refrigerator. I close my eyes and bring Tommy close to my chest and my heart.

We await the arrival of Tabitha and Hank, who has been loving and strong for our family during this awful period. I repeated to him last night what I told him when we learned of the elopement: like it or not, he would be a son to me now.

When he and Tabitha arrive, I embrace them.

"This is my daughter Tabitha and this is my son-in-law Hank, but he is married to our *other* daughter, Hannah," I say to Doug Letter, counsel to the House of Representatives, who has dropped by. My fastidious introduction is quickly the source of much merriment, but I feel vindicated because the confusion is already rampant about who is married to whom.

Tabitha wants to see my speech. I pass her the hard copy and a pen. She crosses out words and makes insertions. At one point, she says, "Dad, you've got to tone this down."

"I did tone it down," I say. "You should have seen it before."

I see Doug Letter smiling as he watches my daughter, one year out of college, edit my speech.

"You're just going to make them madder with all these adjectives," she says. "You don't need so many adjectives."

"You're right. I'll take out some adjectives," I allow. "But Trump's trying to force Pence to help him steal the election. Pence has no power to count electoral votes or decide their validity. That's totally up to Congress. Do you see what they're trying to do?"

"I know, but do it like Tommy," she says. "Let the facts tell the story. This is too divisive, too snarky," she says. "Be conciliatory. You guys won."

Embarrassed by her points when I thought I had already scrubbed my speeches of snark, I quickly incorporate her changes. Later, I will come to see the wisdom in her standpoint. Tabitha graduated from the University of Maryland in 2020 with a degree in psychology. This is where she met Ryan, after she transferred from Amherst College, where she had followed Hannah and Tommy but which she left after finding it too pastoral. She's a city person.

Just before 1 p.m., I collect my papers to head over to the floor. Tabitha and Hank will follow shortly, when an officer comes for them. I glance out the window onto the West Front before leaving the office. Tabitha and Hank look out, too.

On the usually deserted western steps of the Capitol—the place where, four years ago, Trump introduced the ominous image "American carnage" to the world—a surly mob of thousands is now assembling, pressing forward against the fencing, jacked up by Trump's apocalyptic pep talk from a rally he held outside the White House. "If you don't fight like hell," he told them, "you're not going to have a country anymore."

Led up front by fierce-looking Proud Boys, Oath Keepers, Three Percenters, QAnon brawlers, and other shock troops of the extreme right, the crowd, in a matter of minutes, will begin to taunt, push, shove, punch, gouge, scratch, spray, smash, jab, and harass the U.S. Capitol Police force, kicking off four hours of savage, medieval-style violence that will result in eight physical breaches of the Capitol. The insurrectionists and rioters will leave at least five people dead (with several more to come by suicide of officers) and more than 140 officers wounded and injured, many of them hospitalized with traumatic brain injuries, concussions, broken arms, broken legs, broken ribs, broken vertebrae, black eyes, broken noses, lost fingers, broken necks, broken jaws, post-traumatic stress syndrome, and every manner of emotional and psychological damage. Of this unfolding rampage, we members of Congress assembling inside the Capitol still know nothing. Like me, Tabitha and Hank can see only bits and pieces of the advance on the Capitol, but these choppy scenes preceding a wild charge of throngs of fanatics up the Hill are already troubling them.

"Are there enough police to stop them?" Tabitha asks me.

I hesitate because of the size and apparent volcanic energy of the crowd, but assure her it will be fine. From our window, it looks chaotic, but I tell her that our officers are armed and "ready for this kind of thing." I think back to a Black Lives Matter protest in DC on June 2, when masked and helmeted National Guardsmen in camouflage massed on the steps of the Capitol in a formidable and awesome display of force. I have not yet seen the National Guard today, but I tell myself that they must be here *somewhere*, or else they have got to be on the way over. I tell Tabitha that an officer is coming to bring her and Hank up to the visitors' gallery and that I will look up to them from the House floor. I try to communicate a sense of calm that is suddenly beginning to evaporate. *Where is the Guard, actually*, I think, *and what exactly is "this kind of thing" I've said our officers are ready for?*

I give Tabitha and Hank another hug and leave for the floor.

As I enter the House chamber, I feel the eyes of my colleagues on me, searching my haunted countenance and my every gesture for the pain and tragedy I now personify. They don't have to search hard. In the gloomy photographs of that day, from the dark half-circles under my eyes to the deeply etched mask lines across my face, my every pore expresses grief and anguish. It is a whiplash role reversal for me. I am a House member whom others have looked to for lightness and hope through the dread and darkness of the last four years. My Virginia colleague Abigail Spanberger, in an interview on C-SPAN, once called me "the funniest member of Congress," not only the greatest compliment I have received since taking office in 2017 but one I probably secretly aspired to.

But now I am a figure of unthinkable and unspeakable tragedy, someone subsumed by the grim darkness of his days. My family in the course of a week has come to embody the trauma of 2020, a year full of mass despair, individual pain, and wrenching loss. I ardently wish that the people who are staring at me, all my colleagues, could have *met* Tommy, could have *known* him, could have seen that he was the most positive, exuberant, and dynamic life force one could imagine, truly the funniest person you might ever hope to meet (far funnier than me, as proven on family vacations at Stand-up Comedy Night, when he always took top honors in the savage intrafamily competition). The theme of Tommy's life was comprehensively "utilitarian," as Tabitha put it: always advancing the greatest good for the greatest number of people—and the greatest number of animals, maximal happiness and justice all around. But now trauma and anguish are all anyone can see when they look at me, a poor proxy for my handsome and dazzling son. I do not want them to regard Tommy as a figure of tragedy or pity.

A memo from Vice President Pence is passed out. This is a make-or-break moment. Will Pence assert lawless new powers to deny Biden the presidency or will he act according to the Constitution? I get ahold of a copy and speed-read it. It is strikingly good news. Despite lots of genuflecting to the disseminators of the Big Lie, he says he will *not* be exercising unilateral powers he does not have to reject states' presidential electors or return them to the state legislatures for further action. It is not within his constitutional power.

This is big.

I breathe easier now, as any such act would have been a most dangerous flash point for constitutional confusion and political danger. Kevin McCarthy and Donald Trump wanted to block acceptance of Biden's electoral college majority and get Trump named president "immediately" by twenty-six state delegations in a contingent election. Pence has just done democracy a giant favor by not lying and betraying the Constitution, although I am sure the vast majority of the public has no idea how important his little memo is. I do see, just across the central aisle from me, Rep. Liz Cheney scrutinizing the memo, just as I have been doing, and she, too, seems to relax a bit.

At the dais, Vice President Pence is presiding with Speaker Pelosi behind him. No one objects to Alabama or Alaska, but then Arizona is called. A slew of House Republicans, led by Arizona's Paul Gosar, joins Missouri senator Josh Hawley and Texas senator Ted Cruz in objecting to Arizona's electors. So now the joint session recesses, and we move to meet in our respective chambers to consider the merits of this objection, with each side in the debate limited to one hour.

As the senators exit and the House members resume their usual places, I look to my left and see the Democrats and to my right I see the Republicans. I recall a conversation I had with a Republican colleague, Rep. Clay Higgins of Louisiana, once when we were sitting together on the floor. "What do you see when you look out over on our side and then over on your side?" I asked Clay, who is a good-natured man with an awesome Cajun accent.

He looked at me and said, "You tell me first."

"When I look at our side," I said, "I see America today in all its glory: black, white, Hispanic, Asian-American, LGBTQ, straight, women, men, immigrant, native-born. And when I look at your side, I see . . . America in the 1950s."

"When I look at your side," he said. "I see the coasts, New York and California. When I look at my side, I see the heartland."

And so now the debate begins. The GOP objectors to Arizona offer no factual specifics and compensate for their vagueness with right-wing agitprop that sounds tinny to me. "The debate as to the legitimacy of the 2020 presidential election has been suppressed by the left and its propagandists in the media until today," says Rep. Andy Biggs from Arizona, the speaker before me who proceeds to trash his own state's election results—the same results that got him elected—by recycling a judicially demolished claim about the state voter registration deadline. He does not explain why Congress should overturn the legal analysis of all the state and federal courts that have just rejected this precise complaint about how Article II, Section 1, prevents any state actor other than the legislatures from regulating elections or interpreting election laws, even when the state legislatures and state constitutions compel them to do so.

Now it is my turn. I look up and see Tabitha and Hank to my left, in the gallery, sitting next to my friend Rep. Tom Malinowski from New Jersey, who had been a key State Department democracy and human rights official under President Obama. I try to smile at them, but my facial muscles won't cooperate. Tabitha seems fascinated by what is happening; I see her smile, and now I smile the slightest bit.

I rise to say, "Madam Speaker, I claim time in opposition to the objection."

Speaker Pelosi says, "The gentleman from Maryland is recognized for five minutes."

"Madam Speaker," I say, "I first thank you and all my dear beloved colleagues for your love and tenderness, which my family and I will never forget"—I am interrupted by loud applause in the chamber that becomes a thunderous standing ovation.

This is startling. My disorientation returns. Why are they clapping? It is confusing. I hang my head and place my hand over my heart to keep from crying. A passing strange thought crosses my mind: Could the spreading grief over Tommy's death, and the goodwill on display now, somehow magically prevent Congress from dividing bitterly over counting the electoral votes?

The applause subsides, and I am now wishing absurdly that there might be a way to proceed without offending the Republican colleagues who have just tried to console me. I launch into my prepared text, which begins with Lincoln as our bridge figure, although I always think he's going to be a much bigger hit with the Republican side than he actually is:

"Abraham Lincoln, whose name is a comfort to us all, said we have the best government the world ever knew."

I see no reaction over there. Then I work my way back to the three most important words of the Constitution, "We the People," the phrase that anchors our experiment in popular sovereignty. I get to invoke another Republican, former president Gerald Ford, who said, "Here, the people rule." I then casually decide to drop a prepared sentence that I realize could easily boomerang on us. I make this last-second change because I am now haunted by a premonition that, even though our little group has worked and reworked every possible GOP objection, and even though, in a mock session, all of the debates ended up with majorities in both houses defeating the objections on the floor, the day will somehow end badly. Although Senator McConnell has publicly opposed all the backroom games to overthrow the states' lawful electors, something superstitious deep inside tells me not to utter these words, and I don't:

This is the peaceful transfer of power we celebrate and a model for a grateful world.

Instead, I return to the text and use this key passage:

"We're not here to vote for the candidate we want. We are here to recognize the candidate the people voted for in their states. The 2020 election is over, and the people have spoken."

Here I am interrupted by boisterous applause from the Democratic side before going on to make my final points.

And when it is done, I feel relieved to have gotten through it. But as I sit down to lusty Democratic cheers and applause, I am struck by the sensation that I am missing—perhaps we are all missing—what is really happening in or, more precisely, *all around* this proceeding.

None of us can quite express it, but I am definitely feeling it. Strategic violence by extremist elements *outside* the Capitol is fusing with manipulative tactics *inside* the Capitol to try to coerce Vice President Pence and Congress to overthrow the electoral votes in the states and force us into a contingent election. This synchronized coordination of bloodthirsty violence with extraconstitutional strong-arm tactics produces the radically unfamiliar but nonetheless unmistakable feeling of a *coup*. We usually think of a coup as something happening *against* a president, but this would be a coup *by* the president against the vice president and Congress.

Here I have been, focused on strict interpretation of the arcane Electoral Count Act and the quaint Twelfth Amendment, myopic in my grief, polishing my speeches to a fare-thee-well, acting as if this were some kind of moot court argument at the end of which they would hand prizes out to the best oral advocates. These legal and constitutional instruments surely set the table for our present conflict, but in this extraordinary new crucible of struggle, the battle itself could be decided just as well by the deployment of raw power and deadly violence. Gen. Michael Flynn and Trump's other militaristic aides-de-camp have been talking openly over the last many days about imposing martial law to make sure Trump (the "real winner") stays in office; Flynn for weeks has been publicly urging Trump to deploy the military against hostile institutions and to "rerun the election," to conduct a "national re-vote" under direct military control. We have all heard rumors that high-ranking military people are afraid Trump will try to muster and mobilize the National Guard and federal troops to subdue Congress and Vice President Pence, declaring martial law and canceling the 2020 election as fraudulent and corrupt.

Listening to the nonsensical arguments of the next speaker, my new colleague Rep. Lauren Boebert from Colorado (a QAnon sympathizer, brandisher of guns, and rabid Trumpian polemicist), I remember something I once heard Rev. Jesse Jackson say: "Text without context is pretext." The text of all these irrelevant speeches today is just a pretext for executing the president's plans, which involve the strong-arming of the vice president and the disruption of Congress in the counting of votes, the parallel strategies that make brutal violence the context for everything that is happening.

When Boebert completes her bizarre speech, one in which she seems to flatter herself for not challenging anyone to a duel, I receive a worried text from Alyssa Milano, an actress and talented political organizer in California. Alyssa has been checking on me and the family daily since December 31. She found a way to name a star in the sky after Tommy, a gesture that just about destroyed me. She is the first person to alert me that something is seriously wrong.

Alyssa writes: "Are you safe?"

I text back yes, and she writes, "The mob has stormed the building. Please be careful."

I look around: everyone on the floor now is madly talking, texting and pointing as strange and unprecedented sounds ricochet loudly through the building. The mob is on the move within the Capitol, people are saying. Alyssa is right.

It is now after 2 p.m., an hour into these proceedings, and we are still only halfway done with our debate over the objection to Arizona's electoral votes. There are rampant reports circulating of bloody violence being waged against police officers; texts flying in with graphic images of Capitol Police being punched, crushed, mauled and sprayed in the face, speared with American flagpoles; and accounts of the Capitol being "breached" and "overrun." At 2:09 p.m., a text alert from the Capitol Police appears on my phone stating: "All buildings within the Capitol Complex, Capitol: External Security Threat No Entry or Exit," which advises us to "shelter in place" and "stay away from exterior windows or doors."

I see someone from her security team come whisper something to Nancy Pelosi, similar hurried conversations on the GOP side, Steny Hoyer being briefed. As Louisiana congressman Mike Johnson and Arizona's own Raúl Grijalva try to speak, people begin talking out of turn, even shouting openly. When Rep. Paul Gosar tries to speak, one of my Democratic colleagues yells, "Call off your damn Storm-troopers!" and my friend Rep. Dean Phillips from Minnesota stands and shouts, "This is because of *you*." Speaker Pelosi is escorted out, leaving her cell phone behind. Before she leaves, I see her whisper to Jim McGovern, our great Rules Committee chair, a soulful liberal and peace visionary with the deceptively curmudgeonly air of an old-time Boston pol. He soon takes the dais and tries in vain to continue with member speeches. No one is paying attention. All hell is breaking loose, and I see people taking cover behind chairs and desks. Majority Leader Steny Hoyer and Majority Whip Jim Clyburn have been evacuated by security, leaving us with no senior House leaders on the floor.

Someone sends me the now-iconic photo of the invading insurrectionist casually carrying the Confederate battle flag in the Rotunda, a seven-minute walk from where we sit, right in front of the portrait of John C. Calhoun, the South Carolina racist and slavery apologist, the seventh vice president of the United States, who now seems to be smiling a bit from the nineteenth century at this astonishing spectacle taking place in 2021. Across from Calhoun hangs the portrait of Massachusetts Sen. Charles Sumner, the great abolitionist crusader who paid dearly in blood and health in 1856 when South Carolina Rep. Preston Brooks beat him savagely with the golden head of his heavy walking cane as Sumner sat at his desk on the Senate floor. Brooks knocked Sumner out of fulfilling his Congressional duties for three years before he was well enough to come back.

But this is 2021 and I cannot believe my eyes. An insurrectionist rebel has entered the Capitol unlawfully and is brandishing a Confederate battle flag in Congress, something that never happened once during the Civil War.

I look up to see Liz Cheney across the aisle looking equally dumbfounded at the chaos engulfing us.

Liz and I disagree on pretty much all the issues our parties fight over, but in this emerging struggle over democracy, truth and the rule of law, I consider her a critical ally and a real friend. Trump has already been blasting her in public for her political apostasy in not embracing his Big Lie, and she is one of only perhaps a dozen Republicans left in Congress whom I can describe as a constitutional patriot, someone with the mental toughness to place her country and Constitution above party leadership and political calculation.

I like Liz very much and have liked her from the moment we met when we entered Congress together in 2017. She is a mother of five with a great sense of humor and a lawyer with a fine, logical mind. She is tough as nails.

I walk over and show her the picture of the guy with the Confederate battle flag. "Looks like we're under new management," I say.

She looks aghast and shakes her head. "What have they done?" she says solemnly.

Someone gets up and tells us to put on "your gas masks." I didn't even know we *had* gas masks, but there they are, under our seats, wrapped in plastic. Members fumble with them, drop them, ask how to put them on. When people get them working, the masks set off a dizzying buzzing noise that fills the chamber like an elementary school nuclear drill alarm from the last century. Our new House chaplain, Margaret Kibben—our first female cleric, whom I have not yet met and who is in only her third day on the job—goes to the dais to make a surprisingly eloquent plea for "a peace which has more armor than anything we can don" and deliverance from chaos, even in this nauseating din. But, again, there is screaming and panic all around. I hear people phoning their wives, husbands, and children to tell them they love them. I hear someone up in the gallery, maybe our colleague Jason Crow, shouting to everyone to remove their congressional pins so they will not be recognized as members. My colleague and friend Susan Wild, who wrote me a beautiful note about Tommy, seems to have taken ill on the floor, and another colleague tells me it is a panic attack.

All I can consider now is Tabitha and Hank. I am worried sick for them—they may be the only children of members in the Capitol today—and for Julie, my chief of staff. I call Julie. She says my daughter and son-in-law are okay where they are, behind a locked door, barricaded with furniture in Steny Hoyer's office. Tabitha and Hank are hiding under Steny's desk.

"Tell them I love them. Guard them with your life," I tell Julie. I will learn later that Julie finds a fire poker in Steny's fireplace and wields it above the door, vowing "I'm not letting these mother-fuckers get away with this." "She definitely got her Philly on," as Tabitha would explain.

I am picturing them under Steny's desk and I'm about to call Tabitha to hear her voice—

And then—boom!

Boom!

I hear the sound I will never forget, a sound like a battering ram, the sound of a group of people barreling up against the central door with some huge, hard, thick object, hell-bent on entering the House chamber. The members nearby press furniture up against the door, and a number of us farther away run to the door to help protect it, but we are then quickly told to "get back" by Capitol Police officers, who rush in and defend the entranceway with their guns drawn. The pounding at the door accelerates, and we can hear the sound of angry, macho chanting out there, too.

Hang Mike Pence! Hang Mike Pence!

And: *We want Trump! We want Trump!*

Someone official calls upon us to evacuate right away, calmly. Everyone moves, some people run, to the Speaker's lobby, carrying their gas masks. I look up to the gallery again to see our colleagues, who have been frozen in place on the Democratic side, now awkwardly crouching and sliding through the rows to make their way over to the gallery above the Republican side. I see New Hampshire Rep. Annie Custer and California Rep. Sara Jacobs, who is only in her fourth day on the job, crawling their way over. When we escape and are reunited later, a colleague will tell me they decided to cross over to the GOP side because they thought a mass shooter who entered would be less likely to aim at the Republican side of the House. Meantime, the officers up there have locked all the doors to keep the rioters from breaking *in* but will now presumably unlock them to get our colleagues *out*. I feel strange about leaving them up there, but then again, who knows where *we're* going? Where will any of us find safety on January 6?

A bloodthirsty mob of hundreds has entered the building out-side the metal detectors and with no security check. Who knows what weapons they are carrying? It is hard to displace an unbidden thought I have rooted in the images we all carry from the recent Walmart shooting in El Paso; from the 2018 Tree of Life synagogue attack in Pittsburgh; from the 2015 Mother Emanuel AME church massacre in Charleston, South Carolina. What if one of the rioters is carrying an AR-15? Many of us are thinking the same thought.

A handful of officers and young Pelosi staffers—Emma Kaplan, who wrangles members for voting on the floor, is the one I know best and am now looking to for help—leads us higgledy-piggledy down some stairs and into the Capitol tunnels. The members talk madly on their phones. I catch up to Emma and tell her we need to get Tabitha, Julie, and Hank out as quickly as possible. She agrees. She writes a text; she says she will do whatever she can. In the tunnels now, near the tram that runs between the Capitol and the Rayburn House Office Building, we hear shouting, chanting, running, the nauseating buzz of the activated gas masks no one really knows how to work except for some of the military veterans, like Ruben Gallego, Abigail Spanberger, and Jason Crow, who are helping people. Most of the members from the gallery have made it down the stairs and have joined our rapid stream of exit and descent. I hear a southern Republican congressman I know yelling into a phone, "You screwed it up, y'all screwed it all up!" A member tells me that our wonderful colleague Raúl Grijalva is having a hard time moving and will be pushed in a wheeled office chair. I look back but don't see him. I am borne along.

My mind fills with swirling questions as we enter the Longworth Building.

Have we entered a violent power struggle? On our side of the aisle, we have staked everything on popular sovereignty and the mechanisms of democratic election, but Trump and House GOP leaders have been acting on the dictum of the right's favorite philosopher, Carl Schmitt, who said, "Sovereign is he who decides on the exception." We have come to the exception. Trump and his enablers have forced us into a politics in extremis, a place where the rule of law is trampled and violence redefines the terrain of struggle to make an authoritarian deviation from the rule of law possible. If the violence of Trump's own incited mob gets out of hand, he can easily declare a state of siege and impose martial law by activating the Insurrection Act. Will this be his "Reichstag moment," as Gen. Mark Milley, the chair of the Joint Chiefs of Staff, is apparently wondering too?

We Democrats are in love with the rule of law (which is the control of official abuse of power by formal institutional rules) and with voting and the channels of popular participation to ensure democratic self-government. We have thus looked away from a very old threat suddenly staring us in the face again, the threat that Alexander Hamilton warned us of in Federalist No. 1: the threat of an opportunistic demagogue unleashing a violent mob and primitive impulses *against* the Constitution to override the political and constitutional infrastructure of representative democracy. The demagogue panders to the negative emotions of the crowd, pretending to be the champion of the people, only to wage war against the Constitution, the legal order, and the democratic political process, all of which belong to the people. He starts as a "demagogue," one who knows how to whip up the crowd into a mob frenzy, but ends as a "tyrant," a ruler who uses his power to oppress the people, Hamilton said. This scenario is literally in the *first* Federalist Paper.

And when I say *we* have overlooked the potential strategic collaboration of mobs and demagogues to demolish our institutions with insurrectionary violence, I mean *I*. I have overlooked it. How could I have been caught so off-guard? All the clues were staring me in the face. I will face plenty more opportunities in coming days for self-recrimination.

But the question now, as we flee the Capitol from American democracy's old enemy—racist mob violence—is how our constitutional democracy can prevail over Donald Trump's party, which operates like a religious cult and couldn't give a damn about the Constitution or democracy. They are working all the levers of control and violent coercion to keep their leader, his family, and his sycophants in power.

Every part of my past is converging—this struggle will be my life now. As a Maryland congressman who resides in neighboring Montgomery County, I live closer to the Capitol than any other member of Congress (except for my friend Eleanor Holmes Norton, the District's nonvoting delegate) and I know Washington better than all but perhaps a handful of other members of Congress. We must secure the Capitol and defend our government. We must stop the insurrection, arrest the coup, and punish the lawless— from the inside operators in the corridors of power to the street-level brawlers.

As a professor of constitutional law for three decades, an observer of the separation of powers and civil liberties, and someone who reveres liberal democracy and hates fascism and racism, I have the motivation to fight. I have taken big bipartisan groups of incoming House members to the U.S. Holocaust Museum for a tour before their first session begins, to emphasize the essential stakes of liberal democracy. I send everyone to the National Museum of African American History and Culture and the National Museum of the American Indian to learn about the centrality of slavery, racism, and white supremacy in shaping American law and politics and defining the American experience.

But, now, I am driven by the memory and spirit of my lost son, who wanted far *more* from our democracy, not far *less*. He expressed dread of fascism and what it had done to humanity in the last century and horror for Nietzschean politics based only on "force and fraud and the will to power," as he used to mutter every time he read the newspaper. He wanted government to be the active instrument for promoting the general welfare of not just all human beings but all living beings, and he wanted morality to replace violence as the essence of government power.

I wonder where all this chaos is taking us; whether Tabitha, Hank, and Julie are safe and will be rescued soon; whether I should try to turn back and find officers; whether these insurrectionists have firearms; whether Donald Trump's allies plan to escalate the violence; whether Sarah, Hannah, and Tabitha's boyfriend, Ryan, are alright; whether we are facing an insurrection, a coup, or even a civil war; whether we will finally impeach the traitor for setting loose the dogs of war upon us or perhaps invoke, at last, the unsung Twenty-Fifth Amendment; whether dear America will survive this appalling head-first descent into political madness.

I feel curiosity, anger, resolve; but there is one thing I do not feel as we travel down, down, down—"Faster, please hurry," an officer exhorts us—into the dark complex basement passageways of the Capitol, one thing I don't sense as we are jostled this way and shepherded that. There is one emotion I have not experienced at all on this persistently gloomy and objectively terrifying day and that I will not experience all through the night: fear.

I feel no fear. I have felt no fear today at all, for we have lost our Tommy Raskin, and the very worst thing that ever could have happened to us has already happened. But I am still in the land of the living, and Tommy is with me somehow every step of the way. He is occupying my heart and filling my chest with oxygen. He is showing me the way to some kind of safety.

My beautiful son is giving me courage as we flee the U.S. Capitol Building for our lives.

My trauma, my *wound*, has now become my shield of defense and my path of escape, and all I can think of is my son propelling me forward to fight.

PART I

CHAPTER 1

DEMOCRACY SUMMER

———————

It's not incumbent upon you to finish the work, but neither are you free to evade it.

—THE MISHNAH

Tommy was my only son. A middle child and first-born boy born to a middle child and first-born boy, he was my greatest student, one who quickly became my greatest teacher, something like a best friend, too. He transformed me twice politically, once in his young life, when he helped me first run for office, and then again in death, when he filled my broken heart with purpose and strength.

When I first decided to go into electoral politics, it was 2006. I was forty-three, Sarah was forty-four. Hannah was thirteen, Tommy eleven, and Tabitha nine. At the time, all three of our kids were deeply engaged in political questions and all the neighboring puzzles of moral philosophy, but Tommy the most. When it came to politics of either the philosophical or practical kind, Tommy Raskin was a natural, a thoroughbred, bringing a piercing moral and global sensitivity to his interpretation of political problems. The curly-haired ragamuffin running up and down the aisle of the airplane giving everyone high-fives became a moral and political philosopher.

Even as a little boy, Tommy lived his life with a precocious integrity, and always identified with the underdog. He won an award in middle school for explaining to a group of older students why racist jokes were hurtful. He invited a large group of kids to go to the Blair High School senior prom together, regardless of whether they had dates, so that everyone would be included. He stood up in a high school class and objected to the fact that "people are cheating and no one seems to care that students are not learning the material but just copying it," and then threw himself into a peer-to-peer tutoring program called Bliss, to help other kids learn course matter and prepare for their essays and homework. ("Tutoring is real helping," he said. "Cheating helps nobody and just encourages the school bureaucrats to not educate us.")

People think most seriously about morality when they are young and then when they have young children of their own. Having Tommy and two other kids who were serious about ethical action made me think a lot more concretely about my own future. I resolved that it was time for me to stop just writing law review articles, testifying and *opining* about stuff. I had been a constitutional law professor for sixteen years and had always had an essentially academic disposition. I am ardent for reading, writing, teaching, and being with young people, and for this reason, there is nothing in the world greater to me than a college or a university. But, when I was forty-two, I realized that the time had come for me to cross over from scholarship and punditry to fight for the things I believed in: to go *into* politics or resolve just to be a sideline critic.

I had always known that I had a strong political side in me, and politics runs deep in my blood. My maternal grandfather, Samuel Bellman, was the first Jew ever elected to the Minnesota state legislature way back in the 1940s; by the time I got to know him, he was a pillar of the Democratic-Farmer-Labor Party and of the legal establishment of Minneapolis and the Jewish community. During summer visits, we used to watch him solve people's problems all day long while working either as a lawyer in court or a politician on the phone or just as a friend in conversation. When people ask me where I get my tolerance—no, my *love*—for constituent services (such as getting people their Social Security checks, VA benefits, unemployment compensation, or fair treatment by Medicare or some other bureaucracy), I always invoke my Grandpa Sam, who could be curmudgeonly and tough sometimes, but who *always* took the side of the underdog and who never turned away a constituent or a client, no matter how poor—for which he sometimes got paid in chickens and eggs.

My father's side of the family, in Milwaukee, Wisconsin, is also filled with political activists and leaders, like my famous great-uncle Max Raskin, the sainted Milwaukee judge and former city attorney. My paternal grandfather, Benjamin, whom I never met and after whom I was named—my parents turning the name around to make it "Jamin Ben Raskin"—was a plumber, first as a member of the plumbers' union and then as the owner of his own small plumbing business. (Sarah will tell you I'm pretty helpless as a handyman anywhere else in the house, but that I can fix anything in the bathroom without even looking on the internet for help, because it's in my genes.)

One day in 2005, I picked up the newspaper to learn that our state senator, Ida Ruben, who had been in office for thirty-two years and who was the president pro tempore of the Maryland senate, had introduced a pro–Iraq War resolution and a bill to dramatically expand the death penalty in our state. Further research revealed that, although she had definitely done some good things, Ruben had been helping block consideration of marriage equality, which she did not feel the state was ready for. This seemed scandalous to me, because I always took Silver Spring and Takoma to be the progressive heartland of Montgomery County, itself one of the two or three bluest jurisdictions in the state along with Baltimore and Prince George's County. Maryland has a part-time legislature—just ninety days a year in session—and I decided I wanted to run for the state senate from District Twenty, Silver Spring and Takoma Park.

It was very weird going from just analyzing and criticizing things (the life of the law professor) to actually taking responsibility for changing and improving them (the life of the political representative). In the process of my becoming a candidate and political leader, Tommy became my key confidant and adviser—which may sound odd considering that he was eleven years old, but he was a natural-born political visionary and strategist, and he loved politics. His advice to me was soulful. A number of progressives were urging me to run for the state senate—like Mike Tabor, who wrote a column for the *Silver Spring and Takoma Voice*, then a major force in our community; and Marc Elrich, a Takoma Park city councilman and a Bernie-like 1960s radical who had been running unsuccessfully countywide and wanted to try again for the County Council with me on the ballot helping to galvanize a strong progressive network in the eastern county.

But the establishment Democratic politicians were skeptical and worked on me not to run. When Tommy and I would talk to delegates, senators, and power players, they discouraged us. At a countywide Democratic dinner that Tommy accompanied me to, a top official in Montgomery County said to us, "You can't run because you can't beat the Machine."

Tommy asked him in an open and inquisitive way, "Who's the machine?"

And this Democrat proceeded to name three or four people who would certainly be with the incumbent.

"Well, that's four votes," Tommy said. "What about the other one hundred and sixty thousand?"

This precocious lesson in the practical meaning of one-person, one-vote democracy made me smile and made the top power broker wince at the temerity of the eleven-year-old son-of-a-political-upstart.

That amazing lesson became my go-to answer when people asked how we were going to beat the Machine: "By majority vote," I would say, *"one person at a time."*

On a freezing day in January 2006, Tommy introduced me at my campaign kickoff, as did Tabitha. She said, "Please give my dad money; he really needs it." And then Tommy said, "Yeah, he really needs your money." Later in the campaign, he would introduce me by saying, "My dad loves our family a lot, and he loves our community a lot, too."

At my kickoff, he took the photo—of me with the American flag in the background—that we used on our campaign literature, and he was, significantly, at my side after my stem-winder opening speech when a woman came up to me and said, "Great speech, Jamie. Loved your speech. But one thing—take out all the stuff you have in there about gay marriage, because it's not going to happen. It's never going to happen. Even the gay candidates don't talk about it, and it makes you sound like you're really extreme, like you're not in the political center."

I had to swallow hard—I didn't have that many *attendees* at our sub-Arctic kickoff event, and I didn't want to offend her. But then I looked over at my brilliant eleven-year-old son, my son who was taking pictures of me during the kickoff rally. He was looking up at me so hopefully, like I was something far greater than I ever was, and I felt myself becoming a little greater at that moment, because I had this sense that I needed to answer her in a way that was befitting the confidence of this remarkable boy whom fortune and destiny had bestowed upon us.

And I said to her: "Thank you so much for saying that to me, because it makes me realize that it is not my ambition to be in the political center, which blows around with the wind. It is my ambition to be in the *moral center* and to bring the political center to *us*. That is why I'm a progressive; that is why I'm a Democrat."

After we launched that campaign, there was an article in a local newspaper quoting a pundit who described my chances of victory as "impossible"; and nine months later, when we got 67 percent of the vote, there was another article, in the *Washington Post*, quoting a pundit who said my victory was "inevitable." So we went from impossible to inevitable in nine months because the pundits are never wrong, but as I told Tommy, we showed that nothing in politics is impossible, and nothing in politics is inevitable. It is all just *possible*, through the democratic arts of education, organizing, and mobilizing people for change. Tommy must have heard me tell that story a thousand times, but I know he internalized its message from the moment it happened.

Tommy was central to my experience of that first campaign. He helped me create the model that has defined all my campaigns, both the tough campaigns and the easy ones. I have always believed in the grassroots, person-to-person model of campaigning and have veered away from pollsters, TV ads, radio ads, focus groups, high-priced consultants, and negative campaigning—the kind of whole-sale boilerplate campaign tactics that give me a headache.

To me, politics is, at its best, about *education*: educating people about the process, about the issues, about strategies for political change and policy breakthrough. Education about state government was important in our race because so many people in the Maryland suburbs of Washington read the *Washington Post*, watched the national news, worked in federal government or public policy, and focused on federal issues, not state and local ones. So we had the opportunity to teach people about how so many of the problems they were focused on (climate change and air and water pollution; education; the death penalty and criminal justice reform; marriage equality and the rights of people in the LGBTQ community) were profoundly affected by what our state government did.

But how could we compete against a long-term incumbent backed by machine politics, big money, and major organizational endorsements? I knew we would need to create an exciting campaign to move young people and then get them to convince their parents and grandparents to get involved; I had this deep instinct because of a book written by my hero Bob Moses, the philosopher-activist who helped turn the Student Nonviolent Coordinating Committee in the early 1960s into an historic force for sweeping change in Mississippi and throughout the South. Moses later launched the Algebra Project, which has made math literacy a central priority for the civil rights movement in the twenty-first century.

In his book *Radical Equations: Math Literacy and Civil Rights* (coauthored with Charles Cobb), Moses asks the question "How do you organize?" His answer is: you bounce a ball.

You bounce a ball, and some little kids come by to play, and then some bigger kids arrive, and then some high school and college kids, and you begin to talk issues with their parents, and then you organize.

Bob's bouncing ball and remarkable human-scale organizing led to Freedom Summer; the Mississippi Freedom Democratic Party, the great challenge to Dixiecrat politics at the 1964 Democratic National Convention in Atlantic City, New Jersey; the Civil Rights Act of 1964; the Voting Rights Act of 1965; and the transformation of the Democratic Party. The racist backlash to SNCC and the changes it brought spilled a lot of blood, beginning with the murders of James Chaney, Andrew Goodman, and Michael Schwerner and of Medgar Evers and countless others who were sacrificed on the altar of violent white supremacy.

In our campaign, I told Tommy, we already had lots of adult volunteers, but we needed to recruit young people, too.

We needed to bounce a ball.

So Tommy set about to do just that, beginning with his sisters and many cousins. We had to make it fun. We invited all the young people to come to our free events, which were definitely going way outside the Machine, starting with a "Poets and Writers for Raskin" event that drew more than three hundred people in a snowstorm to hear from George Pelecanos, Ethelbert Miller, Judith Viorst, and dozens of others; a rally against the death penalty, with an innocent man freed from death row; a Texas barbecue and square dance with Texas populist Jim Hightower; and a rock concert with our beloved friend Dar Williams.

The key moment for young people was going to be in the summertime, when school was out, right before the September 6 primary. Tommy reported to me that he was having a hard time recruiting older cousins, the generation of teens like Emily, Zachary, and Maggie, and even my younger sister, Eden, to spend the (brutally hot) summer knocking on doors, making phone calls, and organizing events. The problem was, he told me, that "they say 'What are they going to put on their résumé, that they worked for their uncle's losing state senate campaign'?"

I could see the problem he was describing, and said, "Tombo, tell them, first, we're not losing, and second, that they're not really working on my state senate campaign. Tell them they are a fellow in . . . Democracy Summer!"

"What's that?" he said, kind of smiling and bemused.

"We are going to recruit dozens of Democracy Summer fellows, who will get an intensive education from some of the greatest political thinkers, leaders, activists, and campaign managers in America, who live right here in our community and support our campaign. These student fellows are going to become experts on the issues of our times and learn state-of-the-art skills in how to win campaigns—and then we are going to unleash them throughout District Twenty. And I can write a college reference letter for anyone participating in Democracy Summer."

Thus was Democracy Summer born. Our campaign knocked on more than 35,000 doors, and I knocked on 17,000 personally. I still meet people who tell me that I knocked on their door during that first campaign and that they gave me lemonade, or that we threw a football around with the kids.

We signed up more than sixty Democracy Summer fellows that first summer and derived a whole new model for how to run campaigns.

Sixty in itself was a huge number of basically full-time volunteers, but the multiplier effect was unbelievable.

The fellows soon started showing up at canvasses, seminars, and picnics with brothers and sisters, friends, parents, teachers, and other kids who wanted to join. Tommy took twenty-five of my campaign T-shirts to Pine Crest Elementary School to give away, and when he ran out of those, he brought another twenty-five the next day. We were routinely welcoming hundreds of people to our events; one journalist asked me at a rally if I was running for state senate or president, because he had never seen crowds of that size in a state legislative campaign. One young supporter, Lucas Richie, son of my friends Cynthia Terrell and Rob Richie, refused to take off his "Jamie Raskin for Senate" T-shirt until primary Election Day (so we got him an extra one for the purposes of public hygiene). And Tommy learned how to organize: a brilliant young woman who went on to Howard University told me, after we lost Tommy, that he signed her up, two years after my campaign, to cochair the Obama campaign with him in school and changed the trajectory of her life.

The core of it all was constant learning and constant fun. Even the inevitable Machine backlash became a learning moment. When I won the endorsement of Blair High School's *Silver Chips*, the great student paper at the largest high school in the state, the incumbent, Ida Ruben, reacted by calling the school's principal and insisting that he threaten the student editors and reporters with suspension if they did not retract it and endorse her instead. But she picked on the wrong kids. She did not realize that they were endorsing me for my historic defense of student free-speech rights, a topic they knew something about. Pretty soon, the pages of the *Washington Post* were filled with detailed articles about the controversy and supportive editorials ("The Senator vs. Silver Chips") deploring censorship of student political voices and praising the students for their activism. Voters saw TV specials and heard NPR stories about the incident, and the students successfully stood their ground. I still have a framed copy of the endorsement—the first newspaper endorsement I ever got—hanging on my wall.

In that first Democracy Summer, Tommy participated in other key campaign activities, too, including door-to-door canvassing and helping me get ready for the first big debate of my political career. He was like a boxing coach, but one helping me hone my strong arguments on marriage equality, the death penalty, the minimum wage increase, climate change, the state's renewable energy portfolio, and decriminalizing marijuana. He also helped me get ready for what we thought the attacks would be.

Even as a young boy, Tommy was a gifted comedian, prankster, and mimic, and I hope the sadness of his death will never obscure the fact that he was just about the funniest person you could ever hope to meet. One of the first sayings I remember him coming up with, at age five—and repeating incessantly thereafter, pronouncing it like a wisecracking vaudeville comedian after he'd performed a trick or made you laugh at one of his jokes—was *Gotcha on that one! Never to be too sure*, which was invariably followed by a resonant and knowing tongue click against his cheek. Sarah and I had no idea where this comical stage banter came from, and we'd rack our brains trying to figure out how he had invented it. But he deployed it constantly, as nothing pleased him more than being a madcap little trickster who made people laugh hysterically.

During the campaign, he deployed his humor to aid my debate prep and to relax me. The incumbent had been spreading comically false rumors to the press and voters that I had never moved back home to Maryland after law school and was still living in Cambridge, Massachusetts, which was a feeble attempt to make me seem too young. (I was forty-two and had been teaching law school for sixteen years!) Tommy offered me a number of funny rebuttal lines I never got to use, including, "I actually did move down here, but you're confused because my identical twin brother stayed up in Cambridge." He also worked out this retort, to be used if the incumbent claimed I was running because of her age, something our campaign assiduously never mentioned: "I'm not running because the incumbent is getting old. I'm running because *I'm* getting old."

Tommy also took pains to prepare me for the near certainty that the incumbent would brag about the county executive's downtown Silver Spring revitalization project. He showed me a clip of her holding a groundbreaking shovel and saying, "Now you can *shop* until you *drop* and you can *eat* until you're *full*." I will never forget Tommy saying, "Thank you, thank you very much, we really didn't have the ability to eat until we were full before. Thank you for granting us that experience, Senator." The humor of that line at age eleven—and at any age—is just overwhelming, although there was clearly nothing there I could appropriately use in the debate. In fact, Tommy modeled for me and Sarah and his sisters a refreshing generosity of spirit about my opponent; he made sure no one ever uttered a negative public word about her and that the only points of contrast were about issues, never personality.

In 2006, Tommy introduced me when I announced I was running, he campaigned with me on pretty much a daily basis, he helped me at every turn bring Democracy Summer into existence, he came to all my big events, and he was by my side on Election Day and Election Night. He inspired me and made everything funny.

The lessons we learned through Democracy Summer in 2006 became the basis for my public service and the building blocks for my state senate reelection campaigns in 2010 and 2014. Democracy Summer was the center of gravity in my first campaign for Congress in 2016, a nine-way primary fight against a billionaire CEO, a local television celebrity, and six other well-qualified candidates, a race that became the most expensive congressional primary in the history of the United States. We were outspent 9–1, but we had Democracy Summer roaring ahead like a freight train and, by that time, a real legislative record to run on, because we had gotten done in Annapolis everything I had said we were going to do: We abolished the death penalty after centuries. We became the first state south of the Mason-Dixon line to pass marriage equality and then defended it at the polls. We banned the sale of military-style assault weapons and imposed a universal violent criminal background check on all gun purchases. We increased the minimum wage. And we decriminalized marijuana and passed a medical cannabis program. My campaign slogan was "Jamie Raskin, the Effective Progressive."

Once, when I was in the state senate and Tommy wanted to get Chinese food for dinner and I made spaghetti instead, he said I was always making spaghetti, but I held my ground. In response, he told me to give him the phone number of David Moon, my campaign manager, or Marlana Valdez, my campaign chair, because he needed to get in touch with a manufacturer of campaign stuff.

Why? I said.

"Because I want to make a bumper sticker that says, 'Don't Blame Me, I Voted for Ida.'"

I don't know precisely when Tommy came to identify as antiwar, but it was very young. Once it dawned on him that violence was the essence of war, it overthrew his early naïve fascination with military history. I associate this change with his learning the basic facts of the bombings of Hiroshima and Nagasaki and then about the horrors of the Holocaust. These historical events shook him and kept him up at night, sometimes giving him dreadful nightmares, and when he learned about genocidal violence against Native Americans, he stopped playing war and battle games altogether. After he learned what war actually was, and how it destroyed people's lives and communities, his fascination with it ended—or, to be more precise, the *character* of his fascination changed. What he cared about was *stopping* wars and blocking militarism as a cultural and political obsession.

The key problem in Tommy's mind was official state violence, a constant danger to peace and human rights. When George W. Bush ordered the invasion of Iraq in 2003, I was going back and forth over whether to march against the war—it was a question of pure laziness and not wanting to march alone that made me hesitate. But then, a smiling, irrepressible Tommy volunteered to go with me—and that clinched it, because Tommy, of course, for as long as anyone could remember, made everything he did magical. We marched together, my eight-year-old boy and I, down Constitution Avenue, taking turns carrying a sign with a nice Orwellian touch: "Bush's Lies: Weapons of Mass Distraction." Tommy wondered aloud whether President George W. Bush was going to take pictures of us and give them to the police, but I assured him that nothing like that could happen in America after the government got caught spying on the civil rights movement and the antiwar movement in the COINTELPRO scandal. (My boy was obviously already wiser and better grounded in the reality of things than I was.)

When Tommy was at Eastern Middle School, in the magnet program, he became concerned that the magnet students were being set apart from the other students, and this was causing a lot of social tension. He resolved to break out of these arrangements. He'd wander the cafeteria and have lunch with different groups of kids each day. One day, he befriended a boy from El Salvador, and they began to hang out, but when the vice principal in charge of discipline noticed this across-the-tracks relationship, he suspected that something untoward or even criminal (perhaps narcotics dealing) was taking place. We got a call telling us that Tommy had been seen hanging around with a boy whose cousin was reputedly a member of MS-13, and the boy's friends were, too. The school was not disciplining Tommy, we were told, just "warning" him. Among Tommy's numerous run-ins with "illegitimate authority," as my father might have put it, this one frustrated and upset him the most. He experienced in a middle school disciplinarian's reaction to an innocent friendship all the injuries and injustices of race, class, ethnicity, state power, and adult tyranny that he felt in his bones.

That episode reinforced Tommy's powerful libertarian impulses. When he was just in third grade, a kid Tommy knew whom we used to see walking to school was suspended for a week. On the following Monday we were walking to school and I spotted the boy across the street and said, "Look, there's Jimmy, they let him out of jail." To which Tommy replied: "You mean they let him back *into* jail."

In seventh grade, there was an arduous independent research paper requirement for kids in Tommy's magnet program. They had been studying the history of segregation and desegregation, so Tommy, who was at the time slightly obsessed with college fraternities, chose to write an original paper about the response of fraternities and sororities at the University of Maryland to the desegregation of the college and the campus in the 1950s. His research involved interviews with graduates from the period and a scrupulous review of old issues of the *Diamondback*, the college's student newspaper, and *Terrapin*, its yearbook.

Pretty much every Sunday for two months, I would drive him out to the McKeldin Library, on the College Park campus, and he would wade through old newspaper articles on microfiche and hard copies of the yearbook, taking scrupulous notes on oversize note cards. He turned up astounding things, like coed "blackface" parties in dorms, Confederate battle flags flown outside frat houses and brandished inside, portrayals of mock violence perpetrated against frats' African American female cooks in yearbook photos, and sizzling excitement over segregationist Alabama governor George Wallace's visit to campus.

Before Tommy finished working on his paper, there was one more person he wanted to interview, Thomas V. Michael "Mike" Miller, the president of the Maryland State Senate and the longest-serving state senate president both in Maryland and in American history. Tommy kept coming across Miller's name and photograph in his research because Miller was president of the College Park student body during the period in question and also head of one of the lead Greek fraternities on campus at the time of desegregation.

I knew Miller well and had an interesting relationship with this charming old-school politician. A conservative Democrat and consummate inside-moves politician with great personal magnetism, he had strongly backed the incumbent against me when I ran for Senate in District Twenty, putting huge amounts of money into her race while refusing to meet with me. When I defeated her, he was courtly and accommodating, but he fundamentally distrusted my insurgent progressive politics. Yet the Maryland State Senate is an intimate body, with only forty-seven senators, at that time, thirty-three of them Democrats; we used to meet around a single table in our caucus room, and everyone got to know everyone else well. As I saw it, the fundamental bargain was that the progressives would not try to topple Miller's impressive, decades-long reign of power in the senate and, in return, he would allow us to put on the floor all the progressive legislation he disfavored, like marriage equality, abolition of the death penalty, the ban on the sale of assault rifles, and my National Popular Vote Interstate Compact (which he actually ended up enthusiastically supporting). Mike's creative and liberal chief of staff, Vicki Gruber, with whom I was close, helped him pull off this amazing balancing act.

Mike and I shared a common love of American and Maryland history, and we bonded over that. I asked him if he would be willing to do a quick interview with Tommy, and he said sure. I brought Tommy to the senate with me one day, and we went to Mike's office right off the senate floor. I could tell Tommy was pretty awed by it. The office wasn't big, but like Mike, it exuded Maryland: maps of the Chesapeake Bay, hundreds of books about our state, paintings, photos with governors and presidents at Miller's home in Southern Maryland. At this point, Tommy had been toying with the idea of running for governor of Maryland one day, and I could see he was moved by the immense charm of Mike's abundant collection of Maryland artifacts.

Then the interview began. The fourteen-year-old boy on a mission to find out what life on campus was like after *Brown v. Board of Education* was no match for the political mastermind Miller, who answered any tough question with generalities like "It was just a very, very tough time." I'm not sure Tommy had any desire to pose aggressive follow-up questions, but he was unable to anyway, because he could not write fast enough to capture everything Miller was saying. Tommy did not find out whether the white fraternities had ever discussed admitting African Americans, or whether there were any disagreements over civil rights within the leadership of these Greek organizations. He never got to ask a lot of the probing excellent questions he had prepared.

On our way home on Route 50 that evening, Tommy was subdued, until he started asking questions about Mike Miller. I explained that Mike's family went back hundreds of years in Southern Maryland, with one side sticking with the Union and the other with the Confederacy, as Mike told it. Part of Miller's father's side was reputedly involved in Confederate espionage operations against the Union, keeping close track of movements in and out of the federal city, just fifteen or twenty miles away. One day, I had lunch with Mike at his office in Clinton, Maryland, and he showed me down the street the "Suratt house," where Mary Surratt lived before she became the first woman ever hanged by the U.S. government—in her case, for participating in the conspiracy to assassinate Abraham Lincoln. On the three occasions I visited the Miller estate in Southern Maryland, I always felt the weight of William Faulkner's observation "The past isn't dead. It isn't even past."

Tommy said, "I don't think I could do what you do."

I asked him what he meant.

"Not being a law professor," he said; "maybe I could do that, but I mean being a state senator. I'm not sure I could work with people who I don't agree with on, like, basic values, even if I like them as people. It would be too weird for me."

He wasn't saying it to judge me or even to judge Mike Miller. In fact, he hurried to quote back to me something I had told him I learned from my then-colleague Sen. Jim Rosapepe, who had said to me on the Senate floor my first day of holding public office, "Remember, in politics no one is ever as good as they look or as bad as they look."

Tommy was discovering something fundamental about himself. His epiphany made me feel a lot like a *politician*, which is something I rarely feel like—I usually feel more like a professor, which is what my colleagues in the House tend to call me. But his perception of things was piercingly accurate and, compared to him, I may as well have been Mayor Richard Daley. My son and I shared all the most fundamental and essential values, but he was on a different path from his father's; he was on a path of engagement much closer to that of his grandfather, my dad, Marcus Goodman Raskin, not a politician but a public intellectual.

Tommy's bond with my father grew over the years into an intimate channel of political thought and emotion between lively kindred spirits, two ebullient humanitarians who had profound libertarian questions about society and radical moral answers for the world. Each was, in some deep ways, the other's alter ego. After we lost my father on Christmas Eve 2017, I am certain that Tommy also became to me the closest continuing philosophical voice of my father. My dad, whom the grandchildren called Baba, lived during Tommy's childhood nearby, in Washington, DC, right off Sixteenth Street, with my stepmother, Lynn, another loving force of nature whom Tommy adored.

In the late 1950s, when my dad, a Juilliard-trained piano prodigy, chose a life in politics and philosophy over a life in concert piano, my parents moved to Washington, DC, where my dad went to work on Capitol Hill. After John F. Kennedy was elected president in 1960, my dad went to the White House to work at the National Security Council as McGeorge Bundy's special assistant on national security and disarmament. On my dad's first day of work in the Old Executive Office Building, the Bay of Pigs Invasion took place, and he quickly proclaimed his antiwar position, writing a memo to President Kennedy urging that the U.S. military base at Guantanamo Bay be closed, turned into a health clinic, and gifted to the Cuban people as a goodwill gesture in the wake of the failed and bloody invasion.

With his intensifying opposition to a nuclear arms race—as a White House staff member, he, remarkably, joined a picket line outside it during the Cuban Missile Crisis to address protesters from the National Committee for a Sane Nuclear Policy, or SANE—and his aggressive skepticism about the brewing intervention in Vietnam, Dad's days were numbered in the ranks of "the Best and the Brightest," as David Halberstam labeled, with a tinge of sarcasm, the coterie of Cold War intellectuals who brought us the nightmare of the Vietnam War. The iconoclast whom Bundy called a "young menace" in conversations with President Kennedy, Marcus Raskin soon experienced a demotion and "political excommunication" to the Bureau of the Budget as the "education advisor."

Shortly after I was born (on December 13, 1962) and not long after the Cuban Missile Crisis, my dad—and my mom, Barbara Raskin, a journalist and novelist whom we lost in 1999, when Tommy was only four—threw caution to the wind. Dad quit the White House in 1963 and founded, with his friend Richard Barnet, what became not only the first liberal think tank in Washington but, in many ways, the first Washington policy think tank of any kind at all, the Institute for Policy Studies, which continues to thrive today and where Tommy interned in his proud summer of 2012.

Tommy was powerfully drawn not only to my dad's loving nature and absurdist sense of humor, but also to the remarkable odyssey he had traveled through music, academia, and government to take a sweeping stand against militarism and state violence. From an early age, Tommy was fascinated by my dad's passionate leadership in the movement to stop U.S. intervention in Vietnam. Tommy studied that period carefully, reading numerous books on it and taking Amherst professor Vanessa Walker's course on America and Vietnam and whatever other courses he could find on American foreign policy and Indochina.

My dad had gone to law school at Chicago for the express purpose of working out ways to dismantle what he called "the war system." In 1968, when I was five years old, he was indicted, along with Dr.

Benjamin Spock, Rev. William Sloane Coffin, Mitchell Goodman, and Michael Ferber, in the famous "Boston Five" antiwar conspiracy case for allegedly conspiring to aid and abet draft evasion. Amazingly for a conspiracy case, most of the defendants did not know one another before the trial, but had been carefully selected by prosecutors and by then Attorney General Ramsey Clark from different domains of life (including medicine, academia, the policy universe, and students) to send a chilling political message to American society about the costs of antiwar activism. The defendants faced many years in prison, a threat that became a seminal and formative memory in my life.

My dad's alleged acts were all free political speech, much like the charged misconduct of his colleagues, but he was the only one of the defendants actually acquitted by the jury in the case, a result that stunned him and led to his celebrated statement, quoted on the front page of the *New York Times*, that he felt like "demanding a retrial." The convictions of the other defendants were reversed on appeal, and none of them was ever retried by the government or by Attorney General Clark, who never quite lived down the shame of having brought this prosecution in the first place. My dad intensified his activism against the Vietnam War, publishing more books and working in conjunction with Daniel Ellsberg to secure publication of the Pentagon Papers. All of this led to President Richard Nixon's placing him and Richard Barnet on his infamous Enemies List.

Tommy could not get enough of the Boston Five trial (or the "Spock case"). My dad's effort to mobilize both law and public sentiment against violence and war, the idea that criminal defendants would stand together in the face of unjust prosecution (instead of rushing out to get their own lawyers and pointing fingers at one another, like the right-wingers do), and the collision between freedom of political expression and the criminal law would enthrall Tommy from high school through law school.

Tommy saw in his Baba the promise of a public intellectual's career defined by old-fashioned integrity and courage, a model for acting justly in an unjust world. He was drawn at a young age to

Oxfam and Amnesty International and sent them large chunks of whatever money he earned. In college, he wrote blistering essays for antiwar.com on the folly and lies of America's warmakers. He did college summer internships and jobs at the Cato Institute, where he worked with Doug Bandow researching American military entanglements with authoritarian regimes; at J Street, where he worked under Hadar Susskind for peace, human rights, and security for all the people of the Middle East; and the Institute for Policy Studies, where he worked with writer and activist Phyllis Bennis on stopping the Saudi war in Yemen. This last assignment was a special joy for him, as my dad was still alive then, and grandfather and grandson got to see each other at work. My dad constantly bragged to colleagues at the institute about Tommy's impassioned writing and activism.

Tommy felt at home in the world of think tanks, and he experimented with the idea of straightaway becoming a writer and critic; indeed, he became obsessed with the academic pedigree of public intellectuals and writers, wondering just how early one could get off the meritocracy escalator, stop one's advanced schooling, and still be taken seriously as an essayist and intellectual. Before we lost him, we were convinced that he would add a philosophy degree to his JD studies and then spend his career as a professor of law and/or philosophy. But he also could just as well have graduated from law school and become an activist public intellectual, following in my dad's post–law school footsteps and never practicing law. Tommy was born to be a provocative public thinker and a humanitarian activist. As our relative Bob Bergen, who lost his own son, Ben, at a tragically young age, put it, "That young man Tommy Raskin was a game changer."

Tommy's compassion for other people became strikingly visible in high school. He spent many hours a week tutoring fellow students in math and English, using his explanatory and comedic powers to help them understand the material. One day he decided to start mobilizing his fellow magnet students to do the same through his beloved Bliss Project. The hard feelings generated by having divergent and unequal academic pathways in the same building—magnet and non-magnet—were always difficult for Tommy to accept. Tutoring his peers, and befriending them in the process, were the best ways he came up with to resolve some of the tensions created when an advanced magnet program exists at the center of a large public high school, where many students speak English as a second language, others have learning disabilities, and many have not yet connected with an academic purpose.

But issues of class inequality were never far from Tommy's mind. One time, we dropped off one of Tommy's friends at home after school but his friend did not want us to drive up to his apartment building, instructing us to leave him at the corner a block away. This was the second time it had happened, and Tommy spent the rest of the ride home talking about how he was convinced his friend did not want us to see his building and that the "weight of poverty" was crushing for many of his classmates.

As a world-class empath, Tommy felt other people's pain in a visceral and physical way, which led to his decision at the beginning of his senior year at Amherst to become a vegan. He came to identify the callous indifference to human life and suffering practiced in warfare with our callous indifference to animal life and suffering. For him, as he told me one night, the ruthless "slaughter bench" of history in war, as Hegel called it, was made possible by the shady animal slaughterhouse on the outskirts of town. He was convinced that by numbing ourselves to the agony and suffering of animals, we conditioned ourselves to accept the brutalization of human beings by means of war, torture, political oppression, labor exploitation, and ordinary social cruelty. We became part of a thoroughgoing societal killing machine.

He wrote a remarkable poem, "Where War Begins," to explain to friends and family about his decision to go vegan, and he took it on the road. Deploying his photographic recall to recite his poem flawlessly, he performed it widely whenever asked, astonishing his large audiences with his breadth of historical knowledge, versatility of expression, and intensity of emotion about faraway places and events.

With this poem, Tommy converted dozens of family members and friends, maybe hundreds, during his life. He was never a sanctimonious, guilt-tripping vegan; he admitted he loved a lot of meat dishes that he had to give up. If he ever saw other vegans lecturing meat eaters, he would gently intervene to point out that he, like most other vegans, had eaten meat most of his life and that it was best to avoid vegan sectarianism and one-upmanship. "I'm not interested in a vegan club," he would say. "I want a vegan world."

His gentle solicitude for other beings in the world around him also expressed itself in his unusual refusal to drive or even get his driver's license. Both Hannah and Tabitha could not wait to get their learner's permits and their licenses. When Tommy passed his learner's permit test right away in his junior year of high school, both Sarah and I expected that he, too, would soon become a good, responsible driver. He had a year within which to take the road test, and I drove with him several times and found him to be solid and conscientious behind the wheel—perhaps a bit overscrupulous, but that's a common and not undesirable trait among new drivers. Yet, in his senior year of high school, he never seemed ready to take his road test. And then, one day, we were out driving and I asked him if he wanted to take the wheel, and so he did. (Under state law, young people with a learner's permit can drive when a parent or another licensed family member is present.) But he seemed stiff and uncomfortable. We switched back after only five minutes or so, and then, on the side of the road, he told me he didn't really want to get his license. I asked him why not.

"I just don't want to hurt anybody," he said.

I assured him he would not do that and that he was a fine driver and would never drive recklessly.

He sort of tilted his head and lightly shrugged. Something about the gesture struck me hard.

"It's definitely a lot of responsibility," I told him, rubbing his shoulder. "But you are a very responsible guy, and you're a good driver."

"People get into accidents," he said. "I just don't want to do that."

He looked stoical and a touch melancholy. Something about his tone indicated that he had given this a lot of thought. I did not know what to say, and we drove home in silence.

I felt increasingly responsible and guilty about it. In Annapolis, I was on the Maryland Senate Judicial Proceedings Committee, which oversaw all things motor vehicle, and I had often brought home debates relating to driver's licenses and learner's permits, and horror stories about teen drunk driving and tragic prom nights. I remember once telling the kids about a bad accident caused by an octogenarian driver and subsequent proposals to require senior motorists in Maryland to have periodic road tests starting at age seventy-five, provoking a lively dispute that lasted way past dinner. I was also leading the fight in the state senate to pass "Noah's Law," which would compel the installation of ignition interlock devices in the cars of convicted drunk drivers. The law was named after Noah Leotta, a public-spirited young Montgomery County police officer whose life was taken by a speeding drunk driver in Rockville. The issue of driving and road safety was much on the minds of adolescents in our house, and I feared that I had somehow scared Tommy too much as I tried to implant a conviction in our kids never to drink and drive.

Tommy never did get a driver's license. He became a great walker, public transit rider, and early Lyft user. But it seemed as if the existential weight of his decision grew over time, as if a powerful desire not to risk hurting other people required a categorical refusal to enjoy any of the benefits of driving oneself around in our auto-shaped society. There would be no middle path for Tommy. But this refusal, like his extraordinary kindness toward other people and animals in general, struck us as a carefully considered, rational decision—perhaps idiosyncratic but far from unique these days—of a charming and incandescently good boy. Hannah has recently tried to convince me that not driving can be a symptom of OCD as a lot of people who suffer the disorder are freaked out about the possibility of hitting a person or animal and sometimes even imagine that they did so if there is any strange sound on the road. In any event, because Tommy was Tommy, there was never a shortage of friends or girlfriends ready to pick him up when he had somewhere to go.

When I decided to run for Congress in 2016, the "kids" were no longer kids. Hannah was twenty-four and in San Francisco in a training program at Silicon Valley Bank. Tommy was twenty-one, in his junior year at Amherst College, and developing a thesis on the intellectual history of the animal rights movement. Tabitha was nineteen and in her freshman year at Amherst, zeroing in on psychology as a field of interest and about to take her sophomore year abroad in Amsterdam.

As someone who had organized his early political career and campaigns around his children, I found it hard to have them far from home while I was actively campaigning. Democracy Summer at least gave me a way to maintain my interaction with young people and also to keep the Raskin kids engaged with what was happening. But, for me, it was just not the same as having them there, and I told the Democracy Summer fellows stories about them incessantly.

Having some distance from home meant that our children were more focused on the presidential campaign than they were on my congressional ambitions, and at least Tommy and Tabitha were feeling scared and anxious about Donald Trump and the plans he had for America. Tabitha, who had a baseball-playing boyfriend in this period and spent a lot of time with young white males, was warning everyone that Trump was definitely going to win. Tommy saw Trump's stereotyping and scapegoating of immigrants and African Americans as a precursor to fascist policies. From the west coast,

Hannah and Hank did not see things as so dire, and in fact Hank had bet someone that Hillary was going to win.

Although I was overwhelmingly focused on my own race, and therefore not as attuned as they were to the dynamics of the presidential contest, I thought, in my foggy rose-colored-glasses way, that most Americans would recognize Trump as a fraud and a con man, an unstable narcissist, and someone completely unsuited for the most important office in the world. Of course, a majority did recognize that—Hillary beat him by more than three million votes nationally—but his minority of votes were distributed in just the right way for him to eke out a win in the Electoral College.

On November 3, Election Night, the whole family had come with me to the old George Meany Center for Labor Studies, at the National Labor College in Silver Spring, where hundreds of fired-up Democrats joined me, Anthony Brown, and Chris Van Hollen, to celebrate what we hoped was going to be a clean sweep across America. But just minutes after my race was called with a projected 60 percent win (over a Trumpified Republican), and I took the stage to talk about all the important things we were going to do in Congress (such as confront climate change and economic inequality), the networks began to project a Trump victory. It turned into a long, troubled night that ended with a Republican-led House, a Republican-led Senate, and Donald Trump headed for the White House.

On January 20, 2017, the day Trump was sworn in, a few weeks after members of Congress, I could not bring myself to attend his inaugural address. I joined the boycott that was initiated by my late colleague John Lewis and backed by nearly seventy House Democrats. Instead, I invited up to one hundred constituents to join me for a hike from Maryland to DC, through our beloved Rock Creek Park, beginning at the famous Boundary Bridge, in the Eighth District, connecting Maryland to DC. My special guest was Melanie Choukas-Bradley, my friend and constituent and the exquisite biographer of Rock Creek Park. Melanie enchanted us with her lush and microscopic description of the flora and fauna in the park, but also with her rendering of the park as a triumph both for nineteenth-century environmentalism and twentieth-century New Deal democracy.

Our family enjoyed a special feeling on that long hike. Washington now had a strange, quasi-militarized, and occupied quality to it, but we were among friends. As Donald Trump denounced "American carnage" and fantasized that there were millions of people in his inaugural crowd who were never present, we Free Staters walked together as pro-democracy activists hanging tough for the people and the land. When Melanie told us the story of a favorite towering oak tree more than 350 years old, I asked her how trees could survive centuries through hurricanes and thunderstorms, erosion and heat, and every manner of insect and predator, and she said, "It's the part you don't see. It's the roots, which are intertwined so deeply with the roots of the other trees, that make them strong and resilient." When I spoke to the group, I told them we would have to be like those trees and have our roots in the community so strong and intertwined that none of us could be blown over no matter what storms were headed our way.

Tommy's response to Trump's ascent to power was fascinating. Donald Trump was clearly as close to fascism as Tommy ever wanted America to get, and he found it hard to look at him on the screen or even to hear about the things he was doing, certainly not without chanting "He's not the legitimate leader" à la Sacha Baron Cohen in the 2012 film *The Dictator*. Tommy was appalled and crushed by the separation of migrant parents and children at the border, which he immediately declared a human rights violation, and he railed against Trump's racist Muslim ban executive order.

At the same time, though, I heard him patiently but aggressively challenge anyone who called Trump the "worst president in history" in his first few months in office, reminding people of George W. Bush and the lies that led to the Iraq War, which destroyed hundreds of thousands of lives; and of Richard Nixon and his expansion of the Vietnam War and bombing of Cambodia, which also cost hundreds of thousands of lives. Tommy brought up the racism of Andrew Jackson and Woodrow Wilson. He took Truman to task for his decision to drop the atomic bomb on Hiroshima and Nagasaki, and he never let me forget that the revered FDR turned away ships of Jews fleeing Nazi Germany. Tommy wanted everyone to remember that Donald Trump's amorality and indifference to other people's suffering were not close to unique or unsurpassed in presidential history—and Tommy knew his history cold.

Tommy saw violent authoritarianism early on in the Trump movement, but he saw it also in war and other routine exercises of state power. Indeed, he often told me that to a battered wife or an abused child, a man's reign of violent terror in the home *was* fascism. The dropping of an atomic bomb on civilians, the smashing of a nonviolent labor strike, and the subjection of women and girls to genital mutilation were all forms of violent cruelty that should be intolerable to movements that prize freedom.

Tommy was proud of me, I think, in those early years of Trump, for standing as strong as I could, while on the House floor, for political democracy, human rights, and the environment. Still, we had some lively exchanges about the progressive focus on revelations about Vladimir Putin's political and social media interventions to bolster Trump's candidacy. Tommy was skeptical of MSNBC's focus on Donald Trump and the Russian connection. We clashed multiple times on this point. He did not doubt the reality of any of Putin's cyber-machinations to aid Trump, but he wondered how effective they were and, more specifically, whether the growing anti-Putin political rhetoric on the left was setting America up for another Cold War and dangerous nuclear arms race. I tried to convince him that the strongest anti-Putin voices in Congress (including mine) were also the antiwar voices, and that Putin and Trump now constituted a dangerous axis of repressive politics and kleptocratic rule across the globe. For his part, Tommy argued, quite reasonably, that the anti-Putin rhetoric of Democrats risked everyone's losing sight of how organic and homegrown the Trump movement was and that the old habits of Cold War foreign policy died hard. I would say we both moved each other a little bit on the question.

Both of us knew that Donald Trump and his autocratic allies at home and abroad were bringing an authoritarian darkness down across America and all over the world. These traitors to democracy reduced the quality of life—and, indeed, the *quantity* of life, too—for millions and millions of people driven into poverty, disease, and depression. Life expectancy *in America* dropped by two years during the Trump presidency, and deaths "of despair" related to opioids, alcohol, and suicide increased substantially across the population, especially among white men.

The darkness of the era would soon come to envelop us all.

CHAPTER 2

A SEA OF TROUBLES

Forgive me, but it's hard to be human.

—TOMMY RASKIN

At the end of high school, and more seriously in college, depression found Tommy. It entered his life like a thief in the night and became an unremitting beast. The family worked to get him help and to address it, but he otherwise worked hard to keep it to himself. Even Tommy's closest friends and girlfriends knew little of his growing mental health struggles behind the scenes.

When depression arrived, it appeared first as sleeplessness and obsessive anxiety. Tommy always had a difficult relationship with sleep. His room was right next to Sarah's and mine, and we would hear him from a young age talking in his sleep. Sometimes he had nightmares and even the occasional disorienting bout of night terrors, which take nightmares to a different level of intensity. Sometimes he could not fall asleep until late and then had to wake up early, which made him sluggish in the morning and a staunch advocate in high school for any student who got into trouble for falling asleep in class and for later start times. For years, we addressed the sleep problem as a sleep problem and were able to find some natural sleep aids for him, but we did not yet see it as a symptom of a larger threat to his psyche. In college, though, it became something else, as it coupled powerfully with anxiety.

He would worry intensely about whether something he had once said long ago to a classmate in middle school had hurt the boy's feelings, or whether a remark he had made in class the prior week might have been misunderstood by classmates. He would ruminate over past relationships and whether he had said anything unkind or untoward to his girlfriends. When a girl told Tommy she had a crush on him, and he told her he was interested in someone else, he obsessed over whether he had permanently damaged her self-esteem or their friendship.

And if something really went wrong, as it did in his sophomore year in college, it would send him into a full-blown depression filled with obsessive worry and self-doubt, even though we assured him he had done nothing wrong and that no harm could come to him. During his freshman year of college, Tommy had an extraordinary exchange with the right-wing polemicist Dinesh D'Souza, who was paid thousands of dollars to come speak on campus through the Amherst College Republicans and the Amherst Political Union. In his gentlemanly and fair-minded manner, Tommy asked D'Souza whether his defense of the fairness of America's racial hierarchy had properly accounted for the massive government discrimination in distributing veterans' benefits and preferences after World War II to whites but not to African Americans and the accumulated disadvantages attached to such discrimination.

The clarity and moral power of Tommy's questions threw D'Souza for a loop, and he dodged the question. Then he struggled in vain to regain his footing and his audience, finally telling Tommy that if he really supported race-based affirmative action (which Tommy quickly pointed out he had never mentioned), he should walk over to the registrar's office and give up his seat at Amherst, an argument that was an irrelevant distraction and a complete non sequitur on multiple levels that got the audience laughing in amazement.

Tommy's calm and thoughtful response put this ad hominem attack to shame. He told D'Souza that the imputation of hypocrisy to one's adversaries doesn't address their arguments, but rather gins up a counterproductive political discourse. At eighteen years old, Tommy demolished the argument of this propagandist for Donald Trump, reputedly the best rhetorician that the right wing has to offer.

Yet, the following school year, the Amherst College Republicans wrote the dean to insist (absurdly) that Tommy be disciplined for allegedly violating their intellectual property rights by posting his exchange with D'Souza—taken from a totally *public* event—online. The subsequent protracted investigation by the dean of students— apparently meant to show that Amherst would take the grievances of Republicans seriously no matter how spurious they were— dragged on for a year. Tommy had to write long, precocious briefs about why he had violated neither copyright law nor the Republican students' rightful expectations before the dean finally accepted the obvious and dismissed the complaint. It was an outcome that Tommy had hoped would occur in the first week of the complaint's being lodged. Instead, he spent many months in emotional turmoil and was badly distracted from his studies.

It was clear to our whole family—not just to Sarah and me, but also to Hannah and Tabitha—that Tommy was wrestling with a serious health problem. We found him a good psychiatrist from Maryland, who diagnosed his depression and an overlay of obsessive-compulsive disorder, an anxiety disorder that often accompanies and aggravates depression. The combination was dangerous because obsessive and intrusive thoughts can cause a depressed person to replay self-destructive messages on a perpetual feedback loop.

Dr. D got Tommy medicated, which was an agonizingly imprecise process of trial and error that ended up in what we all hoped would be a livable place. Dr. D told Tommy to exercise regularly and to eat well and sleep abundantly. He warned him to avoid alcohol and drugs, which could disrupt his equilibrium and plunge him into depression. We had several family meetings with Dr. D to learn how to help Tommy develop positive routines and schedules. Tommy was fully engaged in working out a health regimen, if not necessarily thrilled to reveal so much to his parents about his personal emotional and psychological life. He was very private about the whole thing.

The development of a diagnosis and treatment plan made things substantially better for him. He resumed his academic and intellectual endeavors with his customary creative energy. Sarah and I were encouraged by where he was now, but we still came to worry about our distance from him while he was away at school, especially in his junior and senior years. There is definitely alcohol consumed at Amherst, as there is at most colleges, and we knew it would be tricky for Tommy, who had just started drinking his freshman year, to refrain entirely. He did not have a drinking "problem," as they say, but given the medication he was on and his underlying condition, *any* amount of drinking would, by definition, be a problem for him, as Dr. D had warned.

To unwind at the end of the week, Tommy would have a few drinks at a party, go to bed late, wake up slightly hungover, and then feel depressed and moody for the early part of the next week. Sometimes, after having had a few drinks, and with his inhibitions loosened, he would cavalierly override his prior decision to have only one or two and would keep on drinking. Several times, he got really drunk. A drunken escapade would lead to a dreadful cycle: he would sleep late, wake up with a hangover and a headache, and sink into depression—alcohol acts as a stimulant when you're drinking it but as a depressant when you wake up and the party's over. Hungover, he would become convinced that he had destroyed a large number of brain cells and had impaired his cognitive capacity permanently. And to prove to himself that he had not squandered all his mental power, he found a way to go online and take, for free, old standardized tests, including the SAT, LSAT, GRE, and GMAT. He took these three-hour, self-graded exams compulsively, sometimes sequentially, with a timer, and if he got a single question wrong, if he placed lower than in the top 99th percentile, he would start all over again. This he was doing *in addition* to his assigned papers, reading, and other homework. It is hard to imagine the stress and anxiety this draconian and self-imposed testing regimen caused him.

He could not stop himself, though. When this behavior first began, I did not understand the OCD nature of the pattern; I thought it was a game of sorts. I tried to argue him out of it, drawing on perceptive things he himself had said about America's testing bureaucracy and mania. Indeed, Tommy had repeatedly challenged the idea that standardized tests have anything to do with *education* because they serve so clearly as an instrument of *social control*. They teach young people nothing other than how to take tests, and treat knowledge like property or trade secrets.

But this matter was not subject to debate. It was in the realm of psychological compulsion, and he was not to be dissuaded. We urged him to talk to Dr. D about it, which he did, and he later found a talking cure therapist at Harvard to come up with other ways of overcoming these urges, including running and doing other exercise, completing crossword puzzles, playing fun games, relaxing, and hanging out with friends. He tamed the beast for the time being, before it got completely out of hand.

When Tommy applied and was accepted to the Friends Committee on National Legislation as one of their young recent college graduate fellows in 2017, he was ecstatic. He would be back home, living in Takoma Park with us, commuting to his office in Washington, and doing precisely the kind of work he wanted to do: defending innocents against war and state violence, promoting human rights, and fighting poverty and hunger. He told me that he thought being out of school would free him from a lot of the anxiety and depression he was experiencing. And so it did: at FCNL, he found a community of likeminded young people, a new girlfriend, a supervisor he adored in Kate Gould, and a great outlet for his writing and speaking on human rights and peace.

This was a time of happiness and emotional strength for Tommy. In the evenings, he brought home hilarious stories of left-wing political correctness run amok in the office and outlandish right-wing conspiracy theories championed by Republican staffers in conversation on Capitol Hill. On weekends, he performed poetry and hung out with his girlfriend at Busboys and Poets, his go-to nightspot; also, a bunch of his friends were back in town.

Although Trump had already launched his ideological rampage against immigrants and federal government workers, he had not yet inflicted the pervasive damage on the country that was coming. But progressive resistance was already building in vivid ways, on the heels of the 2017 Women's March, which Tabitha had accompanied me to, along with Sarah and our close friends Kate Bennis, Katharine Harkins, and Melissa Scanlan. My first major political act of the new administration had been boycotting Trump's inaugural address; the second was calling for my constituents to join me at the Women's March the next day, on January 21, 2017. My campaign rented out the Silver Spring Civic Center and seventeen buses to transport us all to the march. More than two thousand Marylanders showed up to meet me in Silver Spring that day, in a beautiful expression of solidarity and resistance. With my district director Kathleen Connor and her husband, Tom Cove, organizing the masses, we were, from the start, redefining my public office in Congress as an instrument of public organizing and not just as a seat for casting legislative votes.

Tommy was up at Amherst so I called him on the bus ride down to the rally and we talked about how someone needed to write the history of democratic protest in America the way he was writing, for his senior thesis, a history of the ideas of the animal rights movement. With nearly five hundred thousand people in attendance for the DC component alone (and an estimated five million in sister marches in cities across the world), the Women's March was plainly historic: massive, creative, joyful. I saw one sign that read, "Did You Remember to Turn Your Clock Back 400 Years?" Another one read, "Women Are the Wall, and Donald Trump Is Going to Pay for It," a sly allusion to Trump's ludicrous promise to "build the Wall and make Mexico pay for it." We saw a banner that read, "Grab Him by the Midterms." And every few minutes, a call-and-response chant would break out, "Show me what democracy looks like! This is what democracy looks like!"

Tommy said he was impressed by the new anti-Trump movement's having so thoroughly assimilated the ethic of nonviolence, even while Trump clearly tried to pick fights and provoke violence at every turn. It was interesting to me that Tommy had picked up on both sides of that equation. Trump was seeking to activate his most fanatical and unstable followers to do violent things, promising at large rallies, for example, to pay the legal bills of any of his supporters who beat up anti-Trump protesters. Tommy saw these violent appeals for what they were. And he was right that the peaceful self-restraint of the anti-Trump forces was a big victory for the historic movement toward social nonviolence.

Even with all kinds of *valid* questions about foreign interference in our election, no one was falsely claiming that Hillary Clinton had prevailed in the Electoral College vote, even though she carried the popular vote by more than three million. No one was leveling threats against election officials to change results. In their most daring moves, progressives knitted pink "pussy hats," wrote and sang new songs of resistance, organized to help immigrants, and joined Indivisible and other local solidarity groups to keep democracy alive and to support the victims of Trump's policies. But no one ever stormed the Capitol Building, smashed windows, or beat up Capitol

Police officers or tormented DC's Metropolitan Police. No one tried to bomb the DNC and the RNC or threatened to assassinate public officials. Indeed, the Women's March itself came on January 21, the day *after* Trump's inauguration.

Being anchored in this rapidly growing movement for democratic nonviolence—which is considered both the *means* of progress and the *end* of progress—helped Tommy stay in a good place emotionally through the end of college. But that summer, the sickening violence of the "Unite the Right" rally on August 11–12 traumatized the people of Charlottesville, Virginia, and stunned America. Hundreds of unreconstructed Klansmen, white supremacists, neo-Confederates, and neo-Nazis marched on a local synagogue chanting, "Jews will not replace us," invaded the University of Virginia campus carrying torches, killed counterprotester Heather Heyer, and wounded dozens of local people who came to defend their besieged fair city.

For Tommy and everyone in our family, the Charlottesville attack, which we all watched in horror on the news, and President Trump's immediate telling equivocation over it—"There were very fine people on both sides"—shook us badly. My sister Erika, a writer and novelist, and my brother-in-law Keith, an anesthesiologist and professor at the University of Virginia School of Medicine, have lived in Charlottesville for decades, not far from where the violence took place that day. There they raised their children—our beloved Emily, Zachary, and Maggie—minutes away from where these Trump-era fascists marched under the Nazi swastika and the Confederate battle flag. That great city was where we launched a Democracy Summer chapter in 2018 to help the Democrats turn Virginia blue, and it was where dozens of people in our family gathered every year for Thanksgiving. While watching the news, I recognized every spot in downtown Charlottesville that had been struck by this terrorist violence. As Tommy said, as we watched the TV coverage, "It's like they just did this at Auntie Erika's Thanksgiving *dinner table*." Indeed, Erika and Keith joined hundreds of people in the counterprotest against racism and antisemitism and were tear-gassed in the process.

I was afraid the Charlottesville events would haunt Tommy into a dark despondency because of the way historical episodes of fascist violence usually affected him. Yet he seemed to recover quickly when he started his fall fellowship soon thereafter at the Friends Committee on National Legislation. There, he found his professional community exciting and his work on behalf of nonviolent foreign policy profoundly fulfilling.

At FCNL, Tommy lobbied for an end to the war in Yemen and to U.S. complicity with Saudi Arabia's violence against the Yemeni civilian population. He advocated for regional peace and social justice in the Middle East and for a sharp reduction in our bloated military budget. He gave stirring lectures and produced powerful writings on U.S. foreign policy.

It was a pity when Tommy's fellowship came to an end, but FCNL had a hard-and-fast rule that the one-year fellowship ended with no extensions. A good team player, Tommy helped recruit and educate the next class of fellows, and then he chose to take one more year to do independent writing and creative work before applying and going to law school. He produced some extraordinary poetry in that time and some remarkable essays elaborating a favorite theme of his during the Trump years: that Donald Trump was indeed the nightmare that liberals said he was, but that he embodied a deep social arrogance and tribal obliviousness that infects American politics generally, including even, sometimes, American liberals and progressives. He wrote that "Americans of Trump's ilk are not the only ones responsible for lethal inequality," and he wondered whether Trump's cartoonish selfishness allowed liberals to let themselves off the hook too easily. He called for a broader moral and political vision that included the rest of the world, a vision he had been trying to live himself by giving away half his modest salary to groups that handed direct monetary aid to poor people around the world.

"We may object to Trump's condescending rhetoric and galling outbursts about so-called 'shithole' countries, but at the end of the day, we share in his general indifference to the preventable horrors that befall many inhabitants of those countries," he wrote in one essay. "Following our lead, Elizabeth Warren, Bernie Sanders, and other leftist heroes highlight the struggles of American students and low-wage workers—as they should—while saying very little about the millions of afflicted foreigners who will die this year in the absence of wealth transfers from the West."

Tommy simply refused to take the easy way out by equating Trump with unprecedented presidential evil and his progressive critics with moral virtue and political perfection. The world was crying out for redress of savage inequalities and demanding a redirection of resources away from militarism and toward addressing the civilizational emergency of climate change. To be an effective counterweight to homegrown authoritarianism, we should be acting with comprehensive vision on these questions right now, he argued. Opposition to Trumpism was essential, Tommy would say, but smug derision toward Trump was complacent and a political fool's gold.

The radical freshness of Tommy's writings reflected not just his intellectual vitality and moral convictions, but also his growing sense of well-being and emotional self-confidence. Although he was exploring graduate school programs in both philosophy and history, he decided to apply to several law schools in the fall of 2018 and was admitted to Harvard Law School in the spring of 2019 for the fall. By this time, he had a lovely and delightfully earnest girlfriend, a fellow Amherst grad, whom he had gently converted to veganism. The relationship was serious enough for them to discuss what Tommy's moving up to Cambridge might mean for the future. She was going to stay in Washington, but they chose to remain involved on a long-distance basis and to visit each other every other weekend.

However, beginning in the fall of 2019, law school turned out to be a time of some heartbreak and turmoil for Tommy. His intellectual originality and ethical passion erupted in full view of his classmates and professors, and he became a social force in his part of the Harvard Law School community, but his private mental health struggle also returned with a vengeance. The immediate trigger was an encounter with an instructor, an upper-level student, who admonished Tommy, in a private meeting, for not giving his pass-fail class enough attention.

Most students might have been able to roll their eyes and absorb the shock of such a dressing down by a fellow student, but Tommy experienced it as a huge psychological blow. Until then, he was loving law school and totally engaged with the material and the classes. Now, thrown badly off-course, he became convinced he was going to fail this pass-fail course, racked with scholastic self-doubt, and haunted by his old nemesis, sleeplessness.

The episode drove Tommy into a downward spiral of unself-confidence and pessimism. He began drinking to address the stress, which of course brought on depression. He was in a bad way when, one Friday night in late October, he called us drunk and confused. We were at the home of our friends Ann Shalleck and Jimmy Klein for dinner and were so alarmed by his condition that we phoned our friend Katharine Harkins, who was visiting her parents nearby in Cambridge. She went over to his apartment and gently helped him go to a hospital, where they treated him overnight for ingestion of a lot of alcohol, which had created a bad admixture with the medications in his body. Sarah flew up first thing in the morning to see him, and I joined in the afternoon after session.

With the amazing resiliency of youth, Tommy appeared bright-eyed, funny, and embarrassed over the whole incident. We spent a long time talking about the academic prelude to this episode and assured him he had done nothing wrong , which seemed to improve his mood and outlook on things. He was happy to see us and eager to get back to work in his courses.

But there were obviously deeper things going on. Although he had rallied and "reconstituted" quickly, the hospital doctor emphasized how dangerous the situation had been and raised the possibility of Tommy's withdrawing from school for the rest of the year. Tommy wouldn't hear of it. We told him that nothing was more crucial than his physical and emotional health and that he, of course, was the most important person in our lives, along with Hannah and Tabitha. He could come home and focus exclusively on his health for the year and resume his writing and intellectual work, or he could stay in school, in which case we would be checking in on him constantly. Also, we said we needed to set up a more intensive regimen with doctors. He did not want to leave school, which he thought would only isolate and depress him. He wanted to be back in his classes with his friends. He did not want to miss his first year of law school. He said he would have a new weekly psychiatrist to meet with in Cambridge, in addition to Dr. D back home.

We were beset by doubt about the right thing to do. Both Sarah and I wanted him to master this cycle of anxiety and saw the benefits of his being back home. At the same time, Tommy was twenty-three now, obviously a mature and intelligent adult, and he wanted to be in school with his peers, learning and writing and growing, and his living context was obviously closely connected to his emotional and mental health. Didn't we have to respect his judgment? The legal, moral, psychological, and emotional questions all seemed intractable, uncertain, and overwhelming. In the end, he decided, and then we decided together, to have him stay at Harvard. We found him a therapist whom he could work with on the stresses and anxieties of law school and life, and we created a support team of people in Cambridge he could draw from and fall back upon.

Sarah and I both made trips up to Cambridge on weekends to be with Tommy before he came home for Thanksgiving. I remember also feeling so grateful when Hannah flew out from Nevada to spend a week at Michael and Donene's house and hang out with Tommy. The semester finished decently for him, and the spring semester started even better. We felt he had surrounded himself with the love, friendship, and psychological and emotional support networks he needed to thrive while going through law school.

And then COVID-19 hit.

• • •

The law school campus cleared out in March. Tommy went to a section good-bye party, where they all posed closely together for a large group photo, unmasked, undistanced, and unaware of how bad things would get; Tommy and a number of his classmates were soon flipped out with anxiety and racked with guilt over the photo, one of those early-days-of-the-pandemic rookie errors. He flew back to Maryland, and though we were thrilled to have him back home, we could see that the psychic alienation and literal disembodiment of Zoom computer classes, combined with the intense social isolation of those days and the mounting fear of catching and spreading the disease, quickly drained the joy from his law school experience.

Tommy hung tough. We'd stay up late watching Will Ferrell and Sacha Baron Cohen movies (*Borat Subsequent Moviefilm* comes to mind) or playing Boggle or chess. We had family breakfast and dinner every day, and often lunch, too, where we'd make Beyond Sausage subs dressed with onions and green peppers. I remember, during the period of food shortages, several nervous trips to Whole Foods in Silver Spring, where, double-masked and bundled up against the threat of transmission, Tommy and I held our breaths and darted from the breakfast cereals to the orange juice section to frozen "meats," to stockpile huge quantities of Beyond Sausage, and then talked on the car ride home about whether the government could constitutionally ban the use of the word *meat* on labels for non-livestock-sourced food, a raging constitutional issue in the animal rights community, which Tommy was talking to some potential summer employers about.

Tommy gave Sarah and me extended post-dinner lectures on whatever legal doctrines he was working on, a self-designed study method that bolstered his understanding. So much of law school is based on a case method whereby the students try to derive general lessons and rules of law from specific court decisions. This cut against Tommy's systematic nature when it came to learning and understanding. He wanted to read treatises—massive, forest-consuming treatises—even if that meant being forced to learn ten times as much material as everyone else in the class, just so he could place particular cases in an overall doctrinal and historical context. He essentially assigned himself huge mounds of additional homework and taught himself a lot of things that, I suspect as a law professor, most law professors just didn't know.

But all of it was a struggle. Like most young people in his generation, Tommy got thrown for a bad loop by COVID-19. With in-person school closed, social life was reduced to a fragile and masked minimum. Travel was a nightmare in a maze of conflicting federal, state, and local rules and guidance. Relationships were strained, forced into a premature and awkward intimacy or, more likely, into a melancholy virtual oblivion. Budding professional careers came to a screeching stop. A lot of young people faced unemployment, economic contraction, and profound uncertainty. Many were forced, like Tommy, back into their parents' basements or their attic bedrooms with their high school yearbooks and baseball mitts. Tommy expressed a constant fear of catching or having COVID-19 and unwittingly infecting his Grammy, Arlene Bloom, Sarah's mom, with whom he was very close and who preferred not to wear a mask around family members. We spent ridiculous amounts of time deciding when masks needed to be worn and for how long, and where, when, and how to get tested. COVID-19 made even the most informal social occasions among family a matter of profound anxiety and endless planning, analysis, and second-guessing. The whole thing was exhausting and demoralizing.

I was spending most of my time as a congressman addressing the crises created by COVID-19. At the start, hundreds of my constituents were strewn, and stuck, all over the world as air travel ground to a halt, and my district director, Kathleen Connor, and our amazing staff of Erica Fuentes, Nina Weisbrodt, Jennie Foont, Joseph Eyong, and Christa Burton worked around the clock to get planes in the air and our people back home. No sooner did that crisis subside than we were thrust into trying to get ventilators and personal protective equipment, or PPE, for Holy Cross Hospital, Suburban Hospital, and other hospitals in my district. Then, quickly, the economic calamity of all the shutdowns overcame us, and we needed to address the impending tidal wave of unemployment insurance claims, rental evictions, mortgage foreclosures, and business collapses.

Speaker Pelosi named me to the new Select Subcommittee on Response to the Coronavirus Crisis, under the great Jim Clyburn of South Carolina, whose painstaking decision to endorse Joe Biden in the South Carolina Democratic primary pretty much cleared the field and put Biden on the glide path to victory. My legislative and district staffs worked as a team of patriots on both legislation and individual cases to address COVID-19, and I worked with Clyburn and our team to press real public health measures and accountability with regard to the large sums of money we were being forced to spend to rescue public health and our floundering state and local governments.

But the situation we faced was desperate, not just for working families but also for big and small businesses. Even GOP leaders in Congress acknowledged that, for the economy, we needed to act to pass a series of bipartisan emergency bills, beginning with the CARES Act, to get aid to hospitals and the public health care system, first responders, states and counties, public schools, restaurants and small businesses, and other institutions staggering under the weight of this plague.

But the first rule of virus economics, as everyone learned, is that if you are going to seriously bring back the economy, you have to crush the virus. What America really needed was a plan to stop the spread of the disease and extinguish it, yet nothing like a plan was coming from the executive branch. America simply had no plan.

Even with promoting the bare-bones public health message of wearing masks and maintaining social distance, the House of Representatives could barely function because of the sadistic antics of our most right-wing members, who amazingly carried on with proud indignation rather than abject shame and guilt. (Rousseau in his *Confessions*: "How often audacity and pride are on the side of the guilty, shame and embarrassment on the side of the innocent.") Hard-core Republicans made a point of not wearing their masks to committee meetings, apparently to model to the Fox News audience some kind of dubious statement about tyranny and freedom, prompting partisan brawls over the efficacy of masks and our duty to set a responsible example for the public.

Amazingly, the coronavirus response subcommittee, to which Speaker Pelosi appointed me on April 29, spent large chunks of our precious meeting time simply fighting about whether members of the subcommittee itself should wear masks during our hearings. This kind of surreal and slapstick schoolyard brawl spread through the committee system. Because I served on pretty much every committee with Ohio representative Jim Jordan, who was like the guy who was always assigned to cover me in a weekly pickup football game at recess, I often ended up countering his antimasking and anti–public health talking points du jour. I was never proud of getting into the rhetorical mud with him, but his desire to excuse all of Trump's offenses against the American people, always irksome, drove me up the wall during the COVID-19 crisis. I was determined to answer all his speedy polemical provocations and partisan insults with cold, hard facts and ruthless logic.

Sign of the changing times: I had once thought that Jim Jordan and I might be friends because we enjoyed intellectual sparring and had even introduced legislation together to create a press freedom shield to protect reporters from being forced to turn over their sources and notes. I had bonded with a lot of conservative political leaders in my day, both in Annapolis in the Maryland State Senate and in the U.S. House, where I had made good friends with Reps. Lamar Smith (TX) and Trey Gowdy (SC), both of whom had, alas, left Congress. Jordan and I served on most of the same committees together, and I had always thought that policy debate and intellectual exchange were richer if they were built on a platform of friendship and respect. But it became painfully clear to me—yes, even to me, of the perpetual naïve faith that friendship and community can overcome all obstacles—that Jim Jordan considered us his *enemies* in the most basic sense and that his time in Congress was going to be spent demonstrating his absolute devotion to one person alone: Donald Trump.

On our COVID-19 subcommittee, Jordan acted nimbly to cover up for what I saw as Trump's recklessness, incompetence, and mismanagement of the pandemic response. He ridiculed and heckled Dr. Fauci. He tore into the urgent public health mandates of masking and social distancing, which were pretty much all we had to protect ourselves before the vaccines came out. He ignored or minimized the massive death toll of COVID-19, and blamed the economic problems caused by the COVID public health disaster on "Democrat governors." When he went sailing off on an anti-Chinese polemic and filibuster, I would seek recognition to recount at least nineteen different occasions on which Donald Trump lavishly *praised* the beautiful performance of General Xi and the Chinese autocratic government in the first five months of the COVID-19 crisis. But from a public policy standpoint, all of it was just spinning our wheels.

Against the GOP propaganda machine, the White House, and Fox News, there was little we could do to hold back the flood of disinformation. Amazingly, the party that had invented and railed against mythical "death panels" in the Affordable Care Act was now defending Donald Trump and his monstrous mismanagement of COVID-19 that would come to cause *hundreds of thousands* of unnecessary American deaths, according to Trump's own COVID-19 point person, Dr. Deborah Birx, the White House coronavirus response coordinator. My Republican colleagues went from collaborating with Trump's sinister denial of the threat of COVID-19 to covering up for his bizarre promotion of quack medical cures (like hydroxychloroquine and injecting oneself with bleach) to the outrageous embrace of "herd immunity" as the nation's strategy, which meant just letting the disease wash over the population and kill untold numbers of vulnerable people.

Trump and his collaborators were turning the richest and most scientifically advanced nation on earth into a *failed state*, defined as a nation that cannot deliver the basic goods of existence to its own people. That would include the essential good of protection against deadly diseases. The U.S. response to COVID-19 was a historic debacle that will be studied forever as a test case for public health catastrophes, what *not* to do when a pandemic happens. At a time that called for the best and most enlightened democratic leadership in the world, we simply had the worst.

The summer of 2020 brought little relief from COVID-19, but there was one strange silver lining to the situation. Protracted social isolation gave Americans the chance to witness and then really see the continuing brutal violence wielded by certain police officers and certain police forces against African American citizens. The defining event in our household, as in millions of others across America, was the ghastly murder of George Floyd by Officer Derek Chauvin in Minneapolis on May 25, 2020.

"Dad, watch this," Tommy said. He was in our kitchen when he handed me his phone. I watched the excruciating scene of a sadistic white police officer pressing his knee on the neck of a prone, handcuffed African American man, George Floyd, until he was asphyxiated. Tommy looked heartbroken and astonished, as though his mind and heart could not assimilate the reality of so much viciousness and cruelty being densely concentrated in one man, a lawless agent of the state. He was shell-shocked. I tried to comfort him, and we talked about what we might do in response.

In the House Judiciary Committee, we went to work on the George Floyd Justice in Policing Act, legislation to ban chokeholds and strangleholds, require the use of dashboard and body cameras, establish a nationwide database of disciplined officers, forbid lethal force in all cases except to repel lethal force, and get rid of corrupt judge—made doctrines like "qualified immunity," which made law-less cops essentially untouchable. The effort was led by my friend Congresswoman Karen Bass, a progressive leader and the bill's lead sponsor; Congressman Cedric Richmond, my friend on the Judiciary Committee, who was deeply involved in the Biden campaign; and other members of the Congressional Black Caucus. In the meantime, I attended and spoke at a number of BLM protests in my district through the spring and summer, driving on long, deserted boulevards and roads from Silver Spring, Takoma Park, Kensington, and Bethesda to Glen Echo, Cabin John, Frederick, and Rockville, even stopping for some small neighborhood marches called by elementary school pupils.

BLM protesters invited me to come to Lafayette Square to stand up for the rights of the demonstrators after Donald Trump and Attorney General William Barr, on June 1, organized a cross-departmental paramilitary force and unleashed a police riot against protesters, including a lot of my constituents, complete with pepper spray and mounted officers swinging billy clubs—all for the purpose of clearing a path for Donald Trump, who had been reportedly hiding from protesters in a bunker in the White House, to walk across the street to the St. John's Episcopal Church and wave a borrowed Bible upside-down over his head. I decided to go down to Lafayette Square to argue publicly that Trump had just offended to varying degrees all six rights contained in the First Amendment. He had trampled the right of free speech, the right to assemble, the right to petition government for a redress of grievances, and the right of free press as his military force grabbed cameras from reporters and swept journalists off the street. His crashing of the St. John's Church, whose leaders denounced the sordid spectacle, aggressed against the free exercise of religion, and waving the Bible *over* his head was a typically bumbling and laughable effort to establish whatever corrupt authoritarian religion was floating around *inside* his head.

When I was about to leave for the June 4 BLM protest outside the White House, I called down the basement stairs for Tommy, to see if he wanted to come; he said yes, which made me gleeful. There was no one I ever wanted with me on political protest duty more than Tommy, who had an exceptional eye and ear for good signs, good chants, good slogans, good jokes, and good speeches. He turned everything into a learning adventure, and was obviously still outraged about the grisly killing of George Floyd.

We drove down to Lafayette Square and found lots of elaborate Trumpian military fencing under construction as COVID-safe masked crowds of nonviolent youthful protesters gathered to chant support for the sweeping reforms contained in the George Floyd Justice in Policing Act. The homemade signs were vivid: "If You Don't Like Wearing Your Mask Every Day, Try Breathing While Black," "Black Lives Matter/ Black Votes Matter," "Who Do You Call When a Cop Is the Murderer?" and "George Floyd Mattered."

We wended our way through the crowd in our masks, trying to maintain social distance as best as we could, even as we warmly greeted constituents who were excited to bump into me there. There was no organized program happening when we arrived, so we just milled around and talked to people. Tommy teased me whenever someone recognized me and asked to snap a selfie, saying that they "clearly want to photograph the oldest guy at the rally." Someone asked Tommy if he was going to follow in his father's footsteps into politics, and he smiled and said, "I might if he ever leaves office."

I seized on the moment to ask him if he might ever actually think about going into politics, a question I suppose I asked him every once in a while, which is to say, far too often. He said "maybe," knowing of course that I knew he was not necessarily cut out for a career of constant campaigning, petty bickering, and patient waiting for the right moment to make real political changes. I reminded him that when he was in high school, he was once asked to write an essay projecting himself doing something great in the future, and he imagined winning the Nobel Peace Prize for bringing peace to the Middle East, solving the decades-long struggle between the Israelis and the Palestinians, and establishing a two-state solution with a demilitarized zone administered by the United Nations.

"Well, bringing peace to the Middle East—I would definitely do that," he said.

As we walked back to the car that day, Tommy let me hold his hand for a half minute—a half-minute I won't forget—just like when he was a little boy and we walked to the park.

That was the last protest Tommy and I attended together.

That summer of COVID-19 and BLM, Tommy was a legal intern at Mercy for Animals, a leading animal rights and welfare non-profit based in California. Although he was working virtually on Zoom (from Maryland or, later, Cape Cod), he thrived and flourished in the job, getting his first taste of serious lawyering with a set of fascinating work assignments on everything from efforts by the livestock industry to ban the use of the word *meat* for non-animal-based products to corporate-backed legislation that would make it a crime to conceal parts of your identity in job interviews, which was designed to stop whistle-blowers from going undercover and working at slaughterhouses. Tommy received constant positive feedback from his wonderful boss Daina Bray, the general counsel, who was amazed at his prodigious work ethic and his ability to be an effective lawyer for every situation and problem. She told him that he was managing to complete in a week assignments that she had prepared for someone to work on for a month.

But while Tommy began to feel his oats as a real lawyer, something went wrong in his relationship with his girlfriend, the same thing, I suspect, that had gone wrong with a previous girlfriend and so many other relationships with brilliant and beautiful young women. I believe he could not handle the intensity of a good relationship with a young woman interested in marriage and the corresponding prospect of having children. He had emphatically declared himself an "anti-natalist" because he could not agree to the possibility of committing another human being to a life that would include substantial measures of pain, sadness, and suffering (and, what was unspoken here perhaps but very much in the air, depression). However much Sarah and I (and assorted uncles and aunts) tried to describe the joy of having children as we had experienced it, Tommy would not negotiate on his determination that no one has the right to impose the inevitable experience of pain on other people. The fundamental problem was that he did not want to have kids and risk their having to endure the kinds of pain and suffering he himself knew even as a young person.

In hindsight, this looks to me like some kind of a turning point in Tommy's state of mind. He had closed the door on having kids and thus marrying a woman who might want to have children, despite the fact that we had all tried to convince him that people change their minds on this fundamental question in the context of specific loving relationships. Moreover, I also maintained that his ethical argument about not choosing existence for others who will come to suffer pain, even if correct, applied only to the decision to have *biological* children, not to the decision to *adopt* children. Indeed, with adoption, the logic reverses: if one has the means to embrace, love, adopt, and take care of children, to bring them more love and comfort than they would otherwise receive in the world, to diminish the harm and suffering they will encounter in life, then doesn't that become an ethical imperative, too?

We have a lot of adopted kids in our family. Tommy knew this, of course, and he also knew that when it came to adoption, the equities of having and raising children reversed. With adoption, the prospects for dramatically *increasing* joy and happiness on earth went up. Why didn't this alter his thinking? This is a tough pill for me to swallow. I don't know it to be true, but I am afraid that Tommy had thought it through carefully—the way he thought everything through carefully—and concluded that, as a potential father, whether adoptive or biological, he did not have the constancy of psychological and emotional resources to protect his hypothetical children from the ample pain and suffering of this world. In other words, if he was overwhelmed by all of it now, living on his own (at his parents' house), he couldn't possibly protect children who were his own unlimited existential charge and responsibility. It was cold comfort to learn that huge and growing numbers of young people in Tommy's generation feel the same way about not having children. To this day, it is difficult for me to focus on this thought.

It is hard to reconstruct the exact chronology of what went wrong for Tommy in the desperate and endless months of COVID-19 in 2020, when hundreds of thousands of Americans died and everyone's lives were upended.

In the fall semester, Tommy was enrolled in a full complement of engaging classes, but he also taught a class of his own. Carol Steiker, his revered Criminal Law professor, had recommended him as a section leader and teaching assistant to Harvard College professor Michael Sandel for his famous large undergraduate class on justice. This was a delightful surprise for Tommy and a thrill given that he was a natural-born teacher and synthesizer of information. He went right to work and spent easily as much time preparing for the class he taught as he spent on all the classes he was attending *combined*.

Tommy kept his nose to the grindstone all around and, whatever his mood, always found a way to soar in class for his own students. They adored him, and he was deeply invested in their intellectual development. But because he was masked when outside and isolated when inside, his mood and his emotions plummeted with the spirits of the rest of the country. In Takoma Park—our beloved, charming, eclectic, amazingly supportive, progressive multicultural town, which normally thrives on communal life and public celebrations like the Fourth of July, Halloween, farmers' markets, and folk music and street concerts—everything ground to a grim standstill. Our lives moved indoors.

And in Tommy's case, the beast returned. He was battling the disease again, working to adjust his medication with Dr. D, feeling sometimes a little more up and sometimes a little more down, but always experiencing the draining weight and vertigo of depression. I am afraid that he felt embarrassed about this constant slog through quicksand, and he did not want to talk about it. I know that he felt exhausted by it, too.

He told me one day in the kitchen, looking tall and handsome, that he did not think he would ever be happy. I was aware of responding too quickly and perhaps giving him too many rapid-fire answers. I told him that when I had colon cancer in 2009–10, I felt oppressed and overwhelmed by it, and I could not imagine ever being healthy again. But I came back from it. I told him that when you're healthy, you cannot ever imagine being sick, and when you are sick, you can never imagine being healthy, and yet he would indeed be healthy again. And when you're happy, you can't imagine being sad or depressed, and when you're sad or depressed, you cannot imagine being happy again, but you will be. And when that did not seem to convince him too much, I told him that happiness is not so much a single condition as a complex of activities, like working out, and being with friends and family, and eating good food, and loving well and working well, as Freud said.

He smiled and didn't really answer. And I ran out of things to say. I felt like an idiot immediately, although I didn't know why, and I have felt like an idiot every day since.

I wish I had just said, "Hey, this sounds serious, kiddo. Are you thinking about hurting yourself? Are you having thoughts about suicide? Can we all meet with Dr. D?"

I failed to do that. I failed to break the downward spiral.

That is something I will just have to learn to live with.

As the days darkened and the sky froze, the plague raged out of control, and the economic and social crisis wore on day after day. Trump's lethal recklessness and incompetence in confronting the pandemic took place against the background of an already violent, hateful, and polarized culture he was bringing into existence. Four years of increasing white supremacist violence, all the sneering derision directed at immigrants, all the deranged QAnon conspiracy theories, all the odd lying and disinformation about COVID-19, all the racial and ethnic poison injected into the social media—all of it combined to take a dark toll on public mental and emotional health.

In the Trump period generally, and its last brutal year specifically, Americans reported huge jumps in anxiety, depression, and alcohol and substance abuse, especially the industry-facilitated addictive abuse of opioids, which in turn fueled a heroin addiction epidemic. In 2018, before anyone had heard of COVID-19, 69 percent of people asked already reported that the nation's future was "a significant source of stress" for them and their families; in 2020, 68 percent of Americans said that the presidential election alone was a "significant source of stress" in their lives.

The staggering public mental and emotional health crisis of the Trump years hit young people hard, and it hit people already struggling with a mental illness like a tornado. Tommy was in both groups, and he was in good company. The commanding *majority* of young Americans in their teens and twenties was now reporting symptoms of anxiety, depression, and other serious mental health problems. The *Washington Post* reported on November 23, 2020, twenty days after the election and three days before Thanksgiving— just five weeks before we lost Tommy—that the number of young people between ages eighteen and twenty-four who had seriously considered suicide in the prior thirty days had gone up from one in ten in 2018 to at least *one in four* in 2020.

Suicide has now become the second leading cause of death among young people in Tommy's age group.

The mental and emotional health crisis is a national catastrophe for the young that is exacerbated by huge deficiencies in our health care system. Theoretically, physical and mental health needs are supposed to be treated equally in our health insurance plans, but this is a promise honored in the breach, as mental health services and resources are not remotely adequate in most parts of the country.

The good news is that more and more people are talking about the psychological and emotional health crisis, and it is hard to affix to personal mental health problems a social stigma when it is affecting most people of a generation. If a *stigma* is a branding or mark on the skin in its Latin derivative, and when everywhere you turn, people are walking around scarred by it, then it comes to look a lot less like a mark of shame and a lot more like a mark of courage in the face of suffering.

The "sea of troubles" we faced in Trump's America threatened to submerge people already struggling to stay above water with mental illness or prior emotional and psychic trauma—that is, a huge part of the population. At least 25 percent of Americans, or more than eighty million people, have a chronic mental or emotional health condition, as Tommy did.

In the fall of 2020, I was many times conscious, while at dinner with Sarah and Tommy, of censoring stories on the tip of my tongue about the ordeals of immigrants or refugees that we had heard in the Judiciary Committee, because I knew how much they would upset and pain Tommy. I did not want to further darken his mood. But there was no way I could steer him clear of the barrage of news images of the COVID-19 body count, the corpses piling up outside hospitals, the suffering of children losing parents, and the anguish of parents losing children—the "thousand natural shocks" of those brutal days. I would report to him about our work on the Hill to try to address the pandemic, always providing as positive a spin as possible.

But I knew he was internalizing the sense of loss and mass suffering that had overtaken the country. How could he not? We all were.

Tommy was a regular blood donor, and as the crisis unfolded, severe shortages of blood arose, as people were afraid to go out and donate. One cold day, Tommy called a Lyft, grabbed his mask, and went out to give blood. He told us later how he thought he might faint and had to drink apple juice and eat chocolate chip cookies for about a half hour before they would let him go. What an honor it was to be the father of this young man.

I have never had depression before, but when I had colon cancer, I did have to think, quite a bit, about life. I did it all: chemo, radiation, and surgery twice, including one procedure that required Dr. Jonathan Efron and his team at the Johns Hopkins colorectal surgery unit to work on me for more than ten hours and which left me horizontal and immobile for six weeks and floating in an oxycodone fog for five months.

Never once do I remember wanting to die.

I remember Hannah leaving for Amherst and my not being able to go with Sarah to drive her there and help her unpack, and both Hannah and me crying as she left, knowing so little of what the future held for my darling daughter or for me. I remember seeing Tabitha in the middle of the night watch me through a crack in her door as I crawled to the bathroom to hug the toilet, throwing up all over the floor. I remember Tommy being crushed by a teacher who responded to a *Washington Post* article about my illness by telling him that he should expect no special favors or breaks just because his father was sick.

But I never wanted to die. On the contrary, all I could think about was *living*, living to visit Hannah in college and watch her graduate, living to see Tommy graduate from high school and watching his destiny unfold, living to let Tabitha know everything would be okay and to see her through adolescence, living to see marriage equality happen for our lesbian and gay friends, living to abolish the death penalty, living to get rid of the dangerously unfair Electoral College, living to see if maybe Sarah and I got grandkids one day, living to see another sunset at Race Point Beach, living to see us beat climate change and find a path of climate survival for future generations, living to see all my nieces and nephews grow up and get married or just have kids.

I didn't know that feeling of trapped desperation so unthinkable that it leads to a craving for death as an escape. I could not comprehend it.

Many nights I would stay up wondering whether to tell myself that mental illness took Tommy's life or he took his life in response to mental illness. It is a subtle difference that tormented me. The only exit ramp from endless trips around this track was to remember that, when we debated free will and determinism, Tommy always came down on the side of determinism, insisting that our actions are always the product of biological and environmental forces beyond the force field of free will. Ultimately, mental illness taking his life and him taking it in response to mental illness mean the same thing. He simply felt no choice in the face of such total suffering: "My illness won today."

This is not easy for his parents, his family, or his friends to accept, given how much we loved him and how much we knew him to love us back. I have been buffeted by waves of self-prosecution and self-blame over things we might have done differently. We might have insisted on more therapy or on group therapy; we might have taken over the reins from Tommy with regard to his medical appointments or his prescriptions; we might have made the decision never to leave him alone or to have frequent formal family checkups on any thoughts of self-harm. We might have aggressively inspected his room and belongings, with or without his consent—it was his room, and he was twenty-five, but then again, it was our house. Who knows? We are beset with doubt.

Each set of questions leads to another set. The usually spare tree of causal decision making and choices in life becomes, in death, a dense and dark forest of uncertainty and unclarity. The mind wants to believe that there is something that could have been done and, therefore, that there must be something that we did wrong or failed to do right. If we can replay it in our heads, freeze the frame, and make just the slightest adjustment to what we said or did at this or that point, there will be a different outcome. Tommy will return to us. A slant of mental light enters. He will show up and say, "C'mon, Jamie, let's play some Boggle." But a curtain falls down. This is a hopeless fantasy. There is no way to force a different ending in reality, so why does my mind try to force a different reality in my imagination?

For months, I would lie awake and take any and all counterfactual mental excursions that presented themselves—although, thankfully, these have subsided over time. The wonderful grief therapist I have seen on Zoom every Friday at 8 a.m. since January invites me to accept this reality: Mental illness is real, and Tommy had mental illness. His decision was not impulsive or spontaneous, but a premeditated effort, heavily influenced and controlled by obsessive thoughts, to escape an unbearable and unlivable mental agony. Tommy explained this to us cogently in his farewell note, a message I speak to myself like a prayer, and he did not blame us, but asked us for forgiveness.

It is heartbreaking that Tommy would ask us for forgiveness for the painful mental health struggle he had to live through alone, a struggle none of us ever really knew or could know from the inside, given the "problem of other minds" and so much of other people's experience of life being ultimately opaque and unavailable to us. Of course, we forgive him—*we forgive you, dear boy, wherever you are*—and I can only beg my son's posthumous forgiveness for all the missteps, miscues, and missed opportunities I was responsible for as a father in the course of his incandescent and eternally valuable life.

Of all the many retroactive opportunities for self-blame I have pursued, the one that haunts me most is not that I failed to ask how Tommy was doing and feeling and what he was thinking about—because these questions I asked all the time—but rather, that I failed to ask squarely and loudly whether he was thinking about *suicide* or having *suicidal thoughts*. I can remember only three or four times actually using the word *suicide* explicitly with him in connection with his struggles. Sarah did much better in addressing the risk directly with him. But many parents must make the mistake I believe I did in trying not to speak the word too often, for fear of investing it with too much power and presence, as if uttering it might endow it with a terrifying providential aura that would act like a dark spell on the future.

But the truth, of course, is much closer to the opposite: Words gain strange and mystical powers when they are *not* spoken at the times when they *should* be spoken. Not talking about suicide to a depressed person is like not talking about sex and birth control to a teenager. Verbal taboos create mystery, and people gravitate toward mystery. I don't know, and will never know, whether this change in conversation might have altered the trajectory of things for Tommy, but I at least feel convinced that this hard-earned knowledge may be of some practical use to other families struggling in a similar situation. As uncomfortable and intrusive as it may seem, it is *essential* to use the word *suicide* itself in order to demystify and deflate it, to strip it of its phony pretense to omnipotence and supernatural force. *Suicide* is not a "bad word," as Tommy Raskin might have said, for there is no such thing as a bad word. It is just, in reality, a terrible thing and an irreversible detour from the road we all try to walk down together, the road of life.

CHAPTER 3

THE TROLLEY PROBLEM

It is only with the heart that one can see right; what is essential is invisible to the eye.

—ANTOINE DE SAINT-EXUPÉRY, *THE LITTLE PRINCE*

It was important to me for a long while—less so now—to reconstruct the final days to map out exactly how we let our guard down in the final week of 2020.

I believe it was by taking everything happening with Tommy at face value.

I am convinced—no, now we can say *we know*—that Tommy, who really might have been the most honest and direct person I ever met for every other week of his life up until the last, had a secret plan to do this. Part of the plan, in the final days of his life, was not to reveal his pain or depression to his parents, sisters, cousins, uncles, aunts, friends, or doctors, but to act as normal as possible.

He lured us into a dangerous complacency. He clearly wanted this to be his decision and his alone. He did not want to be talked out of it; he did not want to be stopped, and as our friend Whitney Ellenby, who spent a lot of time analyzing this, told us, he did not want anyone else implicated in it. He did not want us to feel guilt—either the guilt we have gone on to feel over not being able to successfully address his mental health problems or the guilt we might have felt had he involved us in his thinking and we failed to dissuade him from his course of action.

I blame myself now for not catching this clever act, because I should have known that when Tommy acted "normal," placid and go-with-the-flow, he was not acting normal in any way *for him*. Normal, everyday Tommy was explosively funny, loving, unpredictable, outrageous and contentious in his opinions, comically litigious, provocative, carried away by ideas at every turn. He would laugh to the moon. He would call his sisters and cousins to pick gentle fights about who said what to whom and why someone in the family didn't actually *call* someone else in the family to wish them a happy birthday or *send* them a letter and a present but instead just *wrote* "Happy bday" on *Facebook*, which Tommy thought was "the lowest of the low, a cheap way to pretend you have celebrated someone's birthday while investing less than six seconds in the endeavor. It's a close call, but you might just as well do *nothing* at that point. I'd rather have nothing from a family member than a perfunctory one-second Facebook comment."

You could have ten hysterical mock fights a day like that one with Tommy, and everybody would come around to his side of the argument. Arlene reminded me that Tommy wrote a letter and convinced his Takoma Park Elementary School principal, Ms. Dunn, to increase recess by five minutes by arguing that *the teachers needed more time for lunch* and, Tommy generously allowed, "I know how hard it is to be a teacher."

When he was eight years old, he answered an assignment asking for a self-description by saying, "I have infinite energy. I have mental energy (mind): riddles, math, reading, history, law etc. I have energy to play with family, friends and more. I am an energetic 8 year old, little, nice, fun, great, fine, mysterious, kind, wild, small (tall) boy!"

But that rambunctious, loving, exuberant, combative, funny Tommy was absent the last week of the year.

He was cool as a cucumber, patient and removed, inscrutable. He was unusually calm, even serene at points. Looking back, I see it clearly: he was acting logistically, methodically, precisely. What time was I coming home from the Hill, he texted me? What time did I want to have dinner? When was Mom coming home from Connecticut? Sarah had gone to visit her mother in Westport and was away for the last week of the year. I didn't think much of it and then realized after the fact that Tommy was trying to make sure she would *not* be with us in the house.

I believe now he had resolved to do it by the end of 2020, that horrific, insensible, catastrophic, death-filled year. With every night filled with dinner and conversation or fun and games with me or his girlfriend or friends or cousins, it would all come down to December 30–31.

When I told him about Republican plans to object to the electors or Trump's plans to try to get Vice President Pence to reject them, there was some interest evinced on his part but not much. There was languid argumentation about the vice president's lack of power to adjudicate the merits of a state's electors and even some mild outrage over the Trump team's re-creating a claim in Congress that had been dismissed in more than sixty courts. But, in hindsight, it seems clear to me that his heart just wasn't in it, any of it, and his mind was elsewhere. He was humoring me all the way. He would just leave the politics to me, something he never really did before. I would have to deal with it on my own.

He was stage acting. He was acting out the theme of *I am not depressed and everything is fine* because he did not want to alert us or activate us to undo his plans.

A stranger might have met Tommy that week and found him witty, beguiling, and dry. But when I look back on it—and undoubtedly there is some hindsight bias here (the tendency to see prior events as far more predictable and inevitable than they were)—I see him that week as being strangely affectless and not truly present with us. His girlfriend, Sarah, agreed, saying he was "withholding—emotionally. The lights were on, but nobody was home. You couldn't break down what he was really saying. He was detached."

Tommy saw Dr. D on the day before he took his life, and Dr. D noticed nothing amiss. He was as dumbfounded and shaken afterward as all of us were. He came over to our house and sat in our bedroom to talk to Sarah and me. This observant psychiatrist looked deflated and hopeless—something that gave me odd consolation for having had the wool pulled over my own eyes as well.

Our insightful friend Whitney told us that a lot of people who resolve to do it experience a kind of relief in their decision, instantly shedding anxiety and worry, and then a focused determination to execute on their decision. Whatever compulsion drives suicidal ideation gives people, at least in some cases, a strange sense of direction. It is hard to reconcile an act of such shattering and destructive consequence for the people left behind with a commanding clarity of purpose, but it is apparently not unusual.

I wish I had known about this dynamic, which others have subsequently mentioned to me, too. I was obviously looking for the wrong signals. In other times when his depression set in, we had seen Tommy lethargic, anxious, and miserable, and that became the face of depression and danger to me. What was visible in him the last week of December was nothing like that. All the people who saw him that week were as baffled and shell-shocked as we were.

I misled myself by looking at external cues rather than internal ones. Tommy had just finished taking his final exams online and writing his final papers, so I simply assumed that the tremendous stress of the semester was over for him—foolishly ignoring that the ravages of depression don't necessarily follow the rhythms of the academic calendar. School's out—so what? I was deceived further because Tommy had clearly done well on his finals, perhaps even spectacularly so. He showed me that week a sparkling First Amendment essay about freedom of speech and incitement to imminent lawless action that I thought was remarkable (even though, in hindsight again, I can see he took far too little *joy* in writing such a commanding essay in sixty minutes under the stress of a final exam).

In sum, we cavalierly expected that the winter break would be a time for Tommy to relax, free from school obligations, and hang out with Sarah and me, Tabitha, his cousins, and his friends. But the "holiday season"—usually filled with family parties and a trip of some kind to someplace cold, like Vermont, or maybe someplace hot, like Florida—had been pretty much leveled and vanquished across America by COVID-19 and nullified by Donald Trump's last-minute veto of some legislation, which kept me from traveling as Congress stayed in town to override. We were reduced to ordering in meals and watching late-night movies.

Beyond Tommy's own elusiveness that final week, I was also caught flat-footed because he was spending a lot of time with Sarah, his lovely new girlfriend, who lived near Catholic University in DC, just a few moments away. With my Sarah gone to visit Arlene in Connecticut; Tabitha still teaching up in Philly the days in between Christmas and the New Year and staying with Ryan's family in Malvern; and Hannah and Hank out in Lake Tahoe, Sarah and Tommy and "Frank Reynolds" (that's me) actually spent a lot of that sheltered COVID-19 week together.

In his post-college period, Tommy had occasionally addressed me by this nickname, after the middle-aged Danny DeVito character on the TV show *It's Always Sunny in Philadelphia*, who insists upon hanging around his children and their friends all the time, the sum total of his rather pathetic but certainly never dull social life. Tommy loved to call me Frank Reynolds whenever I asked him, Tabitha, and Hannah what they were up to and whether they perhaps wanted to have dinner or hang out or go on a hike, which was pretty much all the time. There was no denying it: although I have in Sarah a wife I love passionately, an outstanding brother and brothers-in-law I love, amazing sisters and sisters-in-law I love, spectacular friends from every aspect and phase of my life whom I adore, beloved teachers always on my mind, and truly splendid congressional colleagues from all over America whose company I revel in and who have so much to teach me, and although I am an extrovert who craves seeing all these people, there is simply no one I would rather be with on any given day than my own children. It's not even a close call, although I should say that a perfect day for me is one when I can have all these people together—*all of them*—and the day lasts a really long time. My sister, Erika, always quotes Daisy Buchanan from *The Great Gatsby*: "And I love large parties because they're so intimate. At small parties there's no privacy at all." So my favorite party would be a large party with everyone I love there. But of all the titles I've carried in my life—congressman, senator, chairman, professor, dean—none makes me smile more than "Frank Reynolds."

And so, on Saturday, December 26, the three of us—Tommy and Sarah and me—had taken a long and beautiful, soulful, chilly hike together through the hills of Rock Creek Park, a hike on which Sarah told us all about her family and what it was like growing up Jewish in Delaware but going to Catholic School and how everyone in Delaware knows the Bidens. When we came back from our hike, I made us all bagels and lentil soup for lunch. We got to talking about what games we might like to play over the next few days, and as we brainstormed, Sarah said she thought chess was a "fun game," at which point, in mock outrage, I likened her statement to someone praising Shakespeare (whom I knew Sarah reads constantly and deeply loves) for being a "good writer." Chastising me for being a "chess snob" to his girlfriend, Tommy pointed out that my implication that chess was more than just a fun game only made sense to people who were already steeped in the intricacies of chess and not to those who were just learning it. Tommy was a great and discerning student of different forms of snobbery and had done family evening lectures on academic snobbery, intellectual snobbery, class snobbery, geographic snobbery, reverse snobbery, radical snobbery, defensive snobbery, and other specimens of the human propensity to make social differences sting. I had to concede the point and lose the argument, of course. I *was* being a chess snob—something I had no right to be anyway, given that I barely ever played anymore and my game was rapidly wasting away.

On Sunday, December 27, I gathered up everyone who was still around, including Tommy, and we went to look at a house on the Western Shore of Maryland that I thought might be perfect for the future. As our family has grown to a basic Thanksgiving crew of around twenty siblings and cousins and children, I have grown more and more enamored of the idea of finding a house for all the nearby Mid-Atlantic Raskins and New England Blooms that could accommodate everyone for all the vacations, especially in the summertime. Tommy seemed as engaged as everyone else in the expedition, though perhaps, again, a bit subdued for someone usually at the bull's-eye center of attention.

The point is that I felt things were normal, not in crisis. Another element of the seductive normalcy that governed that sleepy final week of December: Tommy graded his student papers and wrote long, responsive evaluations to his ten students on their essays. He made generous contributions in the names of students to Oxfam and GiveDirectly and wrote them profuse notes explaining why. Again, in hindsight, we could see that these gestures, although quintessential Tommy, might also have been a sign that he was saying good-bye. He had also suddenly decided not to teach as a section leader in a spring semester class on moral inquiry and artificial intelligence, which I thought he was going to love. But we connected no dots. We were blindsided and bypassed.

On Tuesday, December 29, when we made pasta and (fake) sausage for dinner and, upon Tommy's request, garlic bread (a Tommy favorite), we actually spent a long time discussing Harvard's grading system and whether he should accept the recommended "voluntary" curve that the head section leader was working to enforce on the teachers in Sandel's class. Tommy had given a disproportionate number of A– final grades, and he was being encouraged to revise some of them downward, to push them back over the line between A– and B+.

This disturbed him a lot. "Why couldn't every student *get* an A if every student *got* an A?" he asked rhetorically. "What if my students happen to be great and motivated to learn?"

"What if you happen to be a great teacher and you inspire them?" I asked in agreement.

"Thanks, Dad," he said with utter sincerity that I can still hear in his voice and that I can still hear in my ear and in my broken heart.

"But it seems to me *passing strange*," Tommy said, suddenly striking the imitation Christopher Hitchens accent he sometimes adopted when he was debating something (Hitchens being his favorite debater and someone he remembered meeting several times in childhood), "indeed, *bizarre* to have a college that admits overwhelmingly straight-A students but then acts *surprised* when they all get As on their papers once in college. It seems to me that if you don't want all students getting As in class and you want half of them getting Bs of some type, perhaps the Admissions Office should reserve half the places in the entering class for B *students*. Otherwise, your college is just the place where A students go to *become* B students."

This was a cogent point, and it may also be the last funny thing I can remember Tommy saying—a flash of Tommy really being Tommy.

I have racked my brains, my emails, my texts, to ascertain what might have been the precipitating factor in what happened late into the night of December 30 and the early, dark morning hours of December 31, but Tommy's planning was already under way and had almost certainly been a long time coming.

The die was cast.

On Wednesday, December 30, Tommy walked and talked normally with his beloved friends Isabel Thompson and Sam Dembling. He visited Dr. D for an hour. He sent me a draft of the essay he was working on about the moral dilemma he had long struggled with, known as "the trolley problem." A runaway trolley is poised to crash and kill ninety-nine innocent pedestrians. The only person who can stop the trolley is you, by diverting it to another track, in which case it will kill only one innocent pedestrian.

Is it better to passively accept the deaths of ninety-nine innocent people by doing nothing and thereby save one innocent person? Or is it better to actively choose to cause the death of a single innocent person in order to prevent the deaths of ninety-nine innocent people?

The essay was for a winter term class he hadn't even started, but he wanted me to "have" it. I did call him immediately to ask if everything was okay, and he said yes, he just wanted me to read the essay, and his computer had been breaking down. (Actually, he had recently ordered a new computer, so that is an unlikely story.) I asked him if we could have dinner together and hang out. He said sure, but he might be seeing Sarah later that night.

I read the essay, which, not surprisingly, was filled with the type of logical questioning and empathetic reasoning that existed in all his writing. Should we passively accept the death of ninety-nine people about to be hit by a speeding trolley so as not to have to actively take away one innocent person's life, or should we actively cause the death of that innocent person to prevent the otherwise inescapable deaths of ninety-nine others? If we are torn and riven between the two outcomes, as surely most people are, that in itself, Tommy argued, gives us the right guide to action. It tells us that letting ninety-nine people die is, in our moral estimation, the rough equivalent of murdering someone. It might be a little worse or perhaps not quite as bad, and yet we can see they are in the same ballpark.

But whether it is morally superior to go in one direction or the other, when we extricate ourselves from the fantasy extremism of these circumstances, we can know much better *how to live our actual lives*, and isn't that the real point of philosophy?

The trolley problem is solved when we look at the moral weight of the equation built into it. If letting ninety-nine people die might be equivalent to killing one person, then when times are *not* extreme and we don't have to choose between two indigestible options, we all have an overwhelming moral obligation, both not to kill one person but also to *rescue* the ninety-nine. This was Tommy's point. Through collective public means and through private personal means, we must act to liberate other people from hunger and poverty, to save them from starvation.

Doesn't everyone see that? Tommy was saying to the world.

Can you see that?

Just as we have a moral obligation not to murder people in our path, it follows that we have a moral obligation not to let ninety-nine starving people die for lack of food in another land. Our obligations extend beyond our borders. *All life is interrelated*, said Dr. King; *all humanity is involved in but a single process*. Real morality cannot be just an exercise for the classroom; the classroom must help us discover and exercise morality out in the world.

There is plainly no right answer to the trolley problem, no real "solution" to it. Yet Tommy had managed to solve it in his own way, by completely changing the terms of the question.

Sometimes, when I let my thoughts run away with me these days, I wonder if Tommy's even thinking about the trolley problem led him down a blind alley. Did he think, in his stressed frame of mind, that by taking one life, his own, he could somehow save ninety-nine other lives? Did he think he would redirect people's attention to the necessity of human decency and kindness, or was it just a psychological compulsion he was acting on, his illness speaking?

It is a shocking, unanswerable, and unthinkable question.

When I read Tommy's final essay on the trolley problem, I thought about a conversation I had with him before his college graduation. We were up at Amherst, and I asked him if he wanted a new suit for his graduation, and the minute I asked him, I knew it was the wrong question and that he would politely say no. And he surely said no; he just wanted more mosquito nets to stop children from getting malaria in sub-Saharan Africa. So we bought that instead.

On the night of December 30, he and I did have dinner together. He ordered for me a Smokehouse Beyond Burger and a chocolate soy milkshake from HipCityVeg, while he microwaved and ate two burritos. I don't know why he did not eat what I was eating. That puzzles me. He introduced me to that restaurant and loved it, and we would usually eat together. He had finished his dinner before mine arrived, and he watched me eat.

We watched *Family Guy*.

I gave him high-fives at the funny lines.

I remember saying to him: "I love you so much, dear boy," and him replying, "I love you, dear Dad."

I thought it was just another night. The next day was New Year's Eve.

Tommy was fine, Hannah was fine, Tabitha was fine, Sarah was fine, the dogs were fine.

I asked Tommy how he was feeling, and he said, "Good."

I asked him a few questions about classes in the spring semester and told him I was looking forward to really reading his essay on the trolley problem in depth.

I did not register the desultory quality of his answers at the time, but I should have. I rubbed his shoulder, tousled his hair, asked him what his plans were for New Year's Eve, the next day. He said he was going to have dinner with Sarah at 6 p.m. and would then spend the night over at her apartment. He asked if I would be okay at home by myself on New Year's Eve with nothing to do, and I said, of course, I would: "New Year's Eve is always a letdown to me, so I prefer to reduce expectations." He smiled.

When we got tired of *Family Guy*, I asked Tommy if he wanted to watch a movie and said he could pick it out. Or we could watch Hitchens and Galloway debate the Iraq War. But he said he was tired and wanted to get to bed early.

He got up, and I gave him a big hug and noticed again how much taller he was than me now.

"Love you, dear boy," I said.

"Love you, dear Dad," he said.

And that was the last time I saw my son alive.

PART II

CHAPTER 4

"THERE IS A NORTH"

When everything looks hopeless, you are the hope.

—MARCUS RASKIN

I woke up the morning of January 6 thinking not of Tommy exactly, but of Abraham Lincoln.

Lincoln never slips far out of mind for me and not just because my grandfather's beautiful Lincoln bust sits on my desk. I was raised on weekend trips to the battlefields at Bull Run, Antietam, Fort Stevens, and Gettysburg, and the Civil War permeates the culture of our region. The people of Maryland, a state which is cut through directly by the Mason-Dixon line, expressed strong divided feelings about secession, and Lincoln had to dispatch federal troops to subdue rebellious forces in the Maryland General Assembly to guarantee the state would not secede and isolate the federal city from the north.

Right up until 2021, our official state song, "Maryland, My Maryland," had nasty pro-Confederate lyrics calling Lincoln "the Tyrant"; in the summer of 2020, at the suggestion of my friend Elise Bryant, who runs the D.C. Labor Chorus, I wrote a new, pro-Union version of "Maryland, My Maryland" with my friend Steve Jones, who is a serious musician, a song we knew would have been definitely much more to Lincoln's liking.

I had been thinking a lot about Lincoln over this grim week because, like the majority of human parents who have ever lived, Lincoln saw at least one of his children die. In his case, three of his four sons did not make it to adulthood, and he lost his son Willie to typhoid fever when the boy was eleven, a wrenching and catastrophic event in Lincoln's life that took place in 1862, right in the middle of the Civil War, which meant that Abe and Mary Todd Lincoln could not properly grieve for their beloved boy. Lincoln said of Willie: "My poor boy. He was too good for this earth. God has called him home. I know that he is much better off in heaven, but then we loved him so much. It is hard, hard to have him die!"

I also had been studying some of Lincoln's speeches in preparation for our debate today, not just the "House Divided" speech, but his Lyceum Address, where he praised our political institutions and said that if destruction were ever to come to America, it would come not from abroad, from a "transatlantic military giant," but from within, specifically the vigilante forces of racist mob violence that target the press, innocent people, and democracy itself.

But Lincoln was especially on my mind that morning because I had a vivid dream about him the night before during my fitful, restless efforts at sleep. Although I have always loved him, Lincoln had never made it into my dreams before so this was a poignant occasion, and I wrote it down:

In my dream, Lincoln is trying to make his way to Washington, DC, after his election. I can see the top-hatted Honest Abe going through friendly territory in Ohio and Pennsylvania, crowds waving to him. As he comes to Maryland, he is menaced by angry mobs, sharpshooters. He dons a disguise at the Baltimore train station, to avoid the Plug Uglies and other political gangsters at large. Then I see him get on the Metro, which in the dream connects from Baltimore all the way to Montgomery County. We meet Lincoln at the

Silver Spring Metro—right in my district; I can't get anywhere near him, but I am so proud he has come!

We are with a huge, supportive throng, people waving their hats in the air, red bunting flowing in the wind. But the great man is still bedeviled by his enemies. They dart in and out of the perimeter of my dream, on the outskirts of the throng, whispering and conspiring against him. And then Lincoln, barely escaping these hovering assassins, gets on the Metro and rides incognito into the District. He is there on the plaza in front of the White House, on Pennsylvania Avenue. He sees impressive contingents marching in from Pennsylvania and Massachusetts and Maryland, too.

He looks at them in wonder and awe and appreciation.

"There is a North," Lincoln declares soulfully.

Some of the members of the crowd are fanning themselves, because this is a summer dream for some reason, but everyone in the crowd hears him and turns to a neighbor, like Munchkins in The Wizard of Oz, *and repeats in whispered tones the wise words of the president: "There is a North. There is a North."*

When I woke up, I immediately called Sidney Blumenthal, the former Bill Clinton speechwriter and journalist who has been born again as a captivating Lincoln historian, to tell him my dream. My father's friend who has become my friend, Sid has written a multivolume political biography of Lincoln that is spellbinding, and I can't get it out of my head. Sid is a man with a golden pen and a golden penchant for understanding political intrigue and conspiracy from the past, present, and future. Whenever he reviews for me the right-wing's creepy machinations against democracy, I always think about something I heard Gore Vidal say when I was in college: "No, I'm not a conspiracy *theorist*; I'm a conspiracy *analyst*."

I ran the dream by Sid and asked if he'd ever heard of Lincoln saying, "There is a North." After a little reflection and research, Sid remembered that when Lincoln went to tend to the Union soldiers injured at the First Battle of Bull Run (or the Battle of Manassas, as the rebels called it), he was overwhelmed by the casualties and bloodshed and asked the question "Where is the North?" While Sid did not recognize the sentence uttered in my dream, it lingered as a kind of answer to his question.

I spent the drive to the Capitol turning over the sentence "There is a North" in my head, as though Lincoln had provided me with the assurance I needed to repel the coming attacks on the 2020 election.

By the time January 6 arrived, we had been preparing for the counting of the electors for months. We had spent many hours getting ready for everything that Trump's forces might throw at us on the floor of the House in order to convert his stinging defeat into a last-minute victory, to pull a rabbit out of a hat.

Well, we had gotten ready for *almost* everything.

Our preparations had started as far back as May 2020. At that point, Joe Biden was the presumptive Democratic nominee for president, and I knew in my gut that he was going to win the election—that is, I knew he'd win a big majority in the popular vote and that his popular majority would convert to an Electoral College majority. But could his Electoral College majority survive Donald Trump's tactical mischief in the internal maze of this archaic institution? That was the open question.

The Electoral College is a creaky, shadowy place filled with hidden doors and booby traps galore, the perfect jagged battleground for a lawless demagogue like Donald Trump, king of the deep-inside fix, the low blow, and the late hit, just the kind of place to whip up some kind of madcap switcheroo, an insider coup based on indefensible new "interpretations" of the Constitution.

Months before a single vote was cast, and with no evidence for his claims, Trump had already begun to tell audiences that the election was being stolen *from him*, a perfect case of psychological projection but also a prophetic articulation of his own political designs. Trump was going to steal the election, and nothing was going to stop him—not the Electoral College vote, not the popular vote, not the law or the courts. With that increasingly obvious plan in mind, we began to sketch out different scenarios for how Trump would work to bring about a constitutional crisis that would allow him to prolong his increasingly injurious and authoritarian stay in the White House.

In the late spring of 2020, Speaker Pelosi urged the leadership team to think about defending the integrity of the presidential election in the byzantine architecture of the Electoral College. Not many people threw themselves into the task. Most colleagues assumed that the Biden campaign's excellent legal team was on it—which they were, up to a point. Biden's lawyers knew Trump's team would go to work in swing states to try to nullify Biden's winning vote totals, to fabricate vote totals for Trump, and if necessary, to urge GOP legislative leaders to discard popular election results entirely and impose slates of electors pledged to vote for Trump. Justin Antonipillai had arranged telephone meetings between our team and Biden's superb lawyers—Dana Remus, Bob Bauer, Walter Dellinger, and Marc Elias—and they had prepared rugged ground games in the key state legislatures to counterattack on all these fronts.

But relying on Biden's legal team alone as a strategy left out one giant risk factor that was plainly on our home field in Congress: Wednesday, January 6, 2021, when Congress met in joint session to count the electoral votes sent in by the states.

This might have been a humdrum, perfunctory affair in "normal times," a historical period I only dimly remembered now. But Trump and his operatives had worked to destroy every well-established constitutional rule, legal and ethical norm, and customary tradition that might have been an obstacle to imposing his will on everyone else. It seemed indisputable to me that he and his businesses pocketed huge amounts of money from the U.S. government and foreign governments, in direct violation of the Emoluments Clauses. He refused to release his tax returns after repeatedly promising to do so. He had blasted federal and state judges for doing their jobs and treated them as partisan, racial, or ethnic actors who would either bend to his will or suffer his wrath. He had blatantly and continuously interfered in Department of Justice prosecutions and investigations to save his friends from trouble and to injure his enemies. He had trashed America's allies and embraced and praised authoritarian thugs all over the world, from Erdogan in Turkey and Duterte in the Philippines to Xi in China, Vladimir Putin in Russia, and Kim Jong-un in North Korea. Oh, and yes, he relentlessly challenged the integrity and veracity of any elections that he hoped to win but did not go his way.

What would stop him from exploiting weaknesses in the antique Electoral College to try to work his political will on the nation?

Nothing.

I remember when Speaker Pelosi first raised the question of Trump's potential Electoral College gambits. It was a Thursday afternoon in May, in our leadership meeting in her conference room, at the end of a long, hard week. The subject was much on my mind. A fellow night owl, the Speaker had begun to kick it around with me in some of the expansive, late-night telephone conversations she loves. She had asked me to be prepared to say something about it at the leadership meeting.

Nancy's leadership team was always filled with superb and battle-hardened political leaders: my fellow Marylander and the encyclopedic House majority leader Steny Hoyer; the political genius whip Jim Clyburn; our clear-eyed caucus chairman, Hakeem Jeffries; the ruggedly well-organized assistant to the Speaker, Katherine Clark of Massachusetts; army veteran and enterprising lawyer Ted Lieu; the Democratic Congressional Campaign Committee chair, Cheri Bustos; the relentless progressive David Cicilline; the old-fashioned working-class hero and trial lawyer Matt Cartwright; my wonderful friend and the voice of the Midwest, Debbie Dingell; the always effective and impassioned Rosa DeLauro; the great darling of the left, Barbara Lee; and my Judiciary Committee buddies Eric Swalwell and Joe Neguse, serious lawyers both. These pols were afraid of no one. After four years of Trump, they were what Pelosi calls "foxhole fighters," the kind you want next to you in the heat of battle.

Most politicians avoid confronting the arcane mysteries and brainteasers of the Electoral College like children crossing the street on Halloween to circumnavigate a haunted house. I, however, was not one of them. If I had anything to offer this group of political savants and survivors, it was a *constitutional* perspective on events, and there was no escaping the massive problems presented by the obsolescent Electoral College. In my academic and political careers, I have *run directly into* the problems of the Electoral College—and even tried to work out a comprehensive way to overcome them.

When I was elected to the Maryland State Senate and sworn into office in 2007, my very first bill filed in Annapolis proposed a National Popular Vote Interstate Compact by which states would commit to cast their electors for the winner not of the *state's* popular vote, but of the *national* popular vote. This compact commitment would be activated when enough states entered the agreement possessing electors equal to at least 270, so the compact would control the outcome in every election. This is a way to move America to a real popular vote for president, where every citizen's vote equals every other citizen's vote, regardless of where you live. Maryland became the first state in the Union to adopt it, and as of today, fifteen states and the District of Columbia have joined the agreement, a bloc representing 195 electoral votes, or 72 percent of the number needed to activate the compact. The original concept came to me from conversations with Hendrik Hertzberg, a distinguished journalist and old friend, and John Koza, a computer scientist who taught at Stanford (and an American original who invented the scratch-off lottery ticket), via my friends at FairVote, in Takoma Park, Rob Richie and Cynthia Terrell, two champion democracy reformers.

But the debate over the National Popular Vote campaign has turned as partisan as everything else these days, and the old, rickety, accident-prone Electoral College still looms over America like a haunted house on Constitution Avenue, casting a long shadow over whoever gets to live at 1600 Pennsylvania Avenue.

As a constitutional law professor, I had long ventured into the frightening recesses of this founding-era haunted mansion, starting in the basement with white supremacy. Some of my House colleagues were surprised to learn that, when the nation began, Electoral College arithmetic dramatically favored Southern slave masters. The number of electors assigned to a given state was determined by adding the number of representatives in its U.S. House delegation to two (the number of the state's U.S. senators). But the number of representatives in its House delegation was inflated in the Southern states by the Three-Fifths Compromise, the infamous constitutional formula that meant that 60 percent of enslaved African Americans were counted for the purposes of apportioning House seats, even though enslaved people were of course 100 percent disenfranchised. When two additional electors, corresponding to the U.S. senators in the state, were added to that number, these two bonus statewide electors created another Dixie-accented tilt toward the slave masters, as less populous states, which were disproportionately southern (like South Carolina, Alabama, and Mississippi) received the same two electors for their senators as vastly larger states, which were disproportionately northern (like New York, Pennsylvania, Ohio, and Illinois).

This disproportion was even worse than population numbers alone indicated, because race and gender barriers to voting in the southern states meant that the actual *voting population* in these small southern states was even smaller—tiny, really. It has been estimated that the Three-Fifths Compromise alone added more than a dozen members to slave-state U.S. House delegations after the 1800 redistricting and, thus, more than a dozen presidential electors to the southern states. The constitutional inflation of the political power of the southern states was called the "slave power" in the early days of the republic, and everyone knew that Thomas Jefferson owed his victory in 1800 over John Adams to this sectional bloc, which is why he was called the "Negro President," the title of Garry Wills's outstanding book on the subject. These mathematical-political tricks, tucked deeply into the fabric of the Electoral College, worked like a dream for promoting the "slave power." Four out of our first five presidents were slave masters from Virginia, and seven of our first ten presidents were slaveholders. These facts were not contingent or random but arose from the deep architecture of our political institutions.

Progressives today know that the "solid South" of former Confederate states remains the basic organizing principle of right-wing presidential politics, whether in its Dixiecrat phase in the 1950s and '60s or in its GOP Trumpian incarnation today. This is why Georgia Republicans have worked so hard to overthrow the reconstruction of Georgia politics accomplished in 2018 and 2020 by Stacey Abrams and my House colleagues Lucy McBath, Hank Johnson, Sanford Bishop Jr., Nikema Williams, Carolyn Bourdeaux, and David Scott and by Senators Jon Ossoff and Raphael Warnock. This is why the transformation of the Commonwealth of Virginia (where even more of my family members live and vote than in Maryland or Washington, DC) has played such a key role in the opening up of American politics. Virginia and Georgia show how demographic changes and relentless organizing at the block level can break the stranglehold of political white supremacy in national elections.

But even with all that popular organizing, the Electoral College still threatens democracy because it does not respect majority rule and catapulted popular vote losers to the White House in two of the last six elections; it consistently sidelines more than three-quarters of Americans, those living in "safe" red or blue states, as irrelevant spectators to the real contest in swing states; and it offers constant opportunities and temptations for strategic mischief to nullify popular majorities.

When Speaker Pelosi raised the problem of the 2020 election at that meeting, she and I shared a strong feeling about what Donald Trump was going to do. After watching him for several years manipulate the levers of power to have his way, we could see his 2020 campaign strategy for victory in the Electoral College coming from a hundred miles off.

I explained to my colleagues exactly what I thought would happen.

Trump's plan hinged on the Twelfth Amendment and the word *immediately*. Under the Twelfth, if no candidate wins a majority of electors (270) when the votes from the states are counted up, the contest is "immediately" thrown into the House of Representatives for what has come to be called a "contingent election."

Why would Trump's team want to force the election into the House, where Nancy Pelosi is in charge and where a bunch of liberal Democrats are in the majority? Under the Twelfth Amendment, the members of the House "shall choose immediately, by ballot, the President. But in choosing the President, *the vote shall be taken by states, the representation from each state having one vote . . .*" (emphasis added). Thus, rather than choosing the new president in the House on the basis of one *member*, one vote, which is how we usually vote, we would choose on the basis of one *state*, one vote.

When we met before the election, the Republicans had 26 states, the Democrats had 22, and 2 were tied. After the election, in which the GOP shifted Iowa from the Democratic column to the Republican, they would come to have 27 state delegations, we would have 22, and 1 state, Pennsylvania, would remain tied.

The point was this: if Trump could somehow deny Biden 270 in the Electoral College, the race would be kicked into the House for a contingent election, which the Republicans would proceed to win, with their presumed majority of the state delegations in the bag.

Trump would move heaven and earth to get the election thrown into the House. But that meant he needed to strip away at least 37 of Biden's electoral votes to lower his total from 306 to 269 or something lower.

There would be three different ways to do that: (1) persuade, browbeat, or coerce state election officials to overturn actual popular results for reasons of imagined "fraud" or "corruption" and then "find" new votes and new vote totals, catapulting Trump to certified numerical victory in the states; (2) if that did not work, convince GOP-run state legislatures simply to cancel out and disregard their popular election results for similarly pretextual reasons and then use their plenary powers over award of their electors under Article II to install and certify slates of appointed Trump electors before the date the state electors met, December 14, 2020; or (3) if neither of the preceding plans worked and Biden actually made it to 270 or above on December 14, get Vice President Pence to declare newly discovered powers under the Twelfth Amendment to unilaterally reject electors cast by swing states like Arizona, Georgia, and Pennsylvania, and "return" them to the state capitols for further consideration, thereby thwarting Biden's Electoral College majority and "immediately" triggering the contingent state-by-state election in the House of Representatives that Trump desired.

When I first unveiled this analysis to my colleagues, they asked how I knew all this. The answer was simple: I asked myself what I would do to try and steal the election if I were the lawless and incorrigible Donald Trump. Time has revealed that this three-part prognostication based on Trump's record and character was an eerily close prediction of actual events, and the third part of the strategy has now been particularly well-established with the publication of the so-called "Eastman Memo." Trump legal adviser and former law school dean John Eastman drafted a six-point strategy for Trump's victory in the electoral college on January 6. The crux of his strategy was convincing Vice President Mike Pence to unilaterally reject Electoral College votes from seven swing states and then, after the expected "howls" of protest came from Democrats in the joint session, either to declare victory based on the new denominator or to kick things into the House for a contingent election under the Twelfth Amendment. Either way, Trump wins.

To prevent Trump's designs, we would have to work with the Biden-Harris campaign to defend against Trump's efforts to perpetrate voter fraud and to prevent Electoral College slate substitutions in the states. We would have to work to win majorities in more state delegations so we could win outright in a contingent election, or at least block *them* from winning in such a rarified contest; and we would have to figure out every parliamentary maneuver under the Twelfth Amendment and the Electoral Count Act that might take place on January 6 and refine our own responses in detail so we could keep Trump (and potentially Pence) from denying Congress the power to receive the actual electors chosen by popular majorities in the states.

In essence, we had predicted every maneuver coming our way except for one: the unleashing of mob violence to intimidate the vice president and Congress, overwhelm and stop the counting of votes, and provide a pretext and context for Trump to potentially intervene with military force under the Insurrection Act to put down the uprising he himself had helped to organize.

My closest friends from the Judiciary Committee on the leadership team—Joe Neguse, David Cicilline, Ted Lieu, Eric Swalwell— all got what I was saying right away and had pretty much scoped out the situation before. These lawyer-politicians became the core of the team I would work with on everything related to January 6 and beyond.

Still, precious few members and tiny numbers of Americans were focused on the dangers of Trump forcing the contest into a "contingent election." There had not been one of those in the House since 1824. That was when John Quincy Adams defeated Andrew Jackson in the House after Jackson captured a plurality in the popular vote total but fell short of a majority in the Electoral College. The only other presidential contingent election was the infamous 36-ballot race in the House between then-Vice President Thomas Jefferson and his own unfaithful and devious running mate, Aaron Burr, in 1801, when the two ended up in a tie in the Electoral College because of some of the original bizarre quirks of the system, which the always prescient (if not always courageous) Jefferson had called "an inkblot on the Constitution."

Yet, for those of us who saw in the summer that the 2020 presidential election would not, in Trump's mind, be remotely over on November 3, for those of us who recognized that it would instead be fought out in the Electoral College and the even more intricate Electoral Count Act of 1887—it was time to get to work. When asked by *Playboy*'s Brian Karem, an enterprising Montgomery County reporter and a good friend of mine, whether he would abide by the peaceful transfer of power, Trump protested that he did not trust "the ballots" and that there would not be a "transfer of power but a continuation." When I pressed GOP colleagues about what foul play was afoot, they almost always conceded that the contingent election was foremost on the minds of their caucus.

One House member I tried to scope out was my fellow constitutional lawyer Mike Johnson of Louisiana, who was nobody's fool and who was elucidating for the GOP Caucus an attack on election results in states where election officials or courts had made adjustments based on COVID-19 restrictions without the legislature acting itself. Mike would later turn this effort into a House GOP amicus brief on behalf of Texas attorney general Ken Paxton's lawsuit, *Texas v. Pennsylvania*, in which they tried to get the Supreme Court to throw out the electors of Pennsylvania, Georgia, Wisconsin, and Michigan. Although the Supreme Court rejected the suit on grounds of standing, and although no court ever adopted this theory, Mike was adamant that Congress, or the vice president, could "return to sender" the electors from certain cherry-picked swing states that went for Biden and then let the chips fall where they may in the subsequent House contingent election.

A contingent House election looked like it would be friendly terrain for the Republicans. In the 116th Congress, before the election, they controlled 26 state delegations, we had 22, and 2 of them were tied. If nothing changed, it would appear that they would have enough states to pull a majority of 26 votes. But their 26 was not necessarily 26, because one of those states was Wyoming, where the single at-large representative was Liz Cheney, and based on what I guessed from her increasingly outspoken statements against Trump, she might not cast

Wyoming's single vote for him in the House if he did not win the popular election or even the actual Electoral College vote. That does not necessarily mean she would have voted for Biden, but I could not see her voting to make Trump president under a sequence of illegitimate events trashing the election. If the GOP had 25 votes in their column, that would not be enough; the Constitution required a clean majority. The fight was on to flip state delegations and get ready for the once-unthinkable prospect of a contingent House election determining who would be president.

Beginning in early summer 2020, two strategies unfolded on the Democratic side to stop a contingent election gambit from succeeding. With these strategies, we aimed to block Trump's illegitimate continuation in power after losing the election.

The first way to preempt the contingent election gambit and drain it of any practical meaning was by working in the 2020 U.S. House elections to flip over to the Democratic column (or the "tie" column) at least two of the state delegations in the House that were then controlled in the 116th Congress by the GOP. If we could strip away just one of the delegations, Trump would have only twenty-five votes, short of the constitutional majority necessary to choose a new president. And if we could meet the far more challenging goal of taking majorities in *four* new state delegations, it would be Biden, not Trump, who would prevail in any contingent election, completely neutralizing this GOP strategy.

If neither candidate ended up with a majority in either the Electoral College or a contingent House election, then by the time the bell tolled for a new president on January 20, we would have, under the Constitution and the Succession Act, neither a president nor a vice president chosen. In that situation, the Speaker of the House of Representatives—that is, Nancy Pelosi—would become president on that day, something which I pointed out to leadership colleagues, prompting Matt Cartwright to observe, "That's because the Speaker's third in the line of succession after the vice president." To which Pelosi replied, smiling, "No, I'm *second* in the line of succession.

The president himself is not *in* the line." (And now you see why no one messes with Nancy Pelosi.)

What all this meant was that capturing just a half-dozen pivotally placed House seats—in states with tied delegations or states where Republicans had a one-vote majority—would block the potential reversal of Biden's election in a contingent House contest. But winning even six states is far more complicated than it sounds, because you never know which races you really have a chance to win or what the particular dynamics on the ground are in each district. Furthermore, we needed to play simultaneous *defense* for districts that the GOP would be targeting for the purposes of knocking out *our* state delegation majorities and solidifying *their* overall contingent election majority.

After a meticulous "game theory" analysis, my methodical fundraising director, Aaron Jarboe, the national director of Democracy Summer Maddie Kracov, and I identified twenty-two candidates we needed to help—and fast. When I contacted the candidates, the basic response I got was "Finally!" and "Where have you guys been?" They all believed that their races were vital not just to protect and expand the Democratic majority, but also to succeed in a contingent election scenario, which would be their first major challenge in office. Word had begun to spread that there was a "shadow contingent election" being fought out on the ground, too, and the candidates knew it. Rep. Elissa Slotkin was thrilled that someone had put together a coordinated effort on the "contingent election problem," which she said was "keeping her up at night." Challenger Jon Hoadley in Michigan told me he was talking to potential donors on the phone every day about the possibility of the presidential election's going to the House.

So I resolved to "max out" (contribute the maximum allowable amount, which was $2,700 in the primary and general election phases) to these candidates when they needed it and also to set up a new campaign fund-raising vehicle to mobilize national money for them. I had learned about it from my friend, then-Rep. Max Rose (an army veteran and political wunderkind from Staten Island who had taken a GOP seat in 2018 that would prove tough to defend in 2020). The vehicle, called a Joint Fundraising Agreement (JFA), allows groups of candidates with a common vision and purpose to work together to solicit campaign contributions and apportion money equally to all participating candidates. It meant that if we found donors who got the point about the Twelfth Amendment, they could give to all of us at once. (I would not be a personal recipient of campaign cash because Maryland's vote was not in doubt, but I joined so I could lead the effort.) What I registered with the Federal Election Commission as the "Twelfth Amendment Defenders Fund" allowed me to organize and raise money both for promising Democratic challengers and for precarious incumbents in Alaska, Florida, Iowa, Maine, Michigan, Minnesota, Montana, New Hampshire, and Pennsylvania, the states that we determined would be most crucial to assembling a majority of twenty-six Democratic-controlled state delegations in the House or at least denying the GOP a majority. In a matter of weeks, with the help of Aaron Jarboe and Maddie Kracov, I conducted a dozen Zoom fund-raising sessions and made innumerable calls to get the word out about the "shadow election" or "election within the election," and we raised—apart from fund-raising for my own committee and my beloved Democracy Summer project—a remarkable half million dollars and distributed $493,331 of it to the following twenty-two candidates:

Alaska
Alyse Galvin (at-large*)

Florida
Alan Cohn (15)
Margaret Good (16)
Pam Keith (18)

---

* Parenthetical information indicates candidate's congressional district.

Debbie Mucarsel-Powell (26)
Donna Shalala (27)

Iowa
Cindy Axne (3)
Abby Finkenauer (1)
Rita Hart (2)
J. D. Scholten (4)

Maine
Jared Golden (2)

Michigan
Jon Hoadley (6)
Elissa Slotkin (8)
Haley Stevens (11)

Minnesota
Collin Peterson (7)

Montana
Kathleen Williams (at-large)

New Hampshire
Chris Pappas (1)

Pennsylvania
Matt Cartwright (8)
Eugene DePasquale (10)
Christina Finello (1)
Conor Lamb (17)
Susan Wild (7)

That half million we raised was in addition to hundreds of thousands of dollars I was able to give directly to Democrats from my own campaign account, as my district is dark blue and my supporters are visionary enthusiasts for Democratic victory. All told, with direct contributions from my campaign, the Twelfth Amendment JFA, and contributions I made to national and state party committees, I put well over $1 million into our political ground game in 2020. This is airline peanuts compared to what Speaker Pelosi did, but it was a breakthrough record for me, and it helped me educate people about the importance of stopping Trump from snatching victory from the jaws of defeat by forcing a contingent election.

But November 3, 2020, was a cliffhanger Election Day in swing districts across the country, because a massive nationwide Democratic turnout for Biden–Harris was matched and, in some districts, overwhelmed by the tidal wave turnout of Trump followers who had "fallen off" from voting in 2018, but who were now jacked up to go defend their hero and his movement. In Iowa's second district, Rita Hart, the Democratic nominee we had backed, lost by a heartbreaking 6 votes—not 6,000, or 600, but 6.

In general, 2020 turned out to be a rocky year for Democratic House candidates, and this detracted from the success of our first strategy for dealing with the contingent election gambit. Not only did we not topple any of the GOP incumbents we needed to beat in order to capture more state delegations in the House, but we did not succeed in defending a number of vulnerable Democratic incumbents left hanging in the face of the dramatic return of Trump voters. It was painful to see close friends like Florida Democrats Donna Shalala and Debbie Mucarsel-Powell go down amid all the Trumpian fervor and ridiculous rhetoric linking them to Venezuelan dictator Hugo Chávez. We held on to our House majority, but only by a margin of 222–213, which meant that a change of five votes could mean a loss of the House.

Crucially, we saw one state delegation that had been in the tie column before, Iowa, slip over to the GOP column with the loss of our friend Rep. Abby Finkenauer. The GOP margin in a potential contingent presidential election would now be 27–22, with one tied. Even if Liz Cheney refused to participate in a fraud against the election, the GOP would still have twenty-six states in the bag.

This was sobering, but I should make the obvious point that the effort was not wasted, even in the face of all these odds. On the contrary, intervening in multiple key races allowed me to help shore up in vital ways our imperiled House majority and help save some key members who mean a lot to progress in America. In politics, as in life itself, no good act is ever wasted.

Parallel to our ultimately unsuccessful efforts to knock out the GOP advantage in House delegations, we deployed a second approach to stop a contingent election gambit by Trump: we would defend the actual presidential election results against Trump's inevitable attacks on them.

Sometimes during Trump's presidency, there was a tendency within the pundit class to assume that neither Trump nor anyone on his team was capable of executing anything that resembled a coherent political or legal strategy. They felt that because Trump habitually undermined White House efforts on policy questions like health care or infrastructure, he could never steal an election because it required a level of organization and discipline that he was constitutionally incapable of. But this ignored the fact that reelection was the only thing he really cared about. We had already seen in his Ukraine shakedown the lengths to which he would go to exercise coercive leverage over third parties to win a political contest. With his instinct for going for the jugular and his hunger for the bottom line, it was plainly not too difficult for him to understand that he needed 270 in the Electoral College to win, needed to deny Biden 270 in the

Electoral College not to lose, and needed 26 delegations to prevail in a contingent election in the House. He clearly had focused on that end-game strategy, which he had alluded to publicly several times.

And this is where the inaccessibility and inscrutability of the Electoral College worked to his advantage. Because the vast majority of Americans, including many in the media, did not comprehend the byzantine functioning of the Electoral College and the Twelfth Amendment, the details of Trump's actual maneuvers would matter far less than the propaganda screen of baseless allegations thrown up over them. Trump's strategy did not demand complex managerial skills but simply robotic message discipline: the Democrats and cooperating Republicans in name only, or RINOs, were trying to steal his victory from him. He just needed to spread doubt about the election with his poisonous rhetoric and frivolous lawsuits and then get Republicans—ranging from election officials to secretaries of state to statehouse Speakers to the vice president—to fall in line and do as they were told.

Trump could not force a contingent election in the House by sheer will. He needed to deny Biden 270 votes in the electors ultimately counted and accepted by Congress. That's where the predicted three-part plan that I'd mapped out for the leadership team back in May came into play. Trump's broad strategy to steal victory from the jaws of defeat would unfold in the election boards, the state and federal courts, the state legislatures, and finally in the Joint Session of Congress itself on January 6. In part one, Trump would have to figure out a way to overthrow, in federal or state court or in the boards of election, all results in close swing states that gave Biden a victory. This would either switch certain states' electors over to Trump or, at the very least, prevent pro-Biden electoral votes in those states from being successfully submitted to Congress.

Certainly, if he could move any of those swing states' electors to his own column by invalidating specific county-level results through court challenges, that would be ideal, because then Trump would just "win" the election on January 6. But not turning in any electors could also work to deny Biden his 270, which was the major back-stop goal for getting it thrown into a contingent election.

If this failed to work, then part two of the Trump team's strategy would be convincing GOP-run state legislatures to exercise their "plenary power" over the Electoral College by passing laws directly revoking the award of their electors to the candidate who won the popular vote in their states—Biden—and, instead, assigning them (on whatever public grounds they chose or on none at all) directly to Trump-Pence. It would also work to get legislatures simply to withhold sending in any electors at all, invoking "corruption" or "fraud" or technical uncertainty along the way, to keep Biden from getting to 270.

If neither of the preceding strategies succeeded in lowering Biden's electoral vote total in the election below 270, then part three of the plan involved getting either Congress or the vice president himself to reject enough of the Biden electors from the swing states at the Joint Session of Congress on January 6 to get him under 270. But because the House was still in the hands of the Democrats, and because Senator McConnell thought the election was over and at the time apparently wanted nothing to do with Trump's Big Lie, this meant there was only one figure who could theoretically accomplish what Trump wanted on January 6: Vice President Mike Pence. The problem there, of course, was that the vice president's assigned role under the Twelfth Amendment and the Electoral Count Act is ministerial and administrative, and it is left exclusively to Congress itself to count the electors. It would take a breathtaking act of usurpation and arrogance on the part of the vice president to pull this off in the face of so much history and settled legislative understanding to the contrary. But we lived in an age of constant and shameless demolition of rules and norms, and everyone was getting the sense that everything would come down quickly to Pence.

In the days immediately following November 3, part one of Trump's strategy began to unfold before our eyes largely as we'd anticipated. In the state boards of election and the courts, the Biden campaign's extraordinary legal team, led by Bob Bauer and Dana Remus, protected the election at every turn against Trump's efforts to attack mail-in and absentee balloting, to vaguely claim "fraud" and demand completely different rules, to scare off Democrats by promoting COVID chaos and long lines, and to suppress voting in minority areas. There was just no "fraud" or "corruption"—at least none that any self-respecting, incorruptible court could find and invoke to overturn the results of the election—*anywhere*.

When sixty-one different cases and conflicts landed in state or federal courts, the judges, regardless of political party—and at least eight of the federal judges had been nominated by Donald Trump himself—found that both the facts and the law were overwhelmingly against Trump and the other plaintiffs.

No court ever found any systematic election *fraud* against Trump.

No court ever found any systematic *corruption* against Trump.

In a lot of cases, Trump's lawyers, apparently wanting to avoid being sanctioned by the court either for making up evidence or for proceeding on a frivolous legal theory that they knew had no basis in reality, simply dropped Trump's widely promoted claims of fraud and corruption and instead advanced an esoteric legal theory under Article II of the Constitution that any election rules or procedures approved by state or county boards of election or courts without direct state legislative action were unconstitutional. What this argument meant in practical terms was that any actions taken by state and/or county boards of election or courts to protect voters and voting during a pandemic (e.g., enhancing access to mail-in ballots, expanding ballot drop box locations, or taking other steps to accommodate voters) were unconstitutional unless they'd been specifically enacted by the state legislature. And the Trump lawyers' proposed remedy for these alleged violations was that votes in the affected areas—be they in a county or an entire state—should simply be thrown out, disenfranchising millions of Americans. Of course, no court, including the U.S. Supreme Court, had ever adopted this radical claim, which would essentially have made every prior presidential election in American history invalid. Still, this unheard-of argument became the fallback position for people who wanted to support Trump but who understood that his claims about actual fraud were delusional and, as Attorney General Barr eventually put it in formal legal language, "bullshit."

I could write another book about the sixty-one federal and state courts that called out the vast emptiness and pure cynicism of Trump's claims. But when it was all over, the wall of judicial decisions across the country established a perfect factual and legal record of the falsity of Trump's Big Lie. Of course, the struggle to defend the election against Trump soon swept way beyond the courts and election boards, as he refused to accept these stinging judicial defeats and take his loss like a law-abiding citizen and a leader. The first part of Trump's strategy may have been winding down with the election boards and the courts, but the second part was already well under way.

In what Politico's Anita Kumar and Gabby Orr called on December 21, 2020, a "sweeping campaign to personally cajole Republican Party leaders across the country to reject the will of the voters and hand him the election," Trump "used specious and false claims of widespread voter fraud" to try to persuade state GOP legislators, governors, and election officials to disavow, nullify, repudiate, and overturn popular results and either switch in false Trump totals or set aside the voting process entirely and get the legislature to approve Trump-pledged Electoral College slates. This campaign began in fact right after the election, simultaneous with the campaign to twist the arms of election administration officials.

Trump spoke to "at least 31 Republicans, encompassing mostly local and state officials from four critical battleground states he lost—Michigan, Arizona, Georgia and Pennsylvania." He made at least a dozen such phone calls, convened four White House meetings where he brought in twenty state legislative leaders to convince them to exercise their constitutional powers to zero out the election and appoint Trump electors, and he aggressively lobbied Republican members of Congress to be supportive of these efforts as well. All these totally unjustified rearguard assaults on the election "occurred in parallel to his campaign's quixotic efforts to launch recounts and file lawsuits demanding [that] ballots be tossed."

Kumar and Orr, two fine reporters who fought to get Americans to understand the backroom coup that Trump was trying to orchestrate against the formal procedures of U.S. democracy, stated that "Trump's efforts to cling to power are unprecedented in American history. . . . [N]o incumbent president has ever made such expansive and individualized pleas to the officials who oversee certification of the election results."

Trump strong-armed state election officials to simply "find" enough Trump votes to overthrow certified election results for Joe Biden. In the case that became both illustrative and infamous, Trump called the Georgia secretary of state, Brad Raffensperger, and told him, "The people of Georgia are angry, the people in the country are angry. And there's nothing wrong with saying, you know, um, that you've recalculated" the state's vote count to find more than 10,000 new Trump votes. "All I want to do is this. I just want to find eleven thousand seven hundred and eighty votes, which is more than we have. Because we won the state." Raffensperger, who deserves some kind of prize for public servants who do their job and abide by their Oath of Office under blinding political pressure, calmly responded, "Well, Mr. President, the challenge that you have is, the data you have is wrong." To which Trump kept answering, "There's no way I lost Georgia. There's no way. We won by hundreds of thousands of votes."

When Raffensperger refused to participate in Trump's blatant pleas for election fraud, Trump turned up the heat, bringing pressure down on him from other Republicans and unleashing upon him and his family the fury of the local and nationwide Trump mob. Raffensperger later complained, "This should be something for Georgians to celebrate, whether their favored presidential candidate won or lost. For those wondering, mine lost." Trump then tried to force Republican governor Brian Kemp to endorse his delusional claims and call an extraordinary special legislative session to scrap existing state election law and appoint a pro-Trump electoral slate.

Trump worked along parallel lines to coerce GOP state legislative leaders in several other swing states to exercise their plenary powers over the Electoral College to override a popular Biden victory and substitute a slate of legislatively chosen Trump electors. On Friday, November 20, Trump invited a half-dozen Michigan GOP legislative leaders to the White House to convince them to block certification of Biden's popular vote victory in the state on hyped-up grounds of fraud and corruption, so it could be replaced by Trump electors. He tried the same tactic with Pennsylvania Republican state leaders the following Wednesday, the day before Thanksgiving, when he called in to a state GOP meeting to insist that they overthrow Biden's 80,000-vote victory in the state and appoint a Trump slate instead. On November 30, Trump exploded when Arizona governor Doug Ducey, a Republican, certified Biden's statewide election victory by 10,000 votes, an outcome Trump had been lobbying to block and which he disputes up to this very day.

It is a testament to the culture of voting and election norms—and a triumph of underappreciated civic patriotism on the part of a lot of obscure state election officials, many of them Republicans—that Trump was not able to knock out any of Biden's Electoral College votes in swing states before the electors were cast in the state capitols on Monday, December 14. The refusal of GOP officials to buckle under to the most powerful man in the world helped to save America from what the political scientists call a "self-coup" on January 6 organized by the incumbent president.

But Trump's failure to pluck off the states one-by-one was not his last maneuver. For he had his final card to play, and it turned out to be colossal and epic, the most significant and dangerous move yet in his hell-bent campaign to overthrow the actual election. Trump turned his attention to blocking *recognition* of Biden's electoral vote majority on Wednesday, January 6.

This meant that he and his team had to convince Vice President Pence to announce powers never before claimed for a U.S. vice president: the power to push the Congress aside and unilaterally judge and reject electoral votes submitted by individual states. Pence's staff and his estimable outside counsel, Richard Cullen, the former attorney general of Virginia, reportedly told Pence that there was no basis for usurping the counting of electoral votes anywhere in the text of the Constitution, the procedures of the Electoral Count Act of 1887, or the actual historical practice of Congress and vice presidents.

Of course, the novelty of a brazen new departure from practice had never stopped Trump from doing much of anything, so we had to prepare for what the hypothetical possibility of Pence rejecting Biden's electors on January 6 might mean in practical terms. In November, the Speaker had asked our caucus chair, Hakeem Jeffries, to convene a group of members who were interested in this process, including Joe Neguse and me, to meet and get ready for what was coming. Rep. Mikie Sherrill from New Jersey showed great insight in our meetings and asked some of the most trenchant questions I had heard. We also benefited from the views of Zoe Lofgren, my Judiciary and House administration colleague, who was famous for having worked on the Judiciary Committee's staff during Watergate, serving on the Judiciary Committee as a member during Bill Clinton's impeachment, and then acting as one of the impeachment managers during the first impeachment trial of Donald Trump.

I also had a lot of outside help. I called my favorite First Amendment professor, which is to say, my favorite constitutional law professor, Laurence Tribe. A Harvard Law professor and the preeminent constitutional law authority in the country—well, at least for liberals—Tribe remains deep in my heart because it was in his classroom where I found both my academic calling and my wife. In the fall semester of 1986, I met Sarah in Tribe's Constitutional Law class, where I would go not just to hear his densely intricate lectures but also to gaze at the beautiful, beguiling Sarah Bloom. I thought she was staring at me, too, only to learn at the end of the semester that she wasn't wearing her contact lenses.

For many years, Tribe has stayed a constant presence in my political career and intellectual life. I often called him late at night to trade ideas about what might be coming on January 6. He correctly anticipated that the major focus would come down to the proper scope of the vice president's powers.

In October, I had been called by a former prize pupil of my own, Justin Antonipillai, who had gone on to become an attorney at Arnold & Porter, a Senate chief of staff, and a CEO. Justin offered his help in coordinating a small group of other experienced Hill lawyers, including Irv Nathan, a former counsel to the House of Representatives, to kick around with me different potential scenarios for January 6. Their help was invaluable as we scoped out what to do if and when GOP members objected to particular states' electors; what procedures and rules would govern if and when the two chambers met separately to resolve such objections; and then—what became the source of fascinating and scary discussion—what to do if Vice President Pence attempted to step out of his essentially administrative and ministerial role, as defined in the Constitution and the Electoral Count Act, by refusing to receive, open, or count electors from particular states.

The Twelfth Amendment's language reads:

"The President of the Senate shall, in the presence of the Senate and House of Representatives, open all the certificates, and the votes shall then be counted—the person having the greatest number of votes for President, shall be the President, if such number be a majority of the whole number of electors appointed."

The amendment's switch to the passive voice in stating that "the votes shall then be counted" is what gave Trump's partisans the hope that the vice president could unilaterally refuse a state's electors. This absurd reading would of course be at total odds with two centuries of practice and interpretation. Because the amendment is extremely specific about when the vice president is actively involved, Congress has always assumed that it would itself count the electors, both houses have always participated in the counting of electors, and bicameral *congressional counting* became the defining and organizing principle of the Electoral Count Act of 1887, passed in the wake of the chaos of the 1876 election.

To that end, the act strictly limits the vice president's duties to the procedural and ministerial realm, and for ample good reason. It would be a stark conflict of constitutional interest for one person, the vice president, who is historically so often a candidate for president, to be able to unilaterally count, and decide upon the validity of, electoral votes in an election in which he or she is running. This design would cut against everything the Framers believed in. As Madison put it in Federalist No. 10, "[N]o man is allowed to be a judge in his own cause, because his interest would certainly bias his judgment, and, not improbably, corrupt his integrity."

Furthermore, even if Pence wanted to believe in these breathtaking new powers, there was no reason to exercise them here. Trump's claims about fraud were "bullshit," as even his obsequious Attorney General William Barr had to conclude, and sixty-one federal and state courts had blown them to smithereens. Why should Pence risk his reputation and bring the country to the verge of civil war over these fraudulent claims of election fraud?

We had thus refined strong answers for all the questions that might surface. We were ready to defend the Electoral Count Act and oppose any power grab by the vice president. If he tried to pull a rabbit out of a hat like that, we would object immediately, and not only would we have the votes to uphold the objection in the House, we had ample reason to believe that Senator McConnell had a solid majority in the Senate GOP Caucus ready to join Democrats in countering such a blatant effort to steal the election.

Similarly, if Pence played it straight, we were ready to resist objections coming from GOP members on the floor to specific states' electors, a struggle we could always win, given that we were in firm numerical control in the House and that one chamber alone could stop such an objection from moving forward and the Senate was solid itself with McConnell apparently not having any of the nonsense.

Our preparation was solid. Yet something was missing. As much as we brainstormed and conjured our way to the various scenarios Trump might foist on us, we did not look hard enough at recent history to guide us through the whole situation. This was a president who had no qualms about provoking and praising violent conflict and who often seemed to actively court it, whether the target was the media, immigrants, or anti-Trump demonstrators. The specter of violence was out there as a looming possibility. All we had to do was look back as recently as the first day of June 2020.

On Monday, June 1, Donald Trump and Attorney General William Barr assembled a special forces unit of militarized police from across federal government agencies to unleash a paramilitary riot against nonviolent protesters. The declared object was to clear the streets so Trump could march freely over to the St. John's Episcopal Church on Lafayette Square and then, bizarrely, wave an upside-down borrowed Bible above his head to signify an almost burlesque message of Christian nationalism to his religious followers. Under William Barr's direction, this force of mostly unmarked officers, many of them on horseback, issued its order to clear the square an hour *before* the 7 p.m. curfew that had been called by the mayor. The officers then proceeded to use pepper spray, rubber bullets, and batons to assault, scatter, and force the stampede of hundreds of peaceful protesters.

This sordid episode galvanized huge numbers of Americans against Trump's violent lawlessness, but it did elicit support from Trump's most ardent fundamentalist Christian defenders. I remember thinking that this farce revealed exactly what right-wing religion is all about today: neither love nor charity nor nonviolence nor even respect for law or decency, but rather, submission to political power and glorification of state violence to score tribal victories over political enemies. A lot of my young constituents were there that day:

Bethesda–Chevy Chase and Walt Whitman High School students, Montgomery College students, and dozens of others back home from college and graduate school. Many of them were trampled, hurt, and traumatized by Trump and Barr's customized police riot, which hit them like a baton in the ribs.

The events of June 1 shattered all kinds of taboos. President Trump dragged high-ranking military leaders into breaking up peaceful civil protest and defied the Posse Comitatus Act, which restricts the use of federal military personnel in the enforcement of domestic laws. He personally deployed the power of government to rain down violence and pepper spray on protesters assembling in the capital city.

The brutality released on that day showed America how far Trump was willing to go to maintain and extend his power. If he unleashed the violence of *government against protesters* to prop up his power, why wouldn't he unleash the violence of *protesters against government* to do the same? There was a warning enclosed in the events of June 1 that should have been every bit as much a part of our preparations as the Electoral Count Act. But I suspect that the idea of an American president inciting a violent attack on Congress as we conducted our constitutional duties was so utterly strange and unthinkable that it kept us from incorporating the possibility into our analysis and preparation.

In December, Speaker Pelosi asked me to be in a small group that would prepare every final detail of our work for January 6. Also participating would be my close collaborator and Judiciary sidekick Joe Neguse; Zoe Lofgren, the chair of the Committee on House Administration, who would be one of our two vote tellers (along with the ranking member on that committee, Rep. Rodney Davis from Illinois); and Adam Schiff, who was of course the chair of the House Intelligence Committee and the famous lead House manager of what came to be called "the first impeachment of Donald Trump."

We met repeatedly over Zoom through December, brainstormed, divvied up duties, reviewed and mooted our arguments, and got ready to go to the floor to defend the 2020 election against all comers.

On my last call with Speaker Pelosi before the end of year, after Christmas but before New Year's Eve—before Tommy left us—she reviewed our plans and asked me if I was ready and whether I needed anything. I told her that our parliamentary points were ready to go and that I was preparing my speech and would go over it with Tommy, whose rhetorical and legal judgment was second to none.

"This is going to be tough," Pelosi said. When the Speaker gets ready for something, she leaves nothing to chance, and she encourages her team constantly. "In this business, you have to know how to take a punch and you have to know how to throw a punch. It's a tough business we're in, but you're going to be great."

That seemed a long time ago.

She called me also on the evening of January 5 to express her love for Sarah and me and the girls. She asked about Tommy's family service and remembered that I was going to get Tommy's views on my speech. She told me that the whole caucus was with us, heartbroken and sending love.

She asked if there was anything I needed for the floor. I told her we had answers, arguments, plans, and directions ready for everything, including what to do if the vice president broke out of his role: immediate appeals to the House Parliamentarian, objections, moves to adjourn, and new procedural motions to restore the vote-counting role of Congress.

We had prepared for everything.

Everything, that is, except everything that was actually about to happen.

CHAPTER 5

COMPLETE THE COUNT

If you make yourself a sheep, the wolves will eat you.

—BEN FRANKLIN

The first thing I saw when I entered the House chamber on January 6 were the Republican members huddled together in animated confusion. Before I even had a chance to speculate about what was going on, there was a flurry of activity on the floor, and we were all handed a copy of the letter that Vice President Pence had just released. The moment of truth was upon us—what was Pence going to do?

Speed-reading Pence's letter, I learned that he had indeed rejected Trump's demand for him to refuse the electoral votes cast by the swing states of Arizona, Georgia, and Pennsylvania. After genuflecting to the "significant allegations of voting irregularities" and the "concerns" of millions of Americans, the vice president rejected the view that "I should be able to accept or reject electoral votes unilaterally." He emphasized that it was up to Congress to count the electors and that no vice president had ever departed from this practice under the Electoral Count Act from the time of its passage 130 years ago. "It is my considered judgment that my oath to support and defend the Constitution," he wrote, "constrains me from claiming unilateral authority to determine which electoral votes should be counted and which should not."

This remarkable sentence tells us precisely what Donald Trump was demanding of him. Trump wanted to convert the vice president's ministerial role as moderator of the proceedings into the new role of an omnipotent election judge who had the unreviewable power to pick and choose which states' electors would be granted and which denied.

Pence's revealing letter instilled a sense of security in my mind. He had just dispelled my single-greatest fear for the day, and done it in writing. Although we had been prepared to counter a vice-presidential usurpation of congressional vote-counting powers, the process would have been messy, protracted, and dangerous. Pence had steadily walked everyone back from that cliff. Why he did so is anyone's guess, but I imagined that the extraordinary thing Trump was asking him to do was too terrifying a leap into the constitutional void. All of Pence's obsequious deference to Trump over the last four years had endorsed outrageous actions that Trump himself was taking, but this letter was all about Pence doing something breathtakingly and epically outrageous *in his own name*, at a time when he was trying to establish his own political persona. According to his speeches, Senate minority leader McConnell was eager to stop Trump's irresponsible attack on the results of the election, which meant that Pence understood not only that McConnell would be with him on his decision but that the vast majority of the Senate would be with him as well. Pence was also a former member of the House who had been through this process several times and clearly respected the role of Congress.

Pence's letter caused a dramatic stir over on the Republican side of the House. A few GOP colleagues who wandered over to me shared rumors of ambush and conspiracy afoot, a restive spirit among "the Jim Jordan people," who were infuriated by Pence's inexplicable political apostasy. In the meantime, I assured Democrats near me that because Pence was rebuffing Trump's demands, Trump simply had no more tricks up his sleeve to block Biden's election and inauguration. The rest of the day might be wearying and drawn out, but Biden's victory would be secure. The Democratic House alone could block all objections to the electors even if the Senate, for some extremely unlikely reason, took a walk on us.

Still, I kept trying to convince myself that the air of confidence I was trying to project was actually justified. I looked back to the Freedom Caucus rows, at Jim Jordan. An inexhaustible defender of high-ranking allies who abuse their power, Jordan was a tightly wound bundle of coiled energy on the floor, bouncing from member to member, chewing gum, whispering things of seeming political gravity. His public modus operandi I knew well by now, from committee tussles and floor debates: always ferociously attack the opposition, never defend, forget about the merits of the legislation and the truth, and, above all, keep the nasty polemics and self-righteous talking points flowing, the more ad hominem the better. It was low blows all the way.

I suppose I had never forgiven him after the day in a committee meeting when he gratuitously attacked Sarah, screaming out of turn when he did not even have the floor, "Why did Sarah Raskin unmask Michael Flynn?" This was a meaningless question referring to one of the absurd conspiracy theories he traffics in. When Sarah was deputy secretary of the treasury under President Obama, she was apparently briefed, among hundreds of other things, about the activities of Michael Flynn, who later served as Trump's national security advisor for twenty-two days before he had to resign after his lies were exposed about contacts and conversations he had with the Russian ambassador Sergey Kislyak. I told Jordan to turn off his mike and put on his mask.

The air was thick with menace. The House had taken on a darkly cinematic affect, crackling with portent and seemingly on the verge of scary events. I felt the weirdness. The ominous atmosphere merged with my own sense of isolation and dissociation and my cluelessness about what was truly going on behind the scenes with the GOP.

Speaker Pelosi called us to order and turned the gavel over to Vice President Pence, who then announced the arrival of the senators for this rare meeting of members from both chambers, the kind of bicameral session we have only for the State of the Union Address and a few other events of constitutional dimension and public moment. Everyone clapped.

We knew what was coming. Trump would get one of his many eager-to-ride-Air-Force-One GOP House members to stand and present an objection in writing to a state's electoral votes, stating the grounds for the objection "clearly and concisely, and without argument," under the terms of the Electoral Count Act of 1887. So long as Trump could convince even a single Republican senator to sign the objection along with a House member, the House and Senate would have to meet separately in our respective chambers for two hours of debate to consider the objection. Both the House and Senate would have to vote to uphold the objection to a state's electors in order to exclude those electors. Consideration of objections to the electors being cast by the states of Arizona, Georgia, Pennsylvania, and perhaps Wisconsin was the central showdown for which we were all prepared.

And then some kind of ruckus broke out on the GOP side. A dense crowd of members—many non-masked and oblivious, it seemed, to the current COVID-19 pandemic—gathered like a rugby scrum around Paul Gosar from Arizona, a Trumpified true believer who had been denounced by his own siblings because of his increasingly deranged and reactionary political pronouncements. Gosar shouted and tried to be recognized by Vice President Pence to accept some document purporting to prove something or other, but under the rules, as Gosar undoubtedly well knew, he could not be recognized. There was an approved order of business. I guessed Gosar was trying to get Pence to rethink his letter and unilaterally reject the electoral votes sent in by Arizona on the basis of whatever this document was. Pence ignored him, then gaveled him down, to the guffaws and boos of the Freedom Caucus crowd. It was clear now that dozens of House members planned to object to Arizona electors on the grounds that the election was corrupt or fraudulent, which would of course cast doubt on Gosar's own election—but no one gunning for the overthrow of Biden's victory had solved that little paradox yet.

Pence moved quickly to begin the state roll call, which, under the Electoral Count Act, proceeded alphabetically. Alabama was called first, and the clerk read 9 electoral votes for Donald Trump; then Alaska, 3 electoral votes for Donald Trump. Then, when Arizona was called, it was 11 electoral votes for Joe Biden.

At 1:12 p.m., Representative Gosar, who had been screaming bizarrely for months about voter fraud and election corruption and Dominion Systems voting machines, without even a hint of proof of any wrongdoing, stood up to object to certifying the presidential electors from his home state.

Vice President Pence asked whether a senator had signed his objection. I watched as both Missouri senator Josh Hawley and Texas senator Ted Cruz jumped to their feet to say that they had. Hawley's jack-in-the-box performance did not surprise me much, given that he had clearly been auditioning to become Trump's successor as America's top authoritarian populist.

Although Cruz wanted that job, too, I had thought that he might have some lingering measure of self-respect left keeping him from joining this farce. Cruz had once called Donald Trump a "maniac," a "sniveling coward," a "sore loser," and a "serial philanderer" and had said, "If I were in my car and getting ready to reverse and saw Donald in the backup camera, I'm not confident which pedal I'd push." For his part, Trump mocked the physical appearance of Cruz's wife and asserted that Cruz's father had participated in the assassination of JFK. But when it came to paying court to the extreme right, like other once-and-future GOP White House contenders, Ted Cruz was willing to let bygones be bygones. Ten other GOP senators joined the objection.

Once the objection to Arizona's electors had been raised, members of the House and the Senate resolved separately to our own chambers. The objection could be sustained only if both houses approved it, a dubious outcome given our true-blue majority in the House and the 50–50 tie in the Senate in the wake of the astounding election in Georgia the day before of U.S. Senators Jon Ossoff and Raphael Warnock, not to mention McConnell vehemently opposing the whole venture to upset a presidential election, which could have dire unforeseen consequences.

On the House side, Speaker Pelosi called us to order. The rules called for two hours of debate, one hour to each side, with Republicans and Democrats taking turns in five-minute blocks. I listened carefully to my colleagues, even though most of us on both sides of the aisle had long since stopped listening most of the time to each other's speeches. Louisiana representative Steve Scalise started with an argument that amazingly never even asserted any corruption, fraud, or error in Arizona. Instead, he mirrored the language of Trump's failed legal arguments, offering a general indictment of "states where *the Democrat Party* has gone in and selectively gone around" the process of state legislatures making all decisions regarding elections— which he apparently believed was the constitutional requirement, as urged by our colleague Rep. Mike Johnson of Louisiana.

Zoe Lofgren, the chair of the committee that oversees election matters, rose to demolish the Scalise argument. There had been only three presidents impeached in American history—Andrew Johnson, Bill Clinton, and Donald Trump—and Zoe has played roles in the latter two of the three, and was on the House Judiciary Committee staff when Richard Nixon would have been impeached had he not resigned first. She observed that the real political power grab was the one we were witnessing now and that "if Congress selects the next president instead of American voters, we'll have no need for an Electoral College. We'd have no need for presidential elections at all."

Then arose Jim Jordan, drawing from his bottomless reservoir of outrage to make fervent arguments rooted in emotion, tortured factual claims, and logical fallacy. His demagogic rhetoric, beguiling when you first hear it but now cartoonish to my ears, is what passes for rational argument in the GOP today. It was not long before Jordan returned to some oldies but goodies from our days together in the Judiciary Committee, spraying drive-by rhetorical attacks on Jim Comey, the "Russia hoax," and the decision by "Democrat House members" to "impeach President Trump based on an anonymous whistle-blower with no firsthand knowledge who was biased against the president," and so on, a veritable Top 40 of imaginary grievances from the Trump years. This fact-free, logic-impaired discourse is the daily bread fed to Fox News viewers, but is becoming a lethal threat to critical thinking in our besieged democratic republic.

Adam Schiff, whom I always like listening to, went next, ably invoking principles of democratic self-government, making clear that Donald Trump began this train of lies many months ago with the "dangerous falsehood" that "our election would be marred by massive fraud." Adam turned then to say that an attack on Biden's victory would be only a "party triumph" for the GOP, and he invoked our late colleague Elijah Cummings, who liked to say that "we are better than this."

Because Adam had mentioned him, I found myself thinking about Elijah, my beloved friend and colleague from Maryland, who recruited me to the Oversight Committee when I arrived in the House in the 115th Congress. When Jim McGovern asked me to go on the Rules Committee in the 116th Congress, I was persuaded that I could learn about legislative process in a more authoritative way on Rules than I had during my first term as a freshman in the minority. Even though Rules is a notoriously time-consuming committee, with late nights and early mornings, I thought it was a place where serious legislators trained.

But, if I was going to go work with the great Jim McGovern and his fascinating right-hand man Don Sisson, I thought that I should consider stepping aside from another committee, so I went to sit next to Elijah on the floor to explore what he thought about my situation. I could tell he was not pleased at all, and I immediately regretted asking him the question.

"You just do both," he said.

"I know," I responded, "but I also have Judiciary and Admin."

He just looked at me.

"I'm kind of, um, busy," I said, sheepishly and vanishingly.

He was still looking at me. "Raskin!" he said, calling me up short. "You are never *too busy* to be doing what you *ought to be doing*." He enunciated each syllable as if I barely spoke English.

"Yes, Mr. Chairman," I said, sitting up straight and smiling nervously. That was that.

And the next day, Dave Rapallo, Elijah's superb staff director, called to tell me that Elijah was making me the chairman of the Oversight Committee's new Subcommittee on Civil Rights and Civil Liberties. I couldn't tell if this was a punishment or a reward, but it was the best thing that had happened to me in a committee since I arrived. Oversight subcommittee chairman for civil rights and civil liberties—wow! And that was how I came to be a subcommittee chairman in the Committee on Oversight and Reform *and* a member of the Judiciary, House Administration, and Rules Committees. I had gone to Elijah, foolishly, to see about dropping a committee and had ended up keeping them all and *adding* a subcommittee chairmanship.

Never too busy to be doing what you ought to be doing.

Those are words I would carry with me. I loved and admired Elijah, and whenever he said, "*C'mon now*, we are *better* than this," he meant it. You could hear decades of street struggle against Jim Crow apartheid in the timbre of his voice, you could see his determination to save people from the insolence of power in the bulging veins of his neck, and you could feel in his presence the reality that *history was alive*. Elijah, the son of Baltimore, dared and provoked us to make things right.

And just a few minutes after I thought about Elijah's touchstone rhetorical trope that we were all better than this, our debate on these phony objections was interrupted by the sound of a furious, violent mob barreling against our door, loud chants of "Hang Mike Pence," and the jagged noise of shattering glass reverberating through the U.S. Capitol.

Our downward flight into the darkened basement of the Capitol was chaos. We did not know where we were going. Masked and frantic, our staccato steps bouncing off the walls, we kept bumping into one another, especially with so many phones pressed to ears to call spouses, children, parents, staff. I called Sarah and tried to strike a calm, whispered tone to reassure her as well as I could that we would all be safe. I told her I loved her and promised to call back as soon as I learned anything. I called Emma Kaplan, Pelosi's trusty floor aide, who assured me she was doing everything she could to secure the safety of Tabitha, Hank, and Julie in Steny's office, while arranging for a police escort to get them quickly out of the danger zone. Someone told us, *Shh, keep it down, keep it down.* I overheard bits of news flowing from people's iPhones—"Massive breach . . . outnumbered police trying to regain control . . . pressing to enter the Senate chamber . . . shots fired in the House . . . no sign of the National Guard or army yet."

Shots fired in the House? I suppressed my anxiety for a moment and kept moving.

Down the hallways and exit ramps, I glimpsed in my peripheral vision the "rioters"—the "Confederates," as I heard different colleagues call them, because of the photo of the guy brandishing the Confederate battle flag in the Rotunda. They lunged forward in clustered groups, looking to me like zombies in a horror movie, but whenever they came into view, our group broke into a healthy trot pretty impressive for middle-aged politicians. I was not sure a lot of us had it in us. One colleague was struggling, and we slowed down for a second to allow him to catch his breath. When we arrived at our first destination, the Longworth House Office Building dining room, we massed inside the cafeteria. The words about shots being heard in the House were ringing in my ears; I texted Tabitha, Hank, and Julie at 2:43 p.m.:

Me: Where are you?!?! In Hoyer's office? Do police know you're there?
Julie: Yes, they have locked us in.
Tabitha: We are ok
Me: Who is in there with you?
Tabitha: Us three.
Me: Stay put, the police are coming to get you.
Julie: Tell them I barrides the door.
Hank: Barricaded
Me: Ok. Ok Stay put. Julie, I understand you're in touch with Keith and Emma. Do you have a police contact? Are you ok?
Julie: Yes. Jason. We are all safe.

But our texting was interrupted by someone from the Sergeant at Arms Office, who came to tell me, "This is the wrong destination. Members are exposed in here. We're not supposed to be here, Congressman; we're going upstairs."

I began moving toward the Rayburn Building with a group when someone came to get us, saying, "No, not that way." We were redirected to the stairs in Longworth and went up two flights, at which point we were shepherded into the Ways and Means Committee Hearing Room, a room I knew a bit from the first impeachment of Donald Trump, when the Judiciary Committee met there to vote on the impeachment articles.

The room felt much safer from the mob, but it was also perilously crowded for the COVID-19 era. While on the House floor, we had been separated from one another by two seats; in this room, most everyone was seated right next to someone else. People—well, House members—had taken the seats of the committee members at the dais, but there must have been only thirty-five or forty of those and several hundred people in the room, meaning most people were relegated to the public seating area. It felt pretty stifling in there. It escaped no one's notice as we entered that all the Democrats wore masks, as did pretty much all the staffers, but a lot of the GOP members, including Jim Jordan, did not. They paraded around unmasked and defiant, their body language and facial expressions proclaiming their often-explicit position that the COVID-19 crisis was a "Democrat hoax," as we so often heard the Coronavirus Denier in Chief say. Refusing to wear a mask in this safe room struck me as not just dangerous but bizarrely passive-aggressive. Here we had fled, to a packed indoor space because of their cult hero's recruitment, mobilization, and exhortation of a violent armed mob to storm our workplace, and instead of acting responsibly, they were openly flouting the mask rules while grinning and yucking it up.

I made my way quickly to the back of the committee room, where large doors opened on to a library, which miraculously wasn't packed at all, apparently not having been discovered yet by the members. This long room, with bookshelves surrounding a long mahogany conference table in the center, was undoubtedly more comfortable than the committee room with its loud acoustics and jam-packed lecture hall seating.

Already there were my friends Mike Thompson and Anna Eshoo, from California. Anna was seated at the long table with her iPad open. Something about seeing her there filled me with a sense of hope. When I went over to her, she got up and, with a sad and loving smile in her eyes, gave me a tight hug (very tight for the age of COVID-19) for Tommy, saying, "I'm so sorry," and then: "Look." She motioned to the news on her iPad.

This was the first time I encountered extensive footage of the dense, writhing mob of marauders and bullies breaking windows and beating up Capitol Police officers with sticks, lead pipes, baseball bats, Donald Trump flags, and the poles of Confederate battle flags and, most heartbreaking of all, of *American* flags.

I saw a Capitol officer get speared with a flagpole with the Stars and Stripes still attached.

As harrowing as our experience had been upstairs, I had been imagining as a backstory that the people now trying to barrel their way on to the House floor had somehow slipped into the building undetected. The news reports kept saying there had been a "breach," which made it sound like a momentary glitch or a secret trespass. I imagined a game of cat-and-mouse, with right-wing insurrectionists slipping in through side doors.

But this was no case of side slippage; it was a straightforward *siege*, defined by pitched, *violent battles* thick with people all over the Capitol complex, waves of hand-to-hand, medieval-style combat against the officers who greeted us every morning and bade us good night. Everyone could see a paramilitary storming of the entrances to the Capitol, the windows being smashed, and a savage, bloody battle with thousands of combatants raging everywhere, a struggle that our beleaguered security forces were, from the looks of it, clearly losing. This was terrifying, far worse even than what I had imagined during our escape.

I flashed back immediately to the only two words in Donald Trump's inaugural address that stuck with me when I watched it on the night of January 20 (having skipped the actual daytime ceremony for our joyful "resistance hike" through Rock Creek Park): *American carnage.*

When Donald Trump invoked this bloody phrase in 2017, it conjured up no visual image for me at the time. I didn't know what he was referring to, and I supposed that some right-wing speechwriter like Stephen Miller had taken the image to be eloquent or soaring.

But its meaning was now revealed.

This was American carnage.

American carnage was not actually what Trump was *denouncing* on his first day in the presidency; it was what he was *promising*. And he had delivered.

Something else was seriously bothering me, though. How in the hell had all these people gotten into the building? For all the parliamentary scenarios we had unfurled, I realized then that the possibility of a violent assault on the proceeding had never really occurred to me. I had assumed that anyone who attempted anything so daring and transgressive as to *commit violence* against the Capitol Police in order to breach, much less *storm*, the Capitol would be *shot dead on sight*. I kept thinking that if even only one guy in the middle of a regular workday sprinted past the guards, eluding capture and using violence to get in, he would be shot dead because the officers could not wait around to see what kind of danger he would pose on the floor.

Those cops were armed.

I was dumbfounded: Why hadn't the police shot people for trying to enter the Capitol Building without going through a metal detector or a security background check? I had seen the Black Lives Matter protest in the summer, when a huge, armed National Guard presence on the Capitol steps sent an unambiguous message to people petitioning government for a redress of grievances: *Mess with the Capitol or the Congress at your own peril.* What had given this assemblage of fanatics the idea that they could attack and overrun the U.S. Capitol Building and *not get shot* by hundreds of officers? An important puzzle to ponder.

Meantime, Anna Eshoo was worried. She showed me news reports flashing across the screen: Explosive devices had been discovered at the DNC and RNC headquarters. Some commentators were saying that these were just meant to distract police attention from the Capitol, but it was also possible that those who had planted them meant business. After all, they were real explosives. There were untold hundreds of marauders inside the Capitol right now who had entered by going around the metal detectors. They could have explosives on them, too. Many would be armed.

I had a thought I remembered having on 9/11. On that day in 2001, I picked up Tabitha from the Child Development Center at

American University, just twenty minutes after having dropped her off there on my way to the law school. I had spoken quickly to her teachers, Zakia and Lawrence, about what the kids knew and didn't know about the attacks. Then we were en route to pick up Tommy at Takoma Park Elementary School. From there it was to collect Hannah at Piney Branch Elementary School when Tabitha, who was upset because her playdate with her friend Mely was canceled, said, "A bad man hit the building with an airplane." Tommy said, "Two buildings, they hit! One in New York and one in Washington, DC. Are they going to hit any more buildings?"

And I thought then, on 9/11—as I thought now, on 1/6—*Wait, is this thing over? What other surprises might they have in store for us?* When you are in the middle of a crisis, you simply have no sense of its contours, so the thought occurred to me, fleeting and disturbing: *This attack could last for days.*

I asked Anna about Speaker Pelosi, who had been whisked off the floor and taken to an undisclosed location. I knew them to be extremely close friends.

"She's safe. Nancy's okay," she said. "Thank God."

We watched more violence and mayhem on Anna's iPad, transfixed and dumbfounded.

We saw a report of one rioter "being shot as she tried to enter the House chamber."

Gunfire right outside the chamber? My God.

My mind had slipped into following-the-crisis mode on TV, but Tabitha, Hank, and Julie were still in harm's way. I rushed to call Emma Kaplan again. I had no idea how much time had passed since our last call—maybe twenty minutes. Emma said the Capitol Police had evacuated all the members who were up in the gallery: Diana DeGette, Susan Wild, Pramila Jayapal, Jason Crow, others. They had kept their heads down while crossing over from the majority gallery to the minority gallery and then followed the same path we had.

What was I thinking? The police were going to return now and get Tabitha and Hank and Julie out next. They were in touch with Julie.

I called Julie and repeated what Emma had just told me, which she of course already knew. She did tell me that there were three "bolt locks" on the door to Steny's office and that they had barricaded the furniture up against the door, too.

When I called Tabitha, there was no buoyancy to her voice. I told her the police were going to get them out soon.

"When?" she said.

"I don't know exactly, but they got out a bunch of members stuck in the gallery, and they're coming for you guys next. I think they're getting ahold of the situation."

"Okay." She sounded forlorn.

"Where are you now?" I asked.

"Under the desk," she said.

Silence. I could think of nothing else to say. I thought of Steny's mahogany desk and all the antique-y stuff in that room.

"We saw them all coming up the Hill, Dad. There are *thousands* of them."

After we hung up, I stayed in that room and distracted myself for a few minutes by trying to reassure the texters and emailers overrunning my phone that I was fine; their messages were flooding in from Maryland and every other part of my life as they watched images of the bloody violence on the Hill. My communications director, Samantha Brown, who had been texting me, called to say that C-SPAN really wanted to talk to me and would I talk to them? Not until we got Tabitha, Hank, and Julie out, I replied. But then I would be happy to do it, right after I reached Pelosi or Hoyer.

All I really wanted to say to the world was what I wanted to tell our caucus leaders first: that we needed to get back up to the chamber at all costs and complete the counting of the electoral votes. Along with all the other Democratic members I spoke to that evening, I was determined not to buckle under to fascist-style mob insurrection or criminal confiscation of the electoral votes in their mahogany briefcases. We needed to reassure the public that the republic was solid and continuing.

My thoughts turned to words: How should we even refer to these ongoing events? The word *protest* or even *riot* seemed inappropriately mild. There was no doubt we were watching protesters rioting. But this event was far more violent than a protest run amuck, and it was far more politically focused than even the brutal white supremacist riot we had witnessed in Charlottesville, Virginia, nearly two and half years before. Was it an *insurrection*? Perhaps so. Donald Trump had told the Proud Boys, the fascist paramilitary brawlers who bloodied innocent people in Charlottesville that day in 2017, to "stand back and stand by." The TV footage from outside the Capitol identified the Oath Keepers and the Three Percenters as other active organizational players in the violence. All the rhetoric ("Seventeen seventy-six!" "Our House!" "Take the Capitol!" "Hang Mike Pence!" and "String 'em up!") was consistent with this being a "popular insurrection," even though there was nothing popular about it—the vast majority of the country was already appalled and repulsed by the insurrectionists' violence.

But this was no good old-fashioned, popular American insurrection like the Whiskey Rebellion. Its timing, its political rhetoric, and its political targets all coordinated perfectly with the agenda of President Trump and his most militaristic advisers, like Michael Flynn, who were openly talking about martial law in the days following the June 1 police riot in Lafayette Square. It struck me as extraordinary that tens of thousands of avid pro-Trump protesters somehow knew they should be denouncing and chasing Trump's own vice president. This kind of last-minute, fine-grained parsing of the relationship between POTUS and VPOTUS does not happen in large crowds by accident.

These troubling thoughts kept leading me back to one central worry: *We may have just witnessed a coup attempt*, I thought, *and it may still be going on now.*

In my mind, I was beginning to see three different concentric rings to the action: (1) on the outside ring, a mass protest organized on social media that had turned into mass riots, (2) in the middle ring, a violent insurrection outside the Capitol led by domestic extremist and paramilitary groups that massively breached the Capitol and shut down the vote counting, and (3) at the center, Trump and his team's manipulation of all the chaos to try to execute a coup against his vice president and the Congress by overthrowing the 2020 election results and replacing them with a Trump victory in a House contingent election and adding perhaps a healthy dose of martial law to calm everything down.

My meditations were interrupted by Emma Kaplan, who called to tell me that Jason Gandolph, with the Sergeant at Arms Office, and a group of officers had gotten "Julie and Tabitha and her husband [sic] out" and that the three were on the way over to the Ways and Means Committee hearing room. I breathed deeply . . . only to start worrying about their trek down and what kinds of maniacs might be roaming the halls.

I went back out into the library conference room. Seconds later, Tabitha and Hank came in, followed by Julie. It was an emotional and tearful reunion, to say the least. I told them that Emma, too, had thought that Tabitha was married to Hank, which was the big joke of January 6 (to the extent that January 6 had jokes). The media had reported that no member of Congress had his or her kids in the Capitol today except me and that I was accompanied by "my daughter Tabitha and my son-in-law Hank," an accurate statement that left the erroneous impression that Hank was married to Tabitha when he is, of course, actually married to Hannah. I told myself to turn this into a riddle one day for the little kids in our family—Emmet, Gray, and Tess; Tilly and Bennett; Maddox and Kai. In any event, we were all pretty jubilant about their being together and free from danger—or, at least now as free from danger as the rest of us were.

Sam called to tell me I had to go on C-SPAN. I followed my colleagues Henry Cuellar of Texas and Stacey Plaskett, who represented the Virgin Islands as their nonvoting delegate. Stacey had been my prize law student at American University Washington College of Law, someone I had strongly encouraged to become a prosecutor, which became the foundation of a great career. She sometimes teased me about the fact that she was once my student but now had more seniority in Congress than I did.

On C-SPAN, I spoke to host Greta Brawner and told her that members had evacuated the chamber and were in a safe location on Capitol Hill. I told her we were all resolved to go back in and "count the Electoral College votes as is demanded of us by the Twelfth Amendment." I told her that "any violent insurrection against the United States will be put down." I emphasized the gravity of this violence and entered into a stream-of-consciousness diatribe about the insurrection:

"Attacks on the Capitol didn't even happen during the Civil War," I said. "You have to go back to the War of 1812 to find something like this, and that was a foreign power that attacked us. There was no Confederate attack on the Congress. So we're going to complete the count if we have to stay here all night or even all day tomorrow. We're going to swear in Joe Biden and Kamala Harris on January twentieth. This violence is intolerable, lawless, and unacceptable, so we have to finish the job we were sent to do."

It's fairly amazing that for all its blood and gore, even the wrenching period of the Civil War produced no violent confrontation during the Joint Session of Congress itself. To be sure, there were heated rumors that Southern forces might try to block Congress from counting the electors in February 1861, and some thought that Vice President John Breckenridge, a diehard pro-slavery Southerner, might even be arranging with his confederates to be accosted and robbed of the boxes of actual electoral votes as he carried them from the Senate to the House.

But no one in February 1861 ever actually messed with the congressional counting of the states' Electoral College votes, a process left intact, even sacrosanct. No insurrectionist ever came to wave a Confederate battle flag inside the Capitol. And as Lincoln knew, the Capitol itself was such a crucial symbol to the nation of democratic self-government, far more than the White House, that he deployed precious troops from the battlefield to guard the Capitol through the war so nothing untoward could happen to it. The nation then closed ranks around the peaceful transfer of power.

Greta asked me if Donald Trump should address the nation, and I stumbled around and finally said I did not have enough information to respond. The truth was that I feared if he addressed the nation, Trump might blame everything on Antifa and impose martial law, invoking as justification the chaos he had helped unleash. Who knows what he might have said? "The main thing is to complete the count," I said. This was a "constitutional duty," and we were determined to execute it.

It was after 4 p.m. when I finished the interview. We were called back into the main room. Our caucus chair, Hakeem Jeffries, and the Republican Caucus chair, Liz Cheney, now addressed the assembled members from the front of the room, thanking us for our cooperation and pledging a bipartisan commitment to go back in and complete the count. Both spoke passionately, and I was quite blown away by Liz Cheney's emphatic directness regarding the threat to democracy, all while many of the Republicans acted sheepish and evasive about what we were going through.

Liz had been my casual friend since I got to Congress. We did not have many opportunities to socialize, because we are of different parties, different caucuses, different philosophies, different committees and subcommittees, and different regions in a House of 435 members. Still, in Congress (which is a pretty fine microcosm of society), you meet people whom you take a liking to despite everything else, and I have always felt pretty wonderful about Liz as a person. When we met, we established that both of us came from very close big families. In fact, Liz told me that her father, former vice president Dick Cheney, and my father shared a physician in cardiologist Dr. Jonathan Reiner. Neither of us was part of the infamously debauched night life social scene that seems to drive some members of Congress to distraction (and ruin).

One day, Liz called and left me a message to meet her on the floor, where she introduced me to her daughter, who was in town.

"This is the congressman I was telling you about," she said. "He's the one who wrote the book saying that Grandpa never should have been vice president because the Supreme Court was wrong." I got a kick out of this—Liz loved to tease me about my book *Overruling Democracy: The Supreme Court vs. the American People*, in which I argued that the Supreme Court's intervention in the 2000 presidential election was based on fraudulent and specious constitutional arguments. "It was nothing personal," I said. And the truth was that I believed, and still believe, that the Court in *Bush v. Gore* set a number of land mines in voting and in the Electoral College that were exploding all around us right now. That decision expressed the idea that the individual citizen has no federally protected right to vote for president and that state legislatures always possess the power to override a popular vote system.

But I liked the way that Liz seemed to feel it was possible to have dramatic and profound political differences with colleagues without demonizing and dehumanizing them. She was a tough, brawling, partisan competitor, but I never felt she was someone who would attack the constitutional order. I felt certain that she had not lost sight of "our bonds of affection" or the "mystic chords of memory" that Lincoln spoke of that keep this vast, diverse land of America as one nation, not riven by constant lethal conflict.

I walked up to Liz before she left and said, "Good work, Liz. Hang tough."

She thanked me, extended her arm to me, and said, "Are you okay?"

I said I was all right and paused to acknowledge her sympathy. "You think there's any chance they will drop the objections now? This is getting really dangerous."

She just shrugged and smiled sympathetically, made a face, and left.

Speaker Pelosi arrived to address us not long afterward. She thanked the Capitol officers and our staff and emphasized that we would be going back in as soon as possible. She invoked a message of unity against insurrection and terror and said we would not allow any violence to go unpunished. I saw my friend Jennifer Wexton, a Montgomery County native who represents Northern Virginia; she told me of a rumor that the Republicans were divided between those who wanted to pursue their objections, on the assumption that abandoning them now would leave the impression that they themselves were implicated in the violence, and those who wanted to stop the charade immediately in the interest of national unity. They had apparently decided to split the difference by dropping the objection to Georgia but pursuing the objection to Pennsylvania. Word usually leaks out of what is going on in the other caucus's deliberations, but I was not sure how they had even been able to meet and decide this. Maybe it was all rumor.

A large shipment of pizza arrived in the Ways and Means Committee room, from Domino's, to great bipartisan enthusiasm. I brought Tabitha and Hank out to get in line while Julie came to tell me she was working on a way to get them back home. She had been in touch with Faisal Siddiqui, our computer and tech guy, who lives in my district. He had said he would drop them off in Takoma Park on his way home. Would it be safe? As soon as the Capitol Police gave the all-clear to leave the complex, Julie said, it would be fine. She understood from the Speaker's people that the Maryland and Virginia National Guard had arrived and that the DC Guard, mysteriously AWOL for a long time, was finally in the mix too. The news programs were showing participants in the insurrection leaving the Capitol Building—and, astonishingly, not being arrested—with hordes of them leaving the Capitol complex apparently triumphant in the glow of Trump's Tweeted call to "remember this day forever!"

It would be another hour or so before my family members could go. I told Hank that I wanted to call his mom and dad, Peggy and Jay Kronick, to reassure them that he was okay. We reached them near the Boundary Waters of northeast Minnesota, where I had once gone fishing with my brother Noah and my grandfather, and I apologized to Peggy and Jay for putting Hank in danger. They could not have been sweeter and expressed complete support and solidarity with all of us for what we had been through. I thought about how wonderful it would be to have if not a wedding when this was all over, then at least a wedding party.

Shortly before 9:30 p.m., I said good-bye to an exhausted Tabitha and Hank and told them that I'd come home right after we heard the final objections and completed the count. I told them to kiss Hannah and Mommy and Ryan for me and that I'd see them all in the morning.

I hugged Hank good-bye and thanked him for being strong for all of us. I told him to tell Hannah I loved her and would be home soon. And I told him I loved him, too.

I gave Tabitha a tight, fatherly hug. "I promise it won't be like this next time you come to the Capitol," I said, almost casually.

She looked up at me seriously. "Dad, I don't want to come back to the Capitol," she said clearly.

Of all the awful things I had heard and seen so far in this madness, that one cut me to the quick—that and the scene of one of our Capitol Police officers being speared and stabbed by an insurrectionist with an American flagpole. Both those moments, for me, froze and recorded the desecration of something I hold sacred: the idea that democracy is something we will all take care of together as Americans.

I didn't know what to say to my daughter who had said good-bye to her brother the day before and who had just spent the last several hours in fear for her own life. I suddenly felt that everything I did in my work, everything I had done in my career, was a source not of pride or positive good for my family and community but, rather, of violent danger and mad chaos. How did we end up here, with fascists trashing our Capitol Building and killing people? How would we ever return the genie of this political madness to its bottle? Would this be the new face of public elections?

I thought of Tommy and missed him so painfully and sharply—

his radiant goodness, his vast spiritual distance from the cruelties and hypocrisies of normal politics, his gut resistance to the worship of power and demagogues, his daily ethical zeal for the truth and social fairness. And for a second, a thought crossed my mind that I quickly censored: *Thank God he doesn't have to be here to see this and to know what has just happened to Tabitha, Hank, and Julie.* It was the kind of thing that would have kept him up through the night for weeks on end.

As quickly as I'd pushed that thought out of my head, another one just as deep and wrenching came rushing in. When Tommy was in college and reading Nietzsche in one of his classes, he was increasingly haunted by all the darkness and misanthropy he found there but also fearful that Nietzsche was disturbingly perceptive. Tommy was troubled by the Nietzschean idea that public life was all about "force and fraud and the will to power." I told him that my dad, his Baba, disagreed with this and had argued that the vast majority of interactions people had were based on trust, caring, and implied affection. And when people cheated and lied and undermined the social contract, all of us had to act together to stop them. When everything looks hopeless, my dad would say, *you* are the hope.

Maybe some people are driven to care and to help, Tommy allowed. But what about the people who are driven by force and fraud and the will to power? What can we do to stop them?

"That," I said, "is what the rule of law is for."

CHAPTER 6

MIDNIGHT MEDITATIONS AND ORWELLIAN PREPARATIONS

Oceanic society rests ultimately on the belief that Big Brother is omnipotent and that the Party is infallible. But since in reality Big Brother is not omnipotent and the Party is not infallible, there is need for an unwearying, moment-to-moment flexibility in the treatment of facts.

—GEORGE ORWELL, *1984*

I looked around at the bedraggled, confused members milling about in masked or unmasked partisan hubs, and I knew that many hours more would pass before any of us slept. I'd told Sarah they were estimating that we would finish up at around 3 or 3:30 a.m., but even that was seeming optimistic. I went back to the secluded cubbyhole in the back of the room to work on my closing remarks for this fateful night.

But now, having seen these gory scenes of violence, all I could think about was impeachment. If organizing and inciting this assault was not an impeachable offense, then what was? We had all been witness to this atrocity, and neither the House nor the Senate could allow it to stand.

I was not the only one thinking about impeachment. As I prepared my remarks, Julie brought me a call from David Cicilline, Ted Lieu, and Joe Neguse, my judiciary committee brothers-in-arms who wanted to know whether I was on board for us to introduce a resolution to impeach Trump as early as the following Monday, for inciting violent insurrection against the union. I told them: "All aboard, indeed," and "Perhaps we will be able to show by then he not only *incited* the insurrection—or the coup," I quickly added, "but that he *organized* it. He was obviously the key actor all along. Count me in either way."

But I also raised the point that the emergency was far from over and that Trump was an increasingly isolated and delusional desperado who would do anything to hang on to his power. We were all still in grave danger.

We needed to pass a separate resolution calling on the vice president to activate Section 4 of the Twenty-Fifth Amendment, convene the Cabinet immediately, find that the president had proven himself unable to discharge the powers and duties of his office, and remove his powers until his term ended on January 20, two weeks later. They agreed completely.

"Is there any conflict between the Twenty-Fifth and impeachment?" Cicilline asked.

"No way," I said. "Just because you've committed high crimes and misdemeanors doesn't mean you're not incapacitated. And just because you're deranged does not mean you're not also a constitutional criminal."

After hanging up, I found myself thinking about a piece of legislation I'd worked on from my first year in office, a bill whose failure to thrive I lamented now more than ever. The Twenty-Fifth Amendment Resolution, or H.R. 1987, was written to establish a bipartisan, congressionally appointed "body," as provided for in the Twenty-Fifth Amendment, to work with the vice president on determining presidential incapacity in the event of a crisis just like this one. Section 4 of the Twenty-Fifth Amendment leaves this judgment either to the Cabinet or to a special "body" set up by Congress, but my bill would have spelled out more specifically how that body would be composed and how it would act. The commission I proposed would have worked with the vice president to transfer the president's powers to the vice president in the event that the president was rendered, for reasons physical or mental, unable to discharge the powers and duties of his office. Had we passed the bill several years before, that body—I called it "the Commission on Presidential Capacity to Discharge the Powers and Duties of Office"—would have been in place now to respond to this shocking crisis of leadership stability and capacity. (No one held out much hope that Trump's Cabinet—or what was left of it—could ever summon up the independent will to act.) Given Trump's outrageous behavior as president even up to that point in history, there was a lot of idle talk in the air about the Twenty-Fifth Amendment, but I wanted to bring some constitutional and scientific rigor to the table and a commonsense proposal tied not to Trump specifically but to cover breakdowns in all presidencies.

One night at home back then, Tommy asked me what I was doing at work, and we began discussing the bill. He was intrigued by the constitutional architecture supporting the legislation, the legitimacy of defining presidential incapacity under it, and the political wisdom of raising Trump's, or any public official's, mental condition in political dialogue. I had been exploring the Constitution with Tommy since he was a tiny boy, and over time he had become quite the master of complex constitutional doctrine and theory. In the evenings when he was little, he would come into Hannah and Tabitha's room, where I read them all bedtime stories, and sometimes say, "Tell us a case!" I would then set up the facts for them of a Supreme Court decision, and the kids would venture their analysis.

Our friend Paula Kowalczuk always told the story of walking first-grader Tommy to school and listening to him discussing specific Supreme Court decisions and justices' dissenting opinions. By the time of his second year in law school, Tommy's understanding of law was sweeping, so it was a joy that night in 2017 to be able to explain to him a part of the Constitution that he, like pretty much everyone else in the country, knew nothing about.

I had been passionate to learn the history and meaning of the unsung Twenty-Fifth Amendment since the night I was elected to Congress, November 8, 2016, the same night Donald Trump was elected president, because I had this unshakable premonition that we were headed for absolute political and social chaos under this unstable and impetuous man. Six months later, in the spring of 2017, I was still searching for the language to make the case to my colleagues. Tommy was a quick study, so I launched in, and as usual, he helped clarify things for me in a powerful way.

Congress and the states had added the Twenty-Fifth Amendment in 1967. It was designed in response to the assassination of John F. Kennedy and the debilitating illnesses of some other presidents, with the purpose of addressing the Constitution's multiple ambiguities regarding basic questions of succession, political stability, and continuity of government. The effort was bipartisan. Because of the Cold War, a nuclear shadow loomed over every one of the four sections of the Twenty-Fifth Amendment, which was all about guaranteeing clear leadership structures to keep the government going through whatever personal crises came to officeholders.

The Twenty-Fifth Amendment includes careful directions for filling vacancies in the presidency (Section 1) and the vice presidency (Section 2). Section 3 of the amendment empowers the president to transfer his or her powers provisionally to the vice president in the event of a temporary disability, a provision that has been invoked numerous times since 1967, most of them relating to the famous presidential colon. When Ronald Reagan underwent colorectal cancer surgery, he transferred the powers of his office to Vice President

George H. W. Bush and resumed them when the anesthesia wore off. Several other presidents have temporarily transferred their powers over to their VPs prior to a colonoscopy, as when George W. Bush turned them over to Dick Cheney.

Section 4 was the one everyone was talking about in the Trump period, but it was the only one that had never been used. It's in the Constitution because the architects of the Twenty-Fifth Amendment wanted a framework for maintaining stability and coherence in the executive branch if the president suffered a crippling physical or mental disability but was unable, for whatever reason, to transfer their powers to the vice president on their own. Under Section 4, the vice president and a majority of the Cabinet *or* the vice president and a majority of "such other body as Congress may by law provide" have the power to determine that the president is "unable to discharge the powers and duties of his office." This can be for medical reasons, whether physical or mental, such as a president's falling into a coma or suffering a debilitating stroke or experiencing a schizoid break or, indeed, any other functional nonmedical reason—perhaps the president has been kidnapped or vanishes with a lover, as a Republican governor in South Carolina (now a colleague of mine in the House) once did. Once having determined that such a presidential inability exists, the vice president and the Cabinet majority or a majority of the body established by Congress for these purposes then has the power to transfer the duties and powers of the presidency to the vice president until the crisis is over.

Tommy thought Section 4 of the Twenty-Fifth was a stunning find, but he wondered whether this provision was problematic because it could be used "for political reasons to get the president out without using impeachment"—which was the right question. But my answer was no.

To begin with, the key actor under the Twenty-Fifth is the president's own handpicked vice president, who has demonstrated steadfast devotion to the president, at least since Aaron Burr's time in office ended, so there would have to be a serious presidential health crisis for the vice president to act. Moreover, the president's Cabinet—again, handpicked and loyal to a fault for most of our history—would act as a check on a runaway vice president because a majority of its members would have to sign off on any Twenty-Fifth Amendment transfer of power. It is true that the president would not have named members to the "other body" that Congress might appoint to act with the vice president (something Congress has still never done), but here the vice president him- or herself is the backstop check against a breakaway political power grab by Congress.

Furthermore, in any ultimate power struggle under the Twenty-Fifth Amendment, the president has significant structural advantages. If the vice president and the Cabinet or a congressionally appointed body determined that the president was unable to execute the powers and duties of the office, the president would have the right to contradict them and assert his or her capacity. This leads to a final showdown that favors the president's resumption of power, because it takes two thirds of the House and Senate to place the power of the presidency into the hands of the vice president. This means it is far more difficult to invoke Section 4 of the Twenty-Fifth Amendment than it is to impeach and convict a president for high crimes and misdemeanors, which requires only a simple majority vote in the House followed by a two-thirds Senate vote to convict.

All of these checks explain why Section 4 has never been used since 1967. But, as I explained to Tommy, the purpose of my bill, H.R. 1987, was to permanently establish the "other body" called for in Section 4. The Trump Cabinet, already riddled with vacancies and acting secretaries, was comically sycophantic and invertebrate around the president. Tommy agreed that these "toadies and lickspittles" (a classic Tommy phrase) could not be trusted to act even if the president's arrogant behavior sank into complete derangement and dereliction of duty. Therefore, what we needed, I argued, was an independent commission of physicians and psychiatrists who could make a relevant diagnosis about the president's physical and mental health.

Tommy zeroed in immediately on the flaw in my thinking. Composing the commission with only physicians and psychiatrists to judge the president's physical and mental health would "medicalize" what was essentially a practical and *functional* judgment under the Constitution. The language of Section 4, he showed me, invited an inquiry about whether the president is "unable to discharge the powers and duties of office," a condition that could involve numerous different circumstances, including the possibility of the president's going missing or joining a monastic religious cult in another country.

"You don't need a doctor or psychiatrist to make the judgment that the president cannot be located," Tommy pointed out. "Conversely, you could have a president who has a clear medical condition even under the *DSM* but who is nonetheless still able to execute the powers of his or her office. Remember, they say Abe Lincoln was depressed or had bipolar condition, and that's your main *man*. Franklin D. Roosevelt had polio and was in a wheelchair. But those are two of the best presidents of all time. A physical or mental disability is not a constitutional disability. But if the psychiatrists dominate [the commission], they might think that finding a clinical condition is the end of the analysis. You need people who know the *presidency* more than people who know psychiatric jargon."

Tommy was absolutely right about this, and his sensitivity to people being designated, classified, and defined by psychological categories gave him excellent insight on this point. The commission would need disinterested members with executive branch experience (such as former presidents, vice presidents, and Cabinet officers), who could speak to the practical demands and functional requirements of the office. Physicians and psychiatrists could certainly be helpful as commission members, because medical evaluations would be *relevant* in many cases to a president's inability to conduct the duties and responsibilities of the office, but these evaluations must be treated as evidence only and not as any kind of a conclusive and authoritative judgment.

Tommy's insightful questioning led me and my legislative assistant Devon Ombres to revise my initial draft of H.R. 1987, which was top-heavy with physician and psychiatrist members on the commission, to include many more former executive branch officials, whose judgments are more closely tailored to the constitutional language. All his points influenced H.R. 1987, which drew sixty-seven cosponsors in my first session of Congress, all of them, alas, Democrats.

In October 2020, in the wake of President Trump's alarming mental and behavioral deterioration, his personal COVID-19 diagnosis, and revelations that he was far sicker than he was letting on publicly—and after some harsh public words traded between Speaker Pelosi and Trump over his superspreader events and his outlandish irresponsibility related to the pandemic—Speaker Pelosi heard me out. In her inimitable, voracious way, she assimilated in a day or two the history, structure, meaning, and relevance of Section 4 of the Twenty-Fifth Amendment, something that had taken me many months to decode and understand. She quickly grasped its importance for maintaining the stability of the presidency and the continuity of government, especially in the age of COVID-19, with so many elected officials, including Trump, coming down with this serious, crippling, and often deadly disease. The Speaker, who had spoken publicly of the effect of powerful COVID-19 drugs on a person's mental state, welcomed the opportunity to lift this discussion up to a more rigorous and principled constitutional plane.

The Speaker asked me to conduct a press conference with her to explain the Twenty-Fifth Amendment and define my legislative approach to it to the media, Congress, and the country. On October 9, the two of us went down, masked, to the press center in the Capitol. We set forth the terms of my legislation: a bill to establish a permanent bipartisan and bicameral body made up of sixteen members, half appointed by the majority party and half appointed by the minority party, half of them physicians and mental health specialists, half of them former high-ranking government officials such as former secretaries of state and defense, former surgeons general, former presidents and vice presidents, and so on. These people could assess and determine presidential incapacity in the event of a crisis, while the vice president would always remain the controlling force; nothing would happen without him or her.

I considered this a remarkably successful public seminar, and I had wanted to try to move the bill to hearings and passage right then—which would have meant having had the body in place now and ready to act. But ultimately, the Speaker felt the bill would have been deemed way too political then. And in fact, I had received a concerned phone call right after my press conference, from Cedric Richmond, the congressman from New Orleans, a key Biden adviser and as fine a political mind as we have, warning us away from pressing any further on the matter. So, trusting his political judgment and that of the Speaker, I dropped the matter in the 116th Congress. It may or may not have been the right choice. Politically, it probably was, but from the vantage point of January 6, I saw that my decision to do so had left us essentially defenseless against President Trump's rabid derangement in lying about the election results and fomenting violence against Congress. We could impeach him for his high crimes, but there would be no trial in time to remove him. Failing to aggressively build out the architecture of the Twenty-Fifth Amendment, it seemed now, had been a fateful decision; we were completely dependent on the Cabinet and vice president alone to act.

As I waited to return to the floor, I kept revising my remarks in response to the GOP effort to disqualify Pennsylvania's electors. The simpler the better, I said to myself, both because people were exhausted and wanted to go home to sleep, but also because I was getting tired and not necessarily thinking straight. I closed my eyes and drifted off in my chair.

When I woke up, it was after midnight, and Julie was telling me the coast was clear and it was time to go over to the floor to hear the last objections, which would be to Pennsylvania. (Although the GOP had chosen to ditch the Georgia objection—they probably wanted to avoid our speeches about the president's efforts to get Secretary of State Raffensperger to commit election fraud—they continued to rail in public about the votes in that state.) We lumbered into the mass of exiting members and staff and left for the floor. Sleepwalking our way underground with police escorts, through the Cannon Tunnel to the Capitol, we passed the high school artwork hanging there, one winning piece per district. This art usually lifted my spirits. I ordinarily like to stop and admire Maryland District Eight's young artist and some other favorites, but I was exhausted and grieving, and all this art by young people was just making me think about Tommy and all the lost promise in our world and how tawdry and desperate this political life had become.

Arriving on the House floor, I saw Kevin McCarthy and Matt Gaetz. I saw Paul Gosar. It was eerie bumping into so many of our GOP colleagues after this violent interruption, because multiple gnawing mysteries were now settling over us: I wondered if any of them had encouraged or facilitated this assault. I wondered whether all the bloody, vicious violence from the *outside* had been coordinated with the outlandish legislative maneuvers on the *inside*. And I wondered whether GOP members might now refrain from denouncing the insurrection because they shared the same object that night of preventing Biden from being declared the winner.

Most hauntingly of all, I worried about this: Did they actually believe, *really* believe, that the election had been stolen and that Donald Trump had actually won it? And if so, did they countenance all this bloody violence, and would they whitewash it and endorse it into the future?

Some preliminary answers would arrive shortly. Pennsylvania congressman Scott Perry and Missouri senator Josh Hawley, who had given the thumbs-up to the Trump mob while outside the Capitol and had even raised a power fist salute to them, objected to Pennsylvania's electors being accepted by Congress. The two chambers proceeded to re-create in desultory, middle-of-the-night fashion the Arizona debate from before. But, already, you could perceive the outline of an instant Orwellian rewrite of the day's events. The main themes came together in the performance of Matt Gaetz of Florida, who spoke in favor of rejecting Pennsylvania's electors but who never explained how a single act of corruption or fraud took place in Pennsylvania or how a single illegal vote was cast. Rather, he set forth the principal lines of rhetorical attacks Trump's acolytes would rally behind to deny and deflect from the horrors we had just experienced in our workplace.

Gaetz rose and, before getting to the business at hand, invoked the reception I had received when I spoke on the Arizona objection:

> *Madam Speaker, one of the first things we did when the House convened today was to join together to extend our grace and our kindness and our concern for a colleague who has experienced just an insurmountable amount of grief with his family. And I want all of our fellow Americans watching to know that we did that because we care about each other and we don't want bad things to happen to each other, and our heart hurts when they do.*

When I heard this statement, I was touched all over again, and I was prepared to hear a follow-up ringing denunciation of the violence that had just overcome our workplace and threatened our lives. I was excited that he might rise up in defense of Article I and deplore President Trump's whipping up of the mob to attack Congress, even if such a statement were to precede a bunch of false claims about how Pennsylvania had manufactured thousands of false votes (which was what Trump was asking Raffensperger to do in Georgia). Instead, what came out was this:

> *Another important point for the country is that this morning, President Trump explicitly called for demonstrations and protests*

to be peaceful. He was far more—you can moan and groan, but he was far more explicit about his calls for peace than some of the BLM and left-wing rioters were this summer when we saw violence sweep across this nation. Now, we came here today to debate, to follow regular order, to offer an objection, to follow a process that is expressly contemplated in our Constitution; and for doing that, we got called a bunch of seditious traitors. Now, not since 1985 has a Republican president been sworn in absent some Democrat effort to object to the electors; but when we do it, it is the new violation of all norms. And when those things are said, people get angry.

I could feel my blood boiling. Democrats and Republicans alike may have technically objected to electors in the past, but no one in our lifetimes had ever raised insurrectionary mob violence to block the counting of votes or to try to force the vice president to proclaim a power to unilaterally reject states' electoral votes and then force the contest into a contingent election. Meantime, all of us knew that Donald Trump had been encouraging violence and flattering the extreme right as a political strategy ever since he ran for president. Whether it was covering up for the fascists and neo-Nazis in the Charlottesville "Unite the Right" rally or ruthlessly attacking Michigan governor Gretchen Whitmer for her mask mandates and galvanizing far-right extremists in her state, which immediately preceded an extremist plot to kidnap and assassinate her, violence was a way for Trump to mobilize people outside the GOP establishment and terrify his opposition across the political spectrum. We had seen footage of him whipping the January 6 mob into a frenzy during his rally at the Ellipse, telling them that they had to "fight like hell or you won't have a country anymore."

Yet here Gaetz was, already launching the new January 6 corollary to the Big Lie: it wasn't a Trump mob that had stormed Congress; it was Antifa. We were just within an hour or two of having secured the Capitol Building, with hundreds of people badly wounded and traumatized at hospitals throughout DC, Maryland, and Virginia, and with many members of Congress still jittery and waiting for the other shoe to drop. But Matt Gaetz, prominent right-wing spin doctor—this was before his own serious legal problems with the Department of Justice materialized—was already preparing ideological subterfuge and a counterattack, to turn this public relations debacle against shadowy left-wing forces.

> *And I don't know if the reports are true, but the* Washington Times *has just reported some pretty compelling evidence from a facial-recognition company showing that some of the people who breached the Capitol today were not Trump supporters. They were masquerading as Trump supporters and, in fact, were members of the violent terrorist group Antifa. Now, we should seek to build America up, not tear her down and destroy her. And I am sure glad that, at least for one day, I didn't hear my Democrat colleagues calling to defund the police.*

In fact, all those hot-off-the-presses reports about Antifa organizing the attack were a lie, pure propaganda. Antifa had had nothing to do with it, but pumping that concoction into the media bloodstream on the night of the attack gave Republicans a rhetorical antidote to counter the spreading outrage against Trump and his assembled violent insurrectionary forces. In fact, as National Public Radio reported, the new and improved Big Lie—that it was members of Antifa who had stormed the Capitol—was mentioned more than *four hundred thousand* different times online over the next twenty-four hours. There is also evidence that the lie began spreading just as the siege itself began, which strongly suggests that organizers of the insurrection had designed a classic disinformation operation in advance to accompany and camouflage the siege. I saw Texas attorney general Ken Paxton floating the Antifa-did-it theory on January 6 as well.

Planted in these first hours following the siege, the Antifa diversion has continued to thrive in right-wing circles and may never vanish, becoming a case study in how online misinformation aims to rewrite history both in the moment and in perpetuity. When Officer Michael Fanone testified on Tuesday, July 27, 2021, before the newly formed House Select Committee to Investigate the January 6th Attack on the United States Capitol, he spoke of the brutal injuries he suffered at the hands of the insurrectionists, which gave him a heart attack, traumatic brain injury, countless other wounds, and posttraumatic stress disorder. But it was no time at all before right-wingers online wondered whether Fanone had simply been mistaken by protesters for an "Antifa" fighter, even though, as he pointed out at the hearing, he was in his police uniform that day. Of course, the premise of this Bizarro World insinuation is that the "fake" MAGA and extremist protesters, who were actually Antifa fighters, according to right-wing dogma, had attacked a police officer . . . because they thought he was Antifa—which of course makes no sense. In right-wing conspiracy theory and the land of the Big Lie, we had transcended the world not only of fact but of simple logic, too.

By the same token, Gaetz knew that there was no Democratic support for any actual legislative effort in Congress to "defund the police," but he was anticipating the GOP's need to promote this fiction, as the party would be working overtime for months or years to downplay and dismiss the brutal face-to-face violence unleashed against hundreds of our police officers, more than 140 of whom were injured by insurrectionists wielding baseball bats, hockey sticks, steel pipes, flagpoles, bear spray, and other unknown chemical irritants.

When it came to the actual matter at hand, Gaetz never once identified a single act of electoral fraud or corruption in Pennsylvania, much less one that could have altered the outcome of the election that Biden won by more than 80,000 votes. For Trump, Gaetz, and friends, it was enough simply to *assert* that there had been systemic electoral fraud that changed the outcome. No evidence at all was necessary. If Trump told his followers something was true, then it was. He invented his own facts and then communicated them to his worshipful true believers, who not only immediately believed, *knew*, them to be true, but also quickly vilified and ostracized anyone who questioned their dogma.

This system of authoritarian political propaganda was purely Orwellian. Had Gaetz not read *1984* in his Florida high school? Perhaps not. I have kept a copy on my desk since the night Donald Trump and I were first elected. Sitting there, listening to Matt Gaetz's speech, I couldn't help but recognize that there was an alternate Stalinist–Fox News history being written before our very eyes, a record of events designed to recast what we'd all just lived through into something completely different from the reality. Gaetz's speech and the various derivatives it would inspire were the very definition of gaslighting, political revisionism that aimed to rewrite the history of this day before the blood on the floor had even been mopped up.

As we neared the end of this excruciating process, Adam Schiff and I spoke last for the Democrats. Schiff yielded to me to close out, and I chose to focus on the peaceful transfer of power:

> *When you think about it, the peaceful transfer of power is the central condition of maintaining democracy under the rule of law. That is why the famous election of 1801 was such a big deal. When John Adams relinquished the presidency to his passionate adversary and lifelong friend Thomas Jefferson, it was the first peaceful transition of power between parties in a democratic republic in the history of the world. And Adams said, as he rode back to Massachusetts from Washington, that he did this because we are a government of laws and not of men.*
>
> *We will betray this principle if we trade a government of laws for a government of men or, even worse, a single man, or an impressionable and dangerous mob intent on violent sedition and insurrection against our beloved democratic republic.*
>
> *Here is Abraham Lincoln right before the war. "At what point, then, is the approach of danger to be expected? I would answer,*

if it ever reaches us, it must spring up amongst us. It cannot come from abroad. If destruction be our lot, we must, ourselves, be its author and its finisher."

Madam Speaker, my family suffered an unspeakable trauma on New Year's Eve a week ago. But mine was not the only family to suffer such terrible pain in 2020. Hundreds of thousands of families in America are still mourning their family members. Many families represented in the Congress are still mourning their family members who have been taken away from us by COVID-19, by the opioid crisis, by cancer, by gun violence, by the rising fatalities associated with the crisis in mental and emotional health.

Enough, my beloved colleagues. It is time for America to heal. It is time for our families and communities to come together. Let us stop pouring salt in the wounds of America for no reason at all. Let us start healing our beloved land and our wonderful people.

The objection to Pennsylvania's electors was then rejected by a vote of 282–138, with 11 members not voting. We proceeded to declare the counting over. Joe Biden won the presidential election by a vote of 306–232, the margin that was clear two months ago. Amazingly, a majority of the GOP caucus still voted to nullify electoral votes after the rampaging mob violence overran the Capitol.

With cops and National Guard troops everywhere, I staggered back to the Rayburn garage in the middle of the night to meet Julie, who was waiting for me in the car. At 3:37 a.m. on January 7, I had the energy to place just one call to one person I needed to talk to, Jamie Fleet, staff director of the House Administration Committee and the guy I called "the Pelosi whisperer." More than anyone I knew, Jamie kept track of what the Speaker was thinking and feeling and what she needed to hear from members. I wanted to make sure first he was okay and then to tell him that I would be pushing the next day—well, later that day in fact—for impeachment and a new approach to the Twenty-Fifth Amendment to deal with the current crisis. He sounded pleased, a good sign, and we talked schedule for a few moments. Before we got off, I told Jamie how much I treasured a political trip he and I had taken one fine weekend to Pennsylvania in the last election cycle for Democracy Summer. A former member of the Gettysburg Borough Council who got elected while a college student, Jamie knew Pennsylvania like Lincoln knew Illinois and he taught me unstintingly about politics in the Keystone State.

I slept all the rest of the way home. When we got there, I bade Julie good night and thanked her for her heroic and devoted work.

Inside, our dogs Toby and Potter did not even wake up to greet me. I made my way upstairs and collapsed in bed.

"What happened?" Sarah murmured.

"They tried a coup and an insurrection," I said. "But Biden won. He's going to be president."

CHAPTER 7

"THIS IS ABOUT THE FUTURE OF DEMOCRACY"

Duck hunting is a lot of fun until the ducks start shooting back.

—REP. SAMUEL BELLMAN, MY GRANDFATHER

On Thursday, January 7, it wasn't just the Raskin family that woke up in trauma and mourning.

It was all America.

From coast to coast, on the left, center, and maybe even in some quarters of the right, people were aghast, petrified, and spellbound by this brutal violence and the existential enormity of what had just happened. The television footage, drawn from coverage by intrepid reporters and imprudent online postings of cell phone videos by boastful rioters, exposed the nation to scenes of vicious assault against our police officers; of windows being smashed and property being destroyed; of punches being thrown and eyes being gouged; of extremists clubbing officers with steel pipes, baseball bats, hockey sticks, Confederate battle flags, Trump flags, and American flags; of rioters chanting "Hang Mike Pence"; of creepy mob leaders prowling the Speaker's suite; of freakish characters like the "QAnon Shaman" wearing a Daniel Boone–style fur and a hat with protruding horns and red-and-white face paint, carrying a bullhorn ("Ten-to-one odds that's the number one Halloween costume in America in October of 2021," my brother Noah observed when the image came on the news), who casually usurped the desk of the presiding officer of the U.S. Senate; and of the heroic Capitol officer Daniel Hodges screaming in agony in a doorway as he was crushed by Trump rioters in a scene of medieval torture in the twenty-first century.

Disturbing photos rocked America: the iconic-overnight shot of the rioter bearing the Confederate battle flag in the Rotunda; pictures of Proud Boys on the march and Oath Keepers in paramilitary garb snaking their way through the crowd; of the heavyset brawler in the "Camp Auschwitz—Staff" sweatshirt leering at a police officer; and an endless panorama of faces contorted with murderous rage arrayed opposite police officers writhing in pain and anguish. And behind all the destruction was the master of chaos and evasion himself, the president of the United States, who had incited the mob from a dais festooned in POTUS branding, at the rally right before this obscene attack on Congress and the 2020 election.

Safely at home in Takoma Park, Sarah and I were bleary-eyed from lack of sleep. I had gone to bed past 4:30 in the morning and woken up four hours later to do TV interviews. We had hundreds of texts and emails from friends checking on us—it started in the early morning with our friends Jeffrey and Lora Drezner and ran through midnight when I heard from my friend Sally Truitt in Colorado and then Joanne Lichtman in California.

We mourned Tommy throughout the day, comforted Tabitha and Hannah and Hank and Ryan, and were pounded by all the feverish tidings on television of injuries and destruction suffered on the Hill. As Hannah said in the afternoon after reading the news and watching the gory coverage, "The more you learn about it, the worse it gets." What kept us going that day was the bottomless goodwill of family, friends, and neighbors, who piled up on our front porch a fresh abundance of flowers, notes, and other expressions of love, despite COVID-19 and even with Toby and Potter standing guard and demanding their sustained attention. The darkness of our mood was thus broken intermittently by the light brought by so many expressions of concern, and while I had rebelled for several days at even the thought of eating, I now looked forward to the dinners being sent over by friends, family, and neighbors each night through Andrea Dettelbach, who stepped forward to coordinate the crush of kindness from our community. Our house was filled with close friends and family who had come to console us and help manage the interplay of painful dark silence and swirling mad chaos that had suddenly overtaken everything.

From January 7 forward, our emotional baseline of grief and despondency would have to assimilate somehow the additional new trauma of this brutal assault on Congress and the presidential election—trauma that had driven Tabitha and Hank to hide under a desk for hours, trauma that had sent me running for my life, trauma that had scared the daylights out of Sarah and Hannah and everyone else in our family, and trauma that had radically changed the terms of American politics.

Sarah and I were naturally concerned about the girls. Tabitha, who was already reeling and devastated by Tommy's death, was plunged into a new round of shock and grief. Tommy and Tabitha had been inseparable, only two years and one day apart in age, Tommy born on January 30, 1995, and Tabitha on January 31, 1997. In fact, Tabitha's enduring nickname, "Teetah," came from the two-year-old Tommy's inability to pronounce her name properly. In the same way that Tommy had followed his elder sister to Amherst so they could overlap in college, Tabitha had followed Tommy there, although she ended up transferring after sophomore year to the University of Maryland at College Park. The early photos of Tommy and Tabitha together show them intertwined and squealing in laughter and delight; they would have "feet fights" for hours on end; they spoke their own language to the dogs; they knew the movie version of *To Kill a Mockingbird* from start to finish and loved to come to my Constitutional or Criminal Law classes and act out entire scenes between Jem and Scout, something my former students seemed to remember far more vividly than my lectures on *Marbury v. Madison* or *McCulloch v. Maryland*. Now Tabitha was bereft, and overcome by the ugly violence and terror that had washed over our lives.

Hannah, too, was engulfed in agony and sadness. Tommy had looked up to her always: she had taught him how to read when he was little, taught him his first words of French when we went to live abroad; she brought him to parties with her; she babysat him. At Amherst, Tommy ended up rooming his sophomore year with Hannah's boyfriend Nick, and they all had taken good care of one another. Now Hannah Grace was demolished by this sequence of jagged and incomprehensible events. Living on the West Coast was especially hard for her, as most of her friends out there had never met Tommy and they also did not feel the shock and gravity of the insurrection in the same way. On January 6, she had not realized the severity of the violence or the danger we were all in. Meantime, Hank was doubly rocked by the loss of his brother-in-law and by the shattering political violence that had just threatened his own life. We did our best to console and comfort them, but everyone was subdued and numb.

Once more unto the breach: on January 7, I saw the whole Democratic Caucus and progressives across the country shake themselves out of the shock and stupor and resolve to fight for America.

I threw myself back into the political struggle to defend democracy. Trump had two weeks in office left to go, but every minute of those two weeks would pose a clear and present danger to the American people if we failed to arrest his descent into power-mad violence. He had just proven that. Now was the moment to oust him and take down the right-wing authoritarianism he had cultivated and that was spreading like a second deadly plague across the land. In a matter of hours rather than days, I would have to decide which agenda to press and how, and how to respond to the current crisis: Would it be the Twenty-Fifth Amendment or impeachment?

Surely, it would have to be both.

Between the two, impeachment was the more natural option. It was broadly familiar because the House had just voted a year prior, in January 2020, to impeach Trump for abusing his power and obstructing Congress with his "Ukraine shakedown," when he converted foreign military and security aid to Ukraine, extended by Congress to resist Russian aggression, into personal political leverage to extract a promise from President Zelensky to publicly announce a bogus criminal investigation into Joe Biden. The resulting impeachment trial in the Senate ended with votes of 47–53 and 48–52 on the two counts. (The sole Republican defection on one count was Utah's Mitt Romney, who quickly suffered the wrath of Donald Trump and multiple tongue-lashings by the then president.) It seemed pretty obvious that we would have to impeach Trump for his role in organizing and inciting the violent insurrection, which was basically a magnified and intensified version of his attempt to subvert and throw the election with the far more subtle Ukraine shakedown. The press was already predicting that the House would impeach—and that any new House impeachment would result in the exact same partisan stalemate and failure to convict in the Senate.

By contrast, activating a Twenty-Fifth Amendment, Section 4, solution was a complete long shot because it had never been done before and because most Americans—indeed, most members of Congress—could not even describe how the Twenty-Fifth Amendment or its four main provisions worked. Most Americans would instinctively resist the idea that the president's powers could be transferred to the vice president because of an inherently subjective judgment about an "inability" to do the job. That said, the Twenty-Fifth Amendment was slightly less threatening to Republicans than impeachment, because it did not necessarily implicate their party in a program of insurrectionary crimes and constitutional high crimes and misdemeanors. As one GOP southerner put it to me, "Well, if comes to it, and I gotta choose, I'm choosing the Twenty-Fifth, because *anybody* can lose their mind."

Several GOP members and thinkers had actually floated the Twenty-Fifth Amendment as a solution for dealing with the looming dangers of the president's final troubled weeks, when he came to realize that the party was over. This route did not have the fierce partisan sting of impeachment. Even Trump's former political strategist Steve Bannon had raised it demurely at one point, suggesting that the Twenty-Fifth Amendment would become a greater danger to Trump than impeachment.

Nonetheless, all we could do as Democrats was ask Pence and the Cabinet to get together and act, pressuring them to do the right thing. I resolved that I would suggest to Speaker Pelosi and Leader Hoyer that we pass a nonbinding resolution asking Vice President Pence to assemble the Cabinet and activate Section 4 of the Twenty-Fifth Amendment, to save us from any further violence and chaos. At the very least, it would demonstrate that we were acting not only to protect the country from a future Trump presidency, which impeachment and conviction could do through the disqualification procedure, but also to protect the country *right now* against the clear and overhanging danger of more violence in the waning days of the current administration. The appeal to the vice president and the Cabinet to act was not ideal, but it was our best bet from the standpoint of public safety. It would be a historic first in Congress to seek activation of Section 4 of the Twenty-Fifth Amendment, a first that would dramatize the gravity and danger of the moment and allow us to spell out the specific case for the president's overwhelming inability to meet the duties of office. It might even spur Pence into action, as the vice president, of all people, understood the profound danger Trump had placed everyone in the day before. The questions on my mind were where the Cabinet would be on the issue and whether Pence would scramble to appease the bully for his, Pence's, unprecedented disobedience on the sixth.

I decided to do what I always do in a moment of fateful decision making: start writing and, through the formality and intimacy of the writing process, struggle to arrive at the best path forward. I excused myself from the TV noise, the family conversation, all the loving consolation downstairs, and went up to our bedroom for several hours to write a draft of the resolution. This Twenty-Fifth Amendment Resolution would not be to the exclusion of impeachment, and indeed, working on it reaffirmed for me why the impeachment road was central. The Twenty-Fifth Amendment is focused only on the president's demonstrated negligent failure to meet his responsibilities, so it would not require the production of a meticulous historical record of the president's specific active involvement in constitutional crimes.

And that was precisely the record we could assemble in even a truncated impeachment process. Impeachment would offer the country some immediate political catharsis and clarity with regard to the president's leadership in inciting the violent insurrection against the union. We would make an important and definitive factual statement to the nation, to the world, and to history about the president's central role in mounting the January 6 attack. It would form the basis of a broader investigation about the events that befell us.

But what concrete immediate good could impeachment accomplish? Unlike the Twenty-Fifth Amendment, it offered no realistic hope of actually ousting Trump from the presidency in his final two weeks, because Senator McConnell would simply never call the Senate trial before January 20. So the practical effect—even if we actually got two thirds of the members to vote to convict in the Senate, which everyone was saying would be next to impossible—would be to formally declare Trump guilty and then to have a subsequent majority vote to disqualify him from ever holding federal office again. But if disqualification were to be the main payoff, this objective would set us up for the argument that "the whole thing is political and partisan," an early move by the Democrats to remove Trump from the playing field in 2024. That argument could be overcome only by our showing that

Trump was truly a clear and present danger to the American people, and that he had blatantly violated his constitutional oath to preserve, protect, and defend the Constitution. That would indeed become the core of our case.

The Speaker had scheduled multiple House leadership, caucus, and Judiciary meetings through the weekend to discuss what to do. I had spoken by the afternoon to close colleagues and to friends Joe Neguse, Ted Lieu, David Cicilline, Pramila Jayapal, Texas's Lizzy Fletcher and Sheila Jackson Lee, Pennsylvania's Madeleine Dean and Mary Gay Scanlon, Hakeem Jeffries, and Eric Swalwell. The Judiciary Committee progressives were unified: all wanted to impeach. They also backed my Twenty-Fifth Amendment call on Pence to do the right thing and transfer power over the president's manifest inability to govern. Although none of them thought that Pence and the Cabinet would actually respond, they liked the idea of giving Republicans in the House the opportunity to address the current crisis and to underscore the danger we were still in.

Neguse said the Speaker would be taking soundings far and wide through the weekend as to what members wanted to do, and he hoped to make sure we presented a unified front in favor of introducing and passing an Article of Impeachment on incitement the following week. This struck me as right, unless far more compelling evidence surfaced showing that Trump had actually organized the violent assault itself, in which case we could charge him not just with incitement to insurrection but also with a separate charge of either conspiracy to commit seditious insurrection or conspiracy to foment rebellion against the United States. Incitement struck me as the obvious and cleanest way to go, because we could make a compelling case based on already publicly available video both of Trump's incendiary speeches calling on his followers to "fight like hell" to "stop the steal" and of all the sickening violence that followed directly thereafter. An incitement charge would no doubt elicit from Trump a First Amendment defense to impeachment, but I was already forming rebuttals to that self-incriminating strategy, which ultimately relied on the absurd idea that the president of the United States had a free speech right to incite violent insurrection against the union.

Some of the members I spoke with said that Steny Hoyer was hesitant about the whole thing and uncertain that impeachment was the right way to go this late in Trump's term. I had not yet spoken to Steny, but this would not have surprised me. As a slightly more conservative majority leader to a feisty, natural-born-liberal Speaker, Steny rallied the more moderate and conservative forces within the Democratic Caucus and acted as the protector of "front-line" members in swing districts, who often wanted to dampen polarizing partisan conflict that might bring them face-to-face with the wrath of Trump followers. Still, if Steny was a conservative, he was an *institutional* conservative, one who loved the House of Representatives and its customs, procedures, and protocols, and I knew he would be appalled and outraged by the insurrectionist violence that had wrecked our workplace and killed people. I called Steny, and he said he had no doubt that Trump deserved to be impeached, but he wanted to hear from more members about what they thought the consequences of impeachment would be. I felt certain that sentiment in favor of impeachment and invoking the Twenty-Fifth Amendment would be overwhelming.

As I worked on the Twenty-Fifth Amendment Resolution, I spoke twice on January 7 to the Speaker, who called to see how Sarah and I and the girls were doing. In times of distress, Pelosi communicates with an all-enveloping sweetness and compassion that are quite startling for a politician, a job description that usually begins and ends with "toughness" or "hardness." That is not Pelosi. Far from being squeamish or evasive about what had happened with Tommy and why, the Speaker was eager to talk about it. Unlike many of my colleagues, who were reluctant to talk about Tommy directly or even utter his name, she kept asking about Tommy's friends and his dreams and saying that she knew Tommy would be proud of his sister's bravery and that he had been with us through this crisis. She had known a lot of people who dealt with a suicide in the family, and she said that "you can never know someone else's pain and what they are going through." She was quite stoical about events.

The Speaker was very concerned about Sarah and how she was doing; I told her that both Sarah and I felt the tiniest bit better each day but found it hard to talk about Tommy without crying. She said that many people who had never met Tommy were getting to know him through the statement we had written and released immediately after we lost him. She said we had done a lot of people a favor by our openness.

I have always loved Nancy Pelosi, but now I began to feel just how much we all needed her for our most desperate times. Her heart and spirit felt boundless.

The Speaker and I spoke also of our shared political crossroads: "My phone is blowing up with impeach, impeach, impeach," she said. "He has pushed everyone over the edge. We must proceed deliberately and quickly."

It was in moments like this one that the Speaker revealed her true political genius. She had a way of asking for other members' opinions and ideas to help deepen her own understanding while at the same time moving them toward her sense of an emerging consensus, keeping everyone on board in their own fashion while amplifying strong voices and refashioning powerful phrases, words, and arguments she had collected along the way. At every decision point, she loved to quote Lincoln on public opinion: "Public sentiment is everything. With it, nothing can fail; against it, nothing can succeed. Whoever molds public sentiment goes deeper than he who enacts statutes, or pronounces judicial decisions."

The Speaker learned in conversation and fast reading. After a few hours of diving into discussion with colleagues and staff about a subject like impeachment or the Twenty-Fifth Amendment, she resurfaced with not only an authoritative theory of how to proceed but also the strongest arguments, the most vivid analogies, the most resonant turns of phrase to light the path forward. I always saw her speaking in the caucus from the center of moral and political gravity, never explicitly saying what she thought needed to be done until the exact moment of truth arrived to strike forward, and then she was all-in, with everything put out there, an agent of the zeitgeist, of political necessity and historical inevitability. Along the way, she was always testing her arguments, trying out language, articulating and probing the consensus as it was unfolding.

She asked what I thought. I told her that I favored a strategy beginning with a resolution calling on Vice President Pence to assemble the Cabinet and activate the Twenty-Fifth Amendment to strip the president of his powers and transfer them to the vice president. This would make it clear to the country that we were in crisis and would remain unsafe so long as Trump was president, because he had proven himself to be a clear and present danger to public safety and the republic. We would give the vice president twenty-four hours to respond and then separately pass the Article of Impeachment charging Trump with incitement to violent insurrection against the union. I told her that incitement to violence against the union was one of the most serious possible violations of the presidential oath to "preserve, protect, and defend" the Constitution and the government—it is akin to treason—and that the First Amendment could not save your presidency if you breached your Oath of Office in that way.

She was curious about Trump's potential defenses, specifically the idea of a First Amendment defense to impeachment for treason or seditious activity. This, I shared with her, was an essentially comical idea. On that theory, a president could join a revolutionary insurrectionist group and call every day for violent destruction of the union, but Congress would be unable to impeach and convict him for his seditious advocacy or incitement because doing so would offend his free-speech rights. But impeachment is a "political question" in the constitutional sense; it is up to Congress to decide, not the courts, and it is not about throwing people into jail but about removing presidents who violate their Oath of Office to "preserve, protect, and defend" the Constitution.

"Just because a president uses words to commit his constitutional crimes," I said, "does not strip Congress of the power to impeach and convict him for betraying his oath."

The Speaker agreed with the analysis. I told her that Professor Tribe was available for a second opinion, because I knew he would be there for us—he had already texted me to see if I was okay— and I knew how much she trusted his judgment. I also knew he would agree.

"We must move forward fast next week," she said. "Love to Sarah and the girls."

As much as I supported the first impeachment, the Ukraine shake-down that was the basis for it had seemed abstract and impersonal to me. But now, suddenly, the events demanding the impeachment of Trump were, for me, vivid, sharply drawn, bloody red, and con-cretely personal. The violence we'd seen on January 6 was merci-less and unsparing. This new, brewing impeachment awakened deep memories from my childhood of another impeachment drive against a lawless and threatening president, an impeachment that was starkly personal for me.

In 1973, when I was ten years old, John Dean told the Senate Watergate Committee that President Richard Nixon had compiled an "Enemies List" of his critics, people whom he had targeted for hostile government action because they opposed his escalation of the Vietnam War and the political corruption saturating his ad-ministration. We soon learned that my father and his codirector at the Institute for Policy Studies, Richard Barnet, were high up on Nixon's Enemies List. I remember seeing a large, full-page cartoon in the *Washington Post* of a bunch of Nixon's putative enemies that contained caricatures of my dad and Dick Barnet, who were always pictured together. Their friendship even brought a little bit of humor when their FBI files were eventually released, revealing a conversation held between two FBI agents: One of them asked the other how he could tell Raskin from Barnet, and the other agent wrote back, "Barnet is the slightly better dressed one."

Hearing of the Enemies List and seeing the caricatures in the newspaper, and hearing that my parents' taxes were being audited (they were) and maybe even that FBI agents were rummaging through our garbage, brought me back to five years before, 1968, when I learned, at age five, that my dad had been "indicted"—a word I needed to have explained to me—in the Boston Five conspiracy trial for aiding and abetting draft evasion. While this case would be a source of fascination for me many years later, as a five-year-old, I found it terrifying. My dad was facing—as I suddenly heard on TV one evening, at the top of the news after an episode of *The Flintstones*—"many years in prison."

Many years in prison.

The news that my dad would be on trial and could potentially go to jail overwhelmed me. I remember feeling confused and disoriented, unable even to vocalize a question. I remember sitting on a sofa with Noah and our big sister, Erika (whom I called Sissy in those days), and wanting to ask her something but not knowing precisely what to ask. Finally, I asked her whether prison and jail were the same place, and she said yes.

But the iconic story in my family—I have no idea if it's true, but my mom recounted it to everyone, and it has planted itself in my memory as if I remember it directly—had to do with going to see the doctor. Our pediatrician—this much I know is true—was named Dr. Washington, a tall, elegant man with long, silver locks of hair. Dr. Washington was a direct descendant of President George Washington's brother John; his family had lived in DC since the founding of the republic, and he had an office in a stately apartment building in the Adams Morgan neighborhood, near our house, which was on Wyoming Avenue. In my little child's mind,

Dr. Washington *was* George Washington's brother, and I revered him. He was as tall as a tree.

In the story, as I sat up on the examining table, Dr. Washington looked at my throat and checked my eyes and ears, listened to my heart, and tested my knee-jerk reflex. Everything looked fine to him, but the little boy in front of him seemed sad and concerned. His lip quivered.

"Do you have anything you want to tell me or ask me, Jamin?" Dr. Washington said.

"Did you read the newspaper and know my dad is on trial and might have to go to jail?" I asked him.

"Yes," he said. "I saw that."

"Well, is my dad one of the good guys or one of the bad guys?" I said, about to cry.

And Dr. Washington responded, "He is one of the good guys. He is one of the best we have."

Later, according to my mom—and here you must recall that Barbara Raskin was not just a writer and a journalist but also a best-selling novelist who made every story better than it had any right to be, with gorgeous dialogue and striking details, and I loved her for that—Dr. Washington took my mother aside and told her all about our little conversation. And then he said: "Barbara, I am no longer accepting any payment from your family."

Now, when these events (allegedly) occurred, Lyndon B. Johnson was still president, and the attorney general was Ramsey Clark. But Richard Nixon was soon to be president, and when he was, he and Henry Kissinger intensified the brutality of the Vietnam War, ruthlessly expanded its scope, secretly bombed Cambodia, and moved to attack the peace movement and Nixon's critics.

Five years later, the Boston Five was what I thought of when I learned that my father was an "enemy" of the president. When the Watergate hearings blew the thin cover of legality off the Nixon administration and revealed the president's criminal "dirty tricks" and mobilization of the government to persecute his enemies, the need to impeach, convict, and remove this ungovernable president became an article of faith in our family and our community.

I was young then, but not too young to read the newspaper, and I always loved from afar then-Congressman Paul Sarbanes (the father of my colleague and friend in the House today, John Sarbanes), because from his place on the Judiciary Committee, he was the first member of the House to introduce Articles of Impeachment against Richard Nixon. The articles charged Nixon with abuse of power for ordering the Watergate break-in to obtain political intelligence against his opponents; for obstructing justice by unlawfully mobilizing the IRS, the FBI, and the CIA to persecute his adversaries; and for attempting to cover up a string of criminal misdeeds and corrupt actions. I remember going at age ten (unaccompanied by adults) with my friend Jay Spievack to a peaceful march demanding Nixon's impeachment. We chanted, "Jail Nixon, jail Ford / Stop the war and free the poor," and I had my first encounter with police tear gas. I felt in my bones even at age ten that impeachment was the people's weapon of self-defense against a corrupt president who acted like a king.

Perhaps not coincidentally, I became a professor of constitutional law and soaked up every word I could about executive power. For as long as I have studied the Constitution, I have seen, and I have taught, that impeachment is a critical instrument of democratic control over a runaway and lawless president.

I taught my students that our nation was conceived in popular revolution against a mad king. Indeed, the original Articles of Confederation did not even include a president—a role that was added only later, under the Constitution, to improve the efficient administration of the laws, but only under very careful strictures. The organizing democratic idea was representative self-government through Congress, and the Founders armed the legislative branch with ample means of self-defense (impeachment, the power of the purse, oversight) to stop any president from acting arrogantly above the law. Indeed, the whole premise of American constitutionalism and the rule of law is to block people in office from exercising power in unlawful and dangerous ways. Treason and bribery are the essential impeachable offenses because they betray both the national interest and the public interest for the president's private gain.

The impeachment tool reveals that, in the American form of government, Congress must be first among equals. There is a reason that Congress is found in Article I of the Constitution and not in Articles II or III, and there is a reason that Congress can impeach, convict, remove, and disqualify the president but the president cannot impeach, convict, disqualify, or remove *us*. That reason is that Congress is the representative voice of the people—and here, *the people govern*. The president's core job is to implement the laws we develop and to take care that they are faithfully executed. The beautiful and too-often rushed-through Preamble to the Constitution—"We the People of the United States, in Order to form a more perfect Union"—flows immediately into Article I, Section 1, and the emphatic statement that "All legislative Powers herein granted shall be vested in a Congress of the United States . . ." Note what happens in the text there: the sovereign power of the people to form the union, adopt a Constitution, and create a government flows immediately into the power of the Congress of the United States to govern.

Meanwhile, the powers of the president set forth in Article II are few and far between and end up with this forceful warning: "The President, Vice President, and all civil Officers of the United States, shall be removed from Office on Impeachment for, and Conviction of, Treason, Bribery, or other high Crimes and Misdemeanors."

The moment it became clear to me on January 6 that the unleashing of mob violence against Congress was Donald Trump's handiwork, I resolved that this frightful violation of Congress's right and duty to count the electoral votes was about the gravest constitutional violation imaginable and that Trump would have to pay the price for it. He would get his comeuppance for arrogantly and violently dismissing Congress's role in guarding democratic self-government.

• • •

To be clear, and as my right-wing critics liked to point out, I strongly championed impeaching Trump way before January 6 and, indeed, even long before the Ukraine shakedown came to light in 2019. For the entirety of Trump's presidency, I had been focused on an issue I felt was the most open-and-shut case for an impeachable offense that I'd ever seen—until January 6, 2021. It was a constitutional crime so egregious, something that I was convinced of so completely, that it set me apart from the Intelligence Committee–organized focus on the Ukraine foreign interference story and probably made it impossible for me to be a manager in the first trial, which was mildly disappointing at first but which turned out, for me, to be a blessing in disguise.

I was convinced from the start of his time in office that Trump was in plain and continuous violation of the Foreign Emoluments Clause found in Article I, Section 9, Clause 8, of the Constitution. This is America's original ban on foreign bribery and influence peddling. It commands that "no person holding any office of profit or trust under [the United States] shall, *without the consent of Congress*, accept of any present, emolument, office, or title *of any kind whatever*, from any King, Prince, or foreign state" (emphasis added). An emolument is a payment, so this is a prohibition on the president's taking any money at all from foreign governments without obtaining congressional permission for it. Yet Trump, who refused to divest himself of his corporate holdings while in the White House or even to stop doing business while president, was raking in millions of dollars from foreign governments and royals through the hotels, office buildings, golf clubs, and other commercial enterprises he owned.

Moreover, I believed that Trump was also in open and repeated violation of the *Domestic* Emoluments Clause, found in Article II, Section 1, Clause 7, the roughly parallel proscription that categorically prevents a president from receiving *any* compensation or benefit from the U.S. government and taxpayers (or the states) beyond his official salary.

With these provisions, the Founders wanted to establish the undivided loyalty of the president and other federal officeholders to the people of our nation. They knew that service in government would offer officials unlimited opportunities to line their pockets with cash from foreign rulers and wanted to make it impossible for anyone to do that without first obtaining the consent of Congress. They imposed the categorical ban on presidents and other officials taking any cash payments or benefits from the federal government (or the states) beyond their official salaries in order to preserve the integrity of government and to keep it from being converted into a cash cow and capitalist enterprise for the president and his family and friends.

For more than two centuries, American presidents of both parties had followed these strict commands. If they received an occasional gift from foreign officials and wanted to keep it—as when Abe Lincoln received the gift of an elephant tusk he loved from the King of Siam—they quickly came to Congress to seek permission, and Congress promptly granted its approval or, more commonly, refused it. (Lincoln was not allowed to keep the elephant tusk.) Otherwise, presidents steered clear of accepting any gifts or monetary payments "of any kind whatever" from foreign governments, and no president ever even applied to Congress to keep anything that would be worth more than several hundred dollars, much less millions of dollars. For centuries, it also would have been unthinkable for officers in the executive branch of government to be sent to stay at public expense at the president's own hotels, to eat at his restaurants, to play at his golf clubs, or to enter into contracts with his other business entities. It would have been unimaginable that the president's businesses would rent federal property or rake in millions of dollars from federal government disbursements.

All that changed with the inauguration of Trump, who thumbed his nose at both the Foreign and Domestic Emoluments Clauses . . . and then dramatically *expanded* his company's receipt of business from foreign and domestic governments. Whenever foreign

dignitaries checked into the Trump International Hotel in Washington, foreign governments rented space in the New York Trump
office towers, or heads of states or Secret Service agents gathered to
spend time at the president's Mar-a-Lago resort in Florida, Trump
businesses got paid. The meter was running. But he never notified
Congress about his foreign payments, much less asked Congress for
approval to keep them; nor did he return the money he was receiving
from the U.S. government and state patrons. Trump businesses just
pocketed it all—in stupefying contempt of the Constitution of the
United States.

In the face of public protest over his apparent violations of the
Foreign Emoluments Clause, President Trump decided, without
consulting Congress, to begin "donating" hundreds of thousands
of dollars, beginning with a modest payment of $151,470 in 2018,
to the U.S. Treasury to disgorge himself, allegedly, of the "profits"
he had earned from his foreign government business partners. These
periodic unaccounted, undocumented, and unapproved "voluntary
payments" fell comically short of what the law requires and actually gave the game away entirely. The Constitution does not forbid a
president from collecting money from foreign governments without
turning over what he unilaterally deems to be his "profits." It prevents him from receiving *any money at all* (gross or net) from foreign governments without first obtaining *Congress's consent* to take
it. Period. This constitutional bar is there to keep presidents from
selling out the public interest to foreign powers in blatant or sneaky
ways, in scams big or small.

But there is no doubt in my mind that Trump also raided the United
States' federal piggybank directly, regularly collecting certainly
hundreds of thousands and almost certainly millions of dollars in payments made to his businesses by the Secret Service, the Department of
Defense, the Department of Commerce, and numerous other federal
agencies and departments that stayed at or did business with the Trump
Hotels, Trump country clubs, Trump golf courses, and other Trump
enterprises. These payments are strictly forbidden by the Constitution,

and even if Congress *wanted* to assent to them, we could not. Trump was remarkably blasé about what he called the "phony Emoluments Clause." Defending his outrageous plan to host the G-7 Summit at his hotel and golf property in Doral, Florida, in 2019, a plan that he was forced to drop after a torrent of public protest, Trump and his attorneys argued absurdly that the constitutional ban on a president's receipt of foreign government payments does not include "business transactions." Meantime, Trump would often brag that he returned his handsome $400,000 salary to the U.S. Treasury every year. But that is actually the only federal payment he was *allowed* to keep. The rest of it was blatantly unconstitutional, and these ill-gotten gains must be returned to the U.S. Treasury. "Keep your salary, Mr. President," I used to say. "Give the rest of it back."

I fault one person and one person alone for our failure to impeach the president for converting the presidency into a for-profit enter-prise, and that is myself. As one of only a handful of constitutional lawyers in Congress, I was one of a few members who understood how the Emoluments Clauses worked and who had the capacity to move the caucus and the leadership on it. But for the life of me, I was unable to persuade the leadership or move enough of the more conservative Democrats to place this cardinal constitutional sin on the agenda. Although we had seen published reports of lots of both foreign and domestic government payments to the president, Trump sandbagged our investigations and refused to cooperate with any demands for relevant information. Meantime, we failed to come up with an electrifying eyewitness whistle-blower or to identify a single dramatic and memorable example of all the profit-making activity that would stick in the public's mind.

We never made the issue "pop," as the pollsters say. The word *emolument* seemed too academic and archaic to most people, and the violations somehow always struck most of my Democratic col-leagues as too "abstract" and my Republican colleagues as petty and picayune. Yet I was convinced, and I remain so, that the public would have grasped the essential lawlessness of the Trump presidency a lot easier and a lot faster through his self-dealing Emoluments rip-off constitutional-crime spree than by way of the complex and byzantine tale of Rudy Giuliani's galloping foreign political adventures and coercive diplomatic pressure that constituted the Ukraine scandal.

I thought that a few weeks of serious hearings and explanation would make *emoluments* a household term. Everyone can understand the president converting the Oval Office into a moneymaking enterprise, especially considering Trump had come to office with a reputation as a bankruptcy-declaring rip-off artist embroiled in hundreds of lawsuits at any one time. Yet we had not shown how his basic nature found expression in the continuation of Trump enterprises in office, with his adult children exercising "day-to-day management" of his expanding business empire. While I kept writing explanatory op-eds about the issue, sent out Dear Colleague letters, and tried to explain to the Speaker and Steny Hoyer the political necessity of bringing this central constitutional offense to public attention, I failed to change the minds of the Democratic leadership.

Trump's apparently unceasing collection of unlawful foreign government and domestic government payments was not some kind of accidental side grift. To me, it was his *core business model* and *paramount personal commitment* while in office. Indeed, if I am right about that, it may have been the exact reason that he fought so hard to *stay* in office and therefore the possible ultimate motivation both for the Ukraine shakedown and for the explosive insurrectionary violence that overran us on Capitol Hill on January 6. From the outside at least, it looked as though he needed to be in the White House to keep all the money flowing while holding the criminal prosecutors at bay.

My conviction that Trump should have been impeached long before the Ukraine shakedown ever happened created a pretty large distance between me and the 2019 Ukraine investigation, which was centered in the Intelligence Committee and quarterbacked by its very effective chairman, Adam Schiff. As a member of the House

Oversight Committee, I joined in the Ukraine witness interviews in the summer and fall of 2019 and worked hard to master the details of this numbingly complex story with a lot of difficult-to-pronounce and impossible-to-remember names. While I was entirely persuaded that Trump should be impeached for coercively demanding President Zelensky's participation in the 2020 American presidential campaign, I also felt that the public would have an arduous time following the intricacies of this bizarre offense. Its complexity would give the GOP masters of propaganda and obfuscation an opening to easily distort key parts of the narrative. I argued that it would only make sense to bring a separate charge on the Emoluments Clause offenses, which would be simple to follow and would establish in the public's mind Trump's underlying financial motivation for staying in office at all costs, including importuning foreign leaders to become bit players and props in our presidential election.

Through this process, I was repeatedly warned by Amy Rutkin, the Judiciary Committee chief of staff and Chairman Jerry Nadler's able and endearing right-hand woman, that my aggressive focus on Trump's Emoluments Clause violations would cost me a place on the team of Ukraine abuse-of-power House impeachment managers sent over to the Senate. I knew she was probably right about this, and my own chief of staff, Julie Tagen, also kept urging me to dial back my focus on these broader questions of private corruption of public office. Even my senior legislative counsel and good friend Holly Idelson, who was assembling the Emoluments case with me, repeatedly told me that the leadership could not be moved because the more conservative Democrats were already nervous enough about the Ukraine article and did not want us to poke the bear of Trump's outraged followers again with a new round of impeachment charges.

Yet I simply could not stop myself. Of all the extraordinary scams and rip-offs taking place in Trump world, the Ukraine episode seemed to me to be at once gravely serious and terribly remote; I feared there was a good chance the public might not fasten on to the story. Moreover, the central charge of "abuse of power," while again accurate, also set us up for a battle-tested counterattack in the conventional arsenal of presidential impeachment defense arguments: specifically, that the "House has not even alleged an actual crime." This popular argument is a convenient exit ramp for members of the president's party who don't want to defend the president's actual conduct but who cannot bring themselves to vote to convict. It is a handy argument because it exploits the inherent vagueness of the language of the Constitution—"Treason, Bribery, or other high Crimes and Misdemeanors"—to put the focus back on the failings of the prosecution. Nothing is more predictable in a presidential impeachment trial than the rhetorical claim that the House has "not even alleged a crime" or has "simply made one up."

When the Ukraine team (which included the Intelligence, Oversight, and Foreign Affairs Committees) sent over its excellent report on the Ukraine shakedown to the Judiciary Committee, I worked hard to sharpen the factual and legal arguments and move the ball down the field. I also urged Chairman Nadler to add the Emoluments Clause violations to the Articles of Impeachment and, when that (predictably) failed, wrote a thirty-five-page minority report with "additional views," stating that the Emoluments Clauses should be part of the overall articles. It was, I felt, a compelling and cogent argument that really did strengthen the case for impeachment.

But, in the end, I was forced to let it go. A separate report like that would have put progressives in an awkward position and could have driven a wedge in the party right when we needed maximal unity. The scope of the impeachment drive had been defined already by the Intelligence Committee, and leadership would not support expansion to include the financial corruption component. Amy Rutkin again made clear to me that pressing for inclusion of the Emoluments Clause violations in a minority report would sink any remaining chances I had of being part of the impeachment team. I didn't care so much about that, but I agreed not to cause her or the chairman any more problems than they already had.

So I stood down.

And I did fight hard and stand strong for the two Articles of Impeachment we approved in Judiciary, which were Abuse of Power and Obstruction of Congress. Try as I might, I could never articulate a complete and compelling narrative about what had happened with all the various players in the story. It seemed to go on forever and someone was always adding another point: "And then so-and-so did this . . ." And ultimately, whether it was because of the exotic facts of the case or the broad language of the "abuse of power" offense, we could only bring one lone Republican vote to our side and that was just for one of the two charges. Trump was acquitted, and less than two months later, the greatest pandemic in a century hit our shores and Trump's impeachment seemed like a memory from another era.

But now, in January 2021, I felt that we had a real chance to impeach, convict, and disqualify this lawless and corrupt president who had set himself at war against our constitutional order. Trump's efforts to use coercive pressure and incite mob violence to overthrow the results of the election were easily understood and utterly unbearable. This is the point I advanced on the various calls where members spoke: these events were strikingly vivid, and people would never forget them, as much as the Republicans would try to get them to. It would be our job to make this watershed insurrection an imperishable marker in American historical memory and establish that presidents may not unleash mob violence against Congress to overthrow our elections and the constitutional process.

With the broader caucus, the fading argument that impeachment should be avoided because it could polarize the country provoked laughter at one meeting because, as my friend Ayanna Pressley said to me at one point, "What is more polarizing than them trying to kill us?" A lot of people felt as if we owed it "to history," "the Constitution," or "the country" to impeach, even if GOP senators were going to behave like religious cultists asleep in the back of a repainted school bus. There was a Pelosi-style sense of inevitability and clarity gathering about what we needed to do. The die was cast when a member expressed skepticism about impeachment because "it's getting very rough out there, and you should have seen how other members and I were accosted by right-wing militia people on the airplane on the way back home. Both impeachment and the Twenty-Fifth Amendment will be so divisive and will inflame these people."

The Speaker responded, cool as ice: "I'm sorry you had an unpleasant flight. But this is about the future of democracy."

There would be no turning back.

On Tuesday, January 12, I had my hands full.

On behalf of the Rules Committee, and as the principal author, I was bringing to the House floor for debate and a floor vote H.R. 21, our resolution "Calling on Vice President Pence to Convene and Mobilize the Principal Officers of the Executive Departments of the Cabinet to Activate Section 4 of the Twenty-Fifth Amendment to Declare President Donald J. Trump Incapable of Executing the Duties of His Office and to Immediately Exercise Powers as Acting President." The title was a mouthful, but calls on Pence to activate the Twenty-Fifth had been picking up a lot of support over the weekend.

But I was also asked to present the Judiciary Committee's Article of Impeachment before the Rules Committee, a role I'd also played for Chairman Nadler during Impeachment 1.0 (as people were starting to call it), when his wife was taken ill. We had to get Rules to report the impeachment resolution out for a vote the very next day, January 13, 2021.

My plan was to attempt to be on the floor for the Twenty-Fifth Amendment and then in Rules for impeachment in classic Capitol Hill fashion: by being in two places at one time, essentially, darting back and forth between the House floor, on the second story of the Capitol, and the Rules Committee, up on the third. But then, as I was preparing my statements at lunchtime—and eating my daily

Subway sandwich (a veggie pattie with lettuce, tomatoes, and chipotle sauce) that my staffer Maddie Krueger had kindly brought to me at my desk in my still-disheveled new office, where boxes of letters sat on my couch next to unhung family photos—my phone rang.

It was Tabitha, Hannah, and Sarah calling with my cousin Jedd Bellman in the background, too: they did not want me to present the articles in the Rules Committee because they did not want me to become the face and voice of impeachment in the Rules Committee. A year ago, performing this same function had created a lot of new visibility, and that session was practically anonymous compared to what was about to happen in the wake of the insurrection. The Proud Boys and other extreme-right groups were talking about a return engagement on Inauguration Day, just eight days away, and many of them had not left town. There were all kinds of violent threats flying around online and on my office phones.

"It's too dangerous, Daddy," Tabitha said. "I don't want you to do it. You don't have to do everything. Mommy agrees with us."

"I appreciate that," I said, wandering around my office now on my cell phone. "I love you guys so much, and if you really don't want me to do it, I won't do it. But there will be tons of security now in the wake of what happened on the sixth, and there are National Guardsmen and -women all over the place. I just visited the Guard people in from Maryland."

Hannah seemed kind of moved. "Well, I guess we can figure out the safety," Hannah said. "You're clearly the right person to do it."

"Let me just talk to Mommy and see if she feels okay about it," I said.

"No, Daddy," Tabitha said in an urgent voice. "We just lost Tommy, and we can't lose you. It's not worth it."

That argument was pretty much a showstopper. I wanted to tell the girls that Tommy was with me, that I felt him in my heart and I was protected. But they seemed to have everything on their side of the argument: facts, feelings, Sarah, numbers.

I called Sarah, who confirmed that she agreed with the girls: "It's not necessary. Just go talk about the Twenty-Fifth Amendment on the floor. Everyone knows you're trying to keep America safe. No one is going to think you're against impeachment."

That clinched it. What to do? It was Cicilline's impeachment resolution, so I called him and told him the girls were up in arms and beside themselves about the security issue. Was there any way he could come and do it in Rules?

"No problem at all, my friend," he said. "Tell the girls it's all right. Happy to do it. Don't think twice."

He could not have been kinder. The former mayor of Providence, David Cicilline is a great guy and a great legislator, not to mention a one-man melting pot of urban Northeastern political identities: Italian, Jewish, and gay.

Then, before I could call the girls back, the phone rang again: this time it was Nancy Pelosi. I stood with the phone at my ear looking at myself in the mirror in my office—people had been telling me that I was losing weight and needed to eat more. My face looked drawn and haggard.

The Speaker wanted to know how everyone at my house was doing.

"Thank you, Madam Speaker," I said. "Everyone is all right. We're deluged with family now."

"Did you get the flowers?" she asked. She had sent us a beautiful white orchid on behalf of herself and the whole caucus. We had an orchid collection near the piano.

"Yes, thank you so much," I said quietly, for fear of choking up. "It means a lot to us."

"So I wanted to ask you," she said. "How are the girls doing, and Sarah? It's been quite a trauma with Tabitha and January sixth."

"Well, Tabitha is hanging tough; they're all tough, tough as nails," I told her. I suddenly felt weirdly conscious for the first time of being the only male in my nuclear family now, with Tommy gone, one guy surrounded by all these strong, beautiful women. "But they're all a little shaken up about the security situation and all the threats out there," I said. "They asked me not to present the impeachment articles today in Rules, since I'm already managing the Twenty-Fifth Amendment resolution on the floor. So Cicilline is going to do it."

I heard a meditative silence on the other end of the phone. She was absorbing this fragment of information.

"I mean, they're okay," I said. "They just want things to cool down."

Another period of silence as she continued to digest this news.

"Well, listen, Jamie," she said. "I was calling because I wanted to talk to you about the team of impeachment managers and whether you would consider joining the team."

I did not pause: "Of course, Madam Speaker," I said, not having thought about it but not needing to think about it. "Of course, I will. It will be a great honor to serve in this way."

"I wanted to talk to you about serving as the *lead* impeachment manager," she said then.

You could have knocked me over with a feather. I had given such a thing no thought at all. In this moment of profound trauma, so personal and so public, a time of insurrectionary violence and blood and glass and Trump flags strewn on the Capitol steps, I had spoken to no one about this possibility, had given it zero thought. And that meant, of course, that I had never spoken to Sarah about the prospect, much less Hannah or Tabitha and the rest of the family.

And now I intuited what the Speaker had just absorbed in her stratospheric emotional intelligence: if the girls did not want me appearing to talk impeachment for one afternoon in the Rules Committee, how would they feel about my appearing for a week or two in the U.S. Senate, before the eyes of the world? Everything was happening too quickly.

"I will do that," I said. "It would be a great honor." I meant it. That was an understatement. I was a wreck, but I felt something stirring inside me.

"Are you really feeling up to it? There is a lot of stress," she said probingly.

"I feel Tommy with me every minute. He is in my heart, and he is in my chest," I said. "I can do this."

"Talk to Sarah and the girls," she said. "Call me back."

I was despairing already of how that conversation might go. Thank God I had handed off the Rules Committee assignment to Cicilline.

"Will I have Capital Police protection?" I asked her.

"Yes, of course," the Speaker said.

"Can we talk about the team?" I asked.

"Yes, of course, I need to talk to you about that," she said. "But call Sarah and Tabitha and Hannah first."

And so I did. I hung up and dialed Sarah.

"I got out of the Rules Committee," I said.

"Good," she responded. "When are you coming home?"

"Probably around nine or ten. Right after we vote on the Twenty-Fifth Amendment Resolution," I said. "But Nancy just called and asked if I would be the lead impeachment manager."

Ambivalent silence. Then: "What did you say?"

"I said yes but that I had to talk to you—and she wanted me to talk to you."

"That won't be safe," Sarah said.

"I'll have full-time protection, just like Adam had," I said. "I have to do it. It'll be safe. I'm excited to do it. Tommy's with me. I have to do this."

"You'd better call Tabitha and Hannah," she said.

I did. I told them that David Cicilline would be taking my place in Rules but that I would be the lead impeachment manager, with security around the clock. Tabitha just said she missed me and to come home as soon as possible. Hannah said it would be fine with her so long as I was not the only one speaking for the House managers. I told them I loved them.

I called Pelosi back; she picked up right away.

"Did you talk to Sarah and the girls?" she said.

"Yes, they're fine," I offered. "I told them that I'd have full-time security and that I felt Tommy with me every step."

"So they're okay?" she said.

"Yes, they're fine with my doing it. They know how much it means to me to see this through."

"Oh, that's great," she said. "That's great. Please thank Sarah for me. I don't want to bother her."

Julie came in and pointed at the clock, indicating that it was time to start walking to the floor.

"I think I have to head over to the floor for the Twenty-Fifth Amendment Resolution," I told the Speaker.

"Yes, I'm going to join you and speak on it," the Speaker said. "But let's talk about the team for a moment."

She was thinking of a team of seven, as there had been in Impeachment 1.0, but I argued strongly for nine.

"Nine will give us the chance for broader diversity and a bigger statement about America, which is critical given that we were assaulted by white nationalists and Confederate battle flags," I said. "Also, this is an enormously complex factual case with tons of moving parts, and we need a big team of excellent lawyers to master all these details. We want to tell America the story of the attack on America so no one ever forgets it."

She liked these points but said, "I just don't want people out there making a lot of long, boring speeches."

"No long, boring speeches," I said. "That's rule number one. We're telling a single story with a beginning, middle, and end. I will tell them that anyone who thinks this is about making great speeches misunderstands the assignment and will be benched. We're telling one compelling narrative with a lot of video footage about the president committing the worst presidential offense in history by inciting mob violence against our Congress, his vice president, and our election. I promise I will run a tight ship. And we will answer every legal question they have without any lectures."

"And no media!" she said. "I don't want people out there free-lancing on TV without talking to us first."

"Amen," I said. "This is a team of lawyers preparing for trial. Lawyers don't go to the press during a trial except under the most compelling circumstances. I will tell them no media appearances that I don't approve first, and I will run it all through you—"

"And Drew Hammill," she added, referring to her deputy chief of staff. "Just coordinate it all with Drew. I don't want to see people on Sunday news shows unless we approve it. We need absolute message discipline. Everyone thinks this is going to be chaos. The senators do not have high hopes for this, and that's the senators *on our side*."

"No chaos," I said. "Just facts and reason and justice. We will surprise them all. We will have a team of extremely low-ego people, extremely low-ego at least for members of the House, people who know how to be team players."

So we agreed on nine. We agreed on every form of diversity and balance being key. We agreed that lawyers, especially those with serious prosecutorial and trial experience, should be dominant. We agreed that we needed people who had shown that they knew how to put the group before individual ambition or desire for attention.

And then, in about sixty seconds, we agreed on the team itself. I insisted on synergy and chemistry. We needed a team not of rivals but of friends.

The Speaker emphasized that I would be in command of the team. She would be available to me but would not be like "somebody's mother-in-law at Thanksgiving telling you how to roast the turkey."

After thanking her and reminding her I was vegan like Tommy, I told the Speaker I had to leave for the floor and would see her there shortly.

Each side had thirty minutes, and I was (naturally) up against Jim Jordan, the manager of the GOP time. I was amazed and cheered at the number of great Democratic colleagues who had come to the floor to speak on behalf of the Twenty-Fifth Amendment resolution, and I happily gave up most of my own speaking time in one-minute, two-minute, and even forty-five-second blocks to accommodate them all: Ami Bera, Jim Clyburn, Steve Cohen, Danny Davis, Mike Doyle, Adriano Espaillat, Sylvia Garcia, Jimmy Gomez, Hank Johnson, Raja Krishnamoorthi, Sheila Jackson Lee, Zoe Lofgren, Carolyn Maloney, Lucy McBath, Joe Neguse, Jimmy Panetta, Adam Schiff, David Scott, Mike Thompson, Nydia Velázquez, and Maxine Waters.

I knew I would be tempted to rebut every argument Jordan threw at me, so I decided to try to answer his diatribes with a single sentence before turning it over to my colleagues, whose remarks were, to a member, eloquent and compelling. For example, when Jordan railed against "left-wing cancel culture," I responded simply: "The cancel culture of violent white supremacy tried to cancel out all of our lives last Wednesday," and tried to get back on track.

For the Resolution, I recited the "trauma of a violent attack" on Congress and "all of the people who work here" by an "armed, lawless, and enraged mob" that "smashed windows, beat up and crushed Capitol police officers who cried out in agony," "killed or caused the deaths of at least five American citizens," "inflicted serious injuries on dozens of our police officers," "threatened the lives and safety of the three individuals in the line of succession to the president of the United States," erected a gallows outside the Capitol and chanted "Hang Mike Pence," stole and vandalized public property, "terrorized officers, staff, and members," and allowed hundreds of people to "enter the Capitol without metal detectors or any kind of security screening at all . . . to interfere with the counting of Electoral College votes in the 2020 presidential election."

I quoted a shaken Senator Lindsey Graham, who said, "The mob could have blown the building up. They could have killed us all."

Whether or not you believe the president's conduct was a high crime and misdemeanor, I argued, all of us "should be able to agree that this president is not meeting the most minimal duties of office."

The resolution respectfully asked the vice president "to convene the Cabinet and to mobilize the Cabinet" to "articulate what is obvious to the American people—this president is not meeting the duties of office and is clearly not capable of it."

Speaker Pelosi spoke after me and was in her full rhetorical magnificence. She thanked me for working to educate Congress about the Twenty-Fifth Amendment and saluted the full Democratic Caucus for our "love of country, determination to protect our democracy, and the loyalty to our oath that had been so beautifully manifested in this dark past week." She expressed "deepest gratitude to the U.S. Capitol Police for the valor that they showed."

President Trump had "called for this seditious attack," she continued; "he and his family cheered and celebrated the desecration of the Capitol," and as all the dangers escalated, "he ignored and then flat-out rejected the pleas of Congress, including those of his own party, to call off his supporters—the rioters, the terrorists." Then she completed her syllogism: "The president's actions demonstrate his absolute inability to discharge the most basic and fundamental powers and duties of his office. Therefore, the president must be removed immediately."

In the end, the Twenty-Fifth Amendment Resolution passed on a vote of 223–205, but Vice President Pence quickly released a statement that evening as we debated, saying that he wanted us to "avoid actions that would further divide and inflame the passions of the moment." He would not convene the Cabinet, and there was apparently no independent leadership there to declare the president's incapacity to govern, although one thoughtful Trump Cabinet member, National Security Adviser Robert O'Brien, did call me that day to express his condolences about Tommy.

Later disclosures demonstrated the wisdom of our trying to activate the Twenty-Fifth amendment even in the final weeks of Trump's presidency. Bob Woodward and Robert Costa reported in their book *Peril*, about the end of Trump's presidency, that President Trump posed a major national security risk to the United States in his closing days.

According to the authors, the Chairman of the Joint Chiefs of Staff, General Milley, thought that Trump's manifest behavioral instability and desperate search for a way to stay in power created the possibility of an impetuous and undeclared war in his final days. Woodward and Costa reported that General Milley went so far as to call his Chinese counterpart to assure him that the U.S. had no plans to attack China. Milley also apparently instructed officers below him about nuclear launch protocols to ensure that any authorization for a nuclear launch would have to flow through him.

While Republicans immediately pounced on Milley, calling his actions "treason," he was obviously acting to support the foreign policy of the United States—we were not at war with China. If Trump was planning to attack China without a declaration of war by Congress, it would have been not only unconstitutional but also, as Milley understood, a dangerous last-minute gambit to stay in power. This is precisely the kind of deranged and unstable situation that the Twenty-Fifth Amendment is designed to address, but Pence wanted to leave all the constitutional tools sitting in the garage. And so, when we came to regroup on Wednesday, the GOP left us with one choice and one choice only: to impeach the president of the United States for inciting a violent insurrection against Congress to overthrow the 2020 election.

And this we did.

CHAPTER 8

AN ALL-AMERICAN DEFENSE OF DEMOCRACY

A little patience, and we shall see the reign of witches pass over, their spells dissolve, and the people, recovering their true sight, restore their government to its true principles.

—THOMAS JEFFERSON, "LETTER TO JOHN TAYLOR," JUNE 4, 1798

On Wednesday, January 13, at 4:32 p.m., one week after Trump's insurrection blew the doors off the hinges of American politics, and one week before Joe Biden's eagerly awaited inauguration arrived, the House of Representatives voted to impeach Donald John Trump.

It was surreal to see members gathering for this historic vote on the very floor where we had taken cover behind desks and chairs, awkwardly donned gas masks, listened as our doors were barreled into and pounded by the mob, and ultimately fled for our lives—just a week before. Many Members and staffers still avoided the west steps of the Capitol where the stench and sting of tear gas, bear mace, and the insurrectionists' cocktail of chemical weapons still hung acrid in the air. Rumor had it there were bloodstains right outside the chamber in the Speaker's Lounge, where rioter Ashli Babbitt was shot. Julie told me that our friend Jason in the Sergeant at Arms Office, an officer's officer who had helped move Tabitha and Hank around on January 6, was just feet away from Babbitt when she tried to break into the House chamber and was gunned down. He apparently was the last officer who tried to save her life, with a desperate chest press, an experience I could now relate to.

Given the savagery we had experienced just days before, I was surprised to find, when I came down to the floor to debate and vote, a slightly festive atmosphere on the Democratic side. Members were still traumatized and angry, but there was palpable catharsis in the air as we mobilized. We had been through a Trump impeachment vote once before, and during that first round, everyone was careful to strike a pose of solemn regret about having to render such a grave and momentous constitutional judgment. Now, with Trump having blown up every basic constitutional norm and value, he had substantially loosened our inhibitions and unleashed a jaunty, devil-may-care, the-guy's-getting-his-just-rewards attitude on the floor.

Because the Speaker had announced her appointment of the impeachment managers (with me as lead) the night before, my colleagues, news junkies all, wanted to talk to me when I arrived on the House floor, which is like the congressional water cooler. The first person who approached me when I got there was Rashida Tlaib, the pride of Detroit, the firstborn of fourteen children who knows how to take care of people and of business. Strategic, funny, and comfortable in her own skin, Rashida gave me a mixed high five/hug and told me she was thinking of me and Sarah and the girls, and then she said enthusiastically, as if reading my mind, "Raskin, we're going to impeach the—"

"Distinguished gentleman!" I interjected, completing her sentence for her.

"And you're going to go over there and try and convict the—!"

"The fellow!" I interrupted. "The guy!"

There was a moment when the whole country might have understood the comic subtext of this exchange. Infectiously exuberant, progressive to her core, avid for her constituents, the only Palestinian American ever to serve in Congress, Rashida has had numerous episodes of fame and controversy, and one of them arrived when she said at a fund-raiser prior to Impeachment 1.0, "We're going to impeach the motherfucker." Putting down their assault weapons and clutching their pearls, the right-wing pundits had a field day with that one, because how dare someone—a woman, no less!—use profanity to describe someone as innocent, as genteel, as nonprofane as Donald Trump?

The next person to approach was my friend Don Beyer from Virginia, a soulful and gentlemanly member of the House, a former American ambassador to Switzerland, a former lieutenant governor of Virginia, and an extremely successful new- and used-car salesman with Don Beyer Volvo. Like his fellow Virginian Thomas Jefferson, who said, "[I]n matters of style, swim with the current, but in matters of principle, stand like a rock," Don Beyer got along with everyone because of his essential sweetness and exceptional manners, but he was a fountain of nimble and excellent legislative proposals to advance strong democracy and greater fairness in the United States. Don had been my key collaborator on ranked-choice voting ever since I got to Congress and was a champion democracy reformer.

Seeing Don, I asked him if he was ready to impeach the first president who ever tried to stage a coup against America.

"This reminds me," he said, "of a joke we have in the Foreign Service. Why will there never be a coup in America?"

"Why?" I asked.

"Because there's no U.S. embassy there."

"That's funny," I said. "In a macabre sort of way. But I guess you don't need a U.S. embassy if you're running your coup right out of the White House."

As the debate unfolded, the scene on the House floor grew intense as Democrat after Democrat got up to eviscerate Trump for his atrocious misconduct. But some GOP members were also issuing potent statements, verbal and written, in favor of the article, giving rise to lusty cheers on the Democratic side of the aisle.

No one captured the essence of the offense better than Rep. Liz Cheney, my increasingly cherished friend and GOP exile, whose epic and irresistibly quotable statement was read into the record by Chairman Jim McGovern:

> *The President of the United States summoned this mob, assembled the mob, and lit the flame of this attack. Everything that followed was his doing. None of this would have happened without the President. The President could have immediately and forcefully intervened to stop the violence. He did not. There has never been a greater betrayal by a President of the United States of his office and his oath to the Constitution.*

Things looked tense and surly over on the GOP side. You could hear raised voices as Republican functionaries whipped the vote to stop defections and as Cheney's bloc of friends warned their colleagues that Trump's vicious unruliness would spell the end of the Grand Old Party, a point with which I was in profound concurrence. I was moved by the constitutionally insightful framing of Illinois's Adam Kinzinger, an air force veteran whom I did not know but to whom I later made a point of introducing myself, as I found that his office, fortuitously, was right across the hallway from my new one. Trump, he said, "incited this insurrection," and "if these actions— the Article Two branch inciting a deadly insurrection against the Article One branch—are not worthy of impeachment, then what *is* an impeachable offense?" I made a mental note of this powerful formulation: If what Trump had done wasn't impeachable, then what was?

One GOP member whom I was watching closely was Rep. Chip Roy from San Antonio, Texas. Chip was for a while the ranking member of the Oversight Subcommittee on Civil Rights and Civil

Liberties, which I chair, so we'd spent some quality time together over the last few years. Chip had arrived in the 116th Congress a conservative darling because he had been Sen. Ted Cruz's chief of staff and was an exceedingly right-wing guy. But he was nobody's fool and had seemingly wrapped a tight cloak of constitutional principle about himself. I was impressed that he had condemned *Texas v. Pennsylvania*, the November lawsuit brought by Texas attorney general Ken Paxton seeking to overthrow 2020 presidential election results in several states that Joe Biden had won. Chip called the legal argument "a dangerous violation of federalism" and correctly predicted that it would be thrown out by the Supreme Court on the grounds that no state has standing to dispute the legitimacy of another state's presidential election process. He then followed through on his resistance to the propaganda of Trump's Big Lie: On January 3, when most Republicans fell in line behind Trump's disgraceful plan to repudiate electors sent in by Arizona, Georgia, and Pennsylvania, Chip joined other dissidents in the House Caucus to challenge the unfolding GOP strategy to overthrow the 2020 presidential election results. The GOP ploy, he said, "would amount to stealing power from the people and the states. It would, in effect, replace the Electoral College with Congress."

Chip and I had been in touch on the phone the weekend before January 6, and I sensed he was growing anxious over the prospects of a violent attack on members of Congress. After all, our Democratic majority hovered around 5 or 6 votes, depending on who could show up to vote, and there is obviously a spectrum of different ways to make such a slender majority vanish, both peaceful and violent. Chip seemed to me to be genuinely scared of the violent and fanatical extremism growing on the right. He voted against GOP objections to Arizona's and Pennsylvania's electors and was even waiting in line to speak against them when the rioters shut everything down and sent us scrambling for shelter.

Yet, when it came time to vote on impeachment, my spirits sank when I saw the light next to his name on the board flash red: Chip voted no on impeachment and fell back in line with most Republicans. When I searched him out on the floor to see what had happened, he said that the Article of Impeachment had been written too broadly and had First Amendment problems, invoking incitement of insurrection instead of what he saw as the true impeachable offense: Trump's trying to coerce Pence into violating his constitutional Oath of Office.

I could only imagine the political crosswinds blowing in his caucus, and in his district, over impeachment, but I did gently tell him that this argument was just a play on words. The article on incitement to insurrection obviously contained pleadings about the coercive pressure directed against the vice president, which was indeed the entire purpose of the violence brought down on our House. I asked him why, if Trump had a valid First Amendment defense to impeachment for *using words* to incite a violent insurrection to intimidate Pence into violating his constitutional role, he didn't equally have a First Amendment defense for *using words* to try to force Pence to violate his role. I am still waiting for an answer to this question. But I do appreciate the fact that Chip voted against GOP objections to Arizona's and Pennsylvania's electors and that he was blowing the whistle on the utterly mutinous and violent attack on the election, even though he could not find his way to voting for our Article of Impeachment.

Some of my Judiciary colleagues chided me for wasting my time parsing the ideological nuances of different GOP members' positions on Trump and the insurrection. But within these nuances sprang my hope for conviction in the Senate. I had always regarded it as a mistake to focus on political disagreement only within one's own caucus. Democrats who write off all Republicans as robotic right-wing cultists, and Republicans who write off all Democrats as left-wing radical socialists, are missing a lot of opportunities for cross-party coalition in our closely divided chamber. Every House GOP member voting to impeach would make it just a bit easier to woo Republicans in the Senate, especially if they came from the same state, and every statement holding Trump accountable for this monstrous crime would make it that much more difficult for the GOP, in the future, to deny the reality of these events the way they have denied the reality of COVID-19 or climate change.

In the end, the vote was 232–197 to approve a single Article of Impeachment against Trump for inciting violent insurrection against the government as part of a sequence of outrageous and almost certainly unlawful actions he took that were designed to overturn the results of the 2020 presidential election. Ten House Republicans—Liz Cheney, Anthony Gonzalez, Jaime Herrera Beutler, John Katko, Adam Kinzinger, Peter Meijer, Dan Newhouse, Tom Rice, Fred Upton, and David Valadao—nobly joined every House Democrat in voting to pass the article. This group of ten was large enough to create some mutual aid and support and was around 5 percent of the GOP Caucus, sending a chilling signal to the masters of the GOP. Could the party survive defection of 5 percent of its voters? People sometimes forget that politics in America is a game of inches.

Only the third American president ever to be impeached—Andrew Johnson and Bill Clinton came before him—Trump also became the first president in our history to be impeached twice.

With the vote behind us, the matter was turned over to our team of "prosecutors," the nine House members called impeachment managers, who would present the full case to the Senate chamber, where the 100 senators would hear the case at trial and act as juror-judges. They would serve like jurors because they would cast votes on final judgment of the president's guilt (with two thirds, or 67 senators, needed to convict), but they would also act as judges, because they would rule as a group (with 51 constituting the majority) on all constitutional matters and questions of trial procedure that needed to be settled along the way.

As lead manager, I would have the responsibility of overseeing the process of building our nine managers into a strong, cohesive team with an excellent and confidential game plan; hiring our chief counsel, other lawyers, and staff; crafting our "theory of the case," the integration of the evidence, and the legal arguments into a logical, inexorable, bulletproof case for conviction; assembling from a tidal wave of documentary evidence the most relevant video and physical snippets to prove our factual case; coordinating with the Senate and Trump's counsel on critical matters of trial procedure and scheduling; developing a comprehensive trial strategy and assigning appropriate divisions of labor on a trial script to the managers' team; and then leading effective and flexibly adaptive daily presentations of the case in the Senate trial in such a way that would allow us to prevail in the end; all while coordinating with the Speaker and the caucus to maintain an effective public communications strategy to keep America on our side.

This was a daunting agenda, with only two to three weeks to get ready, based on the Speaker's estimate of when the Senate wanted to act. It was time to get to work and to rapidly shift our managers to a prosecutorial frame of mind, to get all of us to stop acting like publicity-seeking politicians and to start acting like painstaking trial lawyers.

The Speaker had invited our team of impeachment managers to meet in her office at 3 p.m. on Wednesday, January 13, shortly after the floor vote, to give us our charge. Like all great leaders, Nancy Pelosi had a finely tuned sense of ceremony and history. We filed into the Speaker's famed conference room, chocolate chip cookies and strawberries waiting for us on the table and the spellbinding portrait of dark-eyed Lincoln behind her at its head. The Speaker showed up in a delightful and convivial mood, surrounded by her able and harried staff. The managers were excited by the novelty of the project and nervous about its immensity. They knew that potentially the whole world could be watching us. It was a humbling moment.

As I looked around the table, I saw that ours was a group defined by the fierce and battle-tested constitutional patriots of the Judiciary

My wife, Sarah, holding Tommy, a baby born smiling.

Baby Tommy's arrival was a big deal for kids in our family.
From left to right: Cousins Emily Blair Littlewood, Zachary
Galyn Littlewood, Tommy's big sister, Hannah Grace Raskin,
and Maggie Ryan Littlewood.

Tommy was a joyful child.

My mom, Barbara Raskin, holds Tommy.

Tommy and I visited my best friend from college and law school, Michael Anderson, and his wife, Donene Williams (our roommate in law school), several months after the birth of their son, Jackson, in 2000.

My dad and Tommy enjoyed strong bonds of love and intellectual affinity. When Tommy was a little boy, my dad taught him to play piano.

Born two years and one day apart, Tommy and Tabitha ("the two Ts") were inseparable and shared a single birthday party each year at the end of January.

The family with Sarah's dad, Herb.

Our family trips have always included grandparents. Here we are with Sarah's mom, Arlene.

Hannah, Tommy, Tabitha, and me along with my friend John Sarbanes, who is holding his son Leo. This was before either John or I was elected to Congress.

Tommy introduced me at my kickoff campaign announcement in 2006 when I declared for the state senate, saying, "My dad loves our family and our community." At ten, he played a surprisingly big role in my campaign and the development of Democracy Summer. (Photo by Angela Davis)

At ten years old, Tommy wrote a speech for my first big state senate campaign event.

Tommy in the center of a crowd of Raskin cousins, uncles, and aunts.

Back row: Sarah's parents, Herb and Arlene Bloom; Sarah's brother Kenneth Bloom and his wife, Abby Meiselman; and Sarah, Tommy, me, and Hannah. *Front row:* Cousins Phoebe and Lily Bloom, and Tabitha.

My successful 2016 race started with the most expensive congressional primary in U.S. history. Pictured here on primary Election Night 2016 are, from left to right, cousin Lily, Tabitha, Hannah, and Sarah; Tommy and cousin Boman are visible in the second row.

Tabitha, Hannah, and Tommy at the wedding of their cousin Jedd Bellman to Sarah Bergen.

The first wave of our family regroups in Charlottesville after a hike. With so many family members living in the DC-Maryland-Virginia area, we have assembled en masse for Sunday brunch for decades.

The Takoma Park Fourth of July Independence Day Parade, a Tommy favorite, coming down Maple Avenue in front of Ann Shalleck and Jimmy Klein's house. (*Photo by Ann Shalleck*)

Tabitha, Tommy, cousin Mariah, Hannah, and Hannah's husband, Hank.

Tommy graduated with honors in History from Amherst College, for which he wrote a thesis on the intellectual history of the animal rights movement. Afterward, he headed off to the Friends Committee on National Legislation in Washington for a fellowship.

Tabitha, Tommy, and Hannah in December 2019, a few months before the arrival of COVID-19 brought a curtain of darkness down on the land.

Tommy, Sarah, and Tabitha in
the summer of 2020.

During a family hike, a favorite outing.

"That boy Tommy Raskin was a game changer," said our relative Bob Bergen. By the time of Tommy's memorial service, on April 3, 2021, the Tommy Raskin Memorial Fund for People and Animals received more than one million dollars in contributions from Tommy's friends and admirers all over the world.

The gallows set the macabre stage at the Capitol on January 6, 2021. *(NurPhoto/ Getty Images)*

"We fight, we fight like hell, and if you don't fight like hell you're not going to have a country anymore." *(Pete Marovich/ New York Times)*

In calling on followers to "stop the steal," Trump never mentioned the sixty-one lawsuits he brought alleging election fraud and corruption that were rejected in federal and state courts across the land. *(Spencer Platt/ Getty Images)*

Many insurrectionists came to Washington armed to fight not only with baseball bats, steel pipes, and flag poles, as well as huge arsenals of guns and knives, but also with mace, bear mace, and other toxic chemical sprays. *(Joseph Prezioso/ Getty Images)*

Racist and extremist groups led the violent assault on the Capitol, beating police officers mercilessly and opening multiple passageways for rioters. *(Jose Luis Magana/Associated Press)*

Capitol officers fought heroically but were vastly outnumbered by rioters. Multiple officers who served in Iraq and Afghanistan said the violence was worse than anything they had seen there. *(Pacific Press/Getty Images)*

Insurrectionists brandishing the Confederate battle flag never made it into the Capitol in 1861 or at any point in the Civil War, but they made it inside in 2021. *(Saul Loeb/Getty Images)*

My chief of staff, Julie Tagen, took this photo on January 6, at 1:30 p.m. from the majority leader's office, while huddled there with Tabitha and Hank. You can see the crowd still streaming in from the March to Save America departing from the front of the White House.

Once inside, the mob moved to invade both the Senate and House chambers in order to shut down the counting of Electoral College votes—that is, to "stop the steal." *(Anadolu Agency/Getty Images)*

After insurrectionists tried to charge into the House chamber with some kind of battering ram, Capitol officers rushed to that central door with their weapons drawn. I was about thirty paces away from here. *(Andrew Harnik/ Associated Press)*

Rep. Jason Crow (D-CO) tends to Rep. Susan Wild (D-PA), who suffered a serious panic attack when the House was stormed.Rep. Lou Correa (D-CA) is at the left. *(Tom Williams/ Getty Images)*

At one point, rioters came within one hundred feet of where the vice president was hiding with his family, as Rep. Stacey Plaskett explained. What if Pence and company had been captured by the insurrectionists chanting "Hang Mike Pence"?

The toxic brew of bear mace, pepper spray, and tear gas used against the police on the lower west front of the Capitol turned that area into what officers called "the gas chamber." Most officers were not wearing gas masks. *(Tom Williams/ Getty Images)*

Speaker Pelosi convened the managers for the first time after the House voted 306–232 to impeach Trump for inciting insurrection against the Congress. At trial we would not give political speeches, I told the team, but tell one tight story. *(Erin Schaff/ New York Times)*

We spent more than seven hours and most of Super Bowl Sunday rehearsing our arguments, a grueling practice session that paid off when the trial began. *(Erin Schaff/ New York Times)*

The dazzling Stacey Plaskett (D-VI), the first delegate from a territory ever to serve in the role of impeachment manager, was my law student at American University's Washington College of Law. She loved to tease me because she had more seniority in Congress than I did. *(Erin Schaff/New York Times)*

A lot of our time was spent before trial talking strategy on Zoom and writing up our arguments and statements of fact for the different "chapters" of our case-in-chief. *(Erin Schaff/New York Times)*

I told the managers that our factual case for Trump's guilt was airtight and our legal arguments unbeatable, but they should not suppress their passionate feelings about the violence unleashed against us. *(Erin Schaff/New York Times)*

Here I caucus with the chief impeachment counsel Barry Berke, Rep. Joe Neguse (D-CO), and the deputy chief impeachment counsel Sarah Istel.
(Erin Schaff/New York Times)

After the 57–43 Senate vote on the last day of trial, we watched Senator Mitch McConnell's astounding speech holding Trump responsible for the attack but then defending his not guilty vote on the basis of the jurisdictional argument the Senate had rejected on the first day of proceedings. *(Erin Schaff/ New York Times)*

The managers were eloquent and powerful in their post-trial statements. I thought the always-compelling Joaquin Castro made a striking case for how Trump had been indicted and convicted in the eyes of history. *(Erin Schaff/New York Times)*

A mother, a grandmother, and a natural-born teacher and lawyer, Rep. Madeleine Dean (D-PA) touched American viewers whenever she spoke during the trial and also showed concern for my emotional state throughout. *(Erin Schaff/New York Times)*

Room 219 off the Senate floor, our "war room." (I tried and failed to get people to call it our "democracy room.") *(Erin Schaff/New York Times)*

America's constitutional dream team after the final vote: Lieu, Cicciline, Plaskett, Neguse, me, DeGette, Dean, Swalwell, and Castro. *(Photo by Julie Tagen)*

The July 27, 2021, testimony of Sergeant Aquilino Gonell and Officers Michael Fanone, Daniel Hodges, and Harry Dunn before our Select Committee on the January 6 insurrection reminded America of the bloody violence visited on hundreds of police officers on the front lines of the battle to defend Congress on January 6. *(Chip Somodevilla/Getty Images)*

The outpouring of love and support I felt in my district during the trial was overwhelming. Even looking back on it now, I am overcome with emotion.

Signs of Democracy
Takoma Park, Maryland 2021

"Tyranny, like hell, is not easily conquered; but we have this consolation: The more difficult the struggle, the more glorious in the end will be our victory." —Thomas Paine

All four of us spoke at Tommy's COVID-safe memorial service on April 3, 2021. As I said that day, I "hardly recognized the world" with Tommy not in it. *Clockwise from top left:* Me, Sarah, Tabitha, and Hannah. *(Photos by Matthew Bergh)*

Committee, but it was enhanced by a few "ringers-with-zingers," as I called them, drawn from other committees. They were a strikingly impressive lot and a beautiful tableau of twenty-first-century America—all of America:

Joaquin Castro (TX): A leading member of the Foreign Affairs Committee, Joaquin had the passion, eloquence, and command of the world situation to make him an important asset in articulating how the assault on Congress during the transfer of power had been registered around the world. He was also blessed with an unflappably calm and steady demeanor.

David Cicilline (RI): A former public defender with the acclaimed federal Public Defender Service in Washington, DC; the former mayor of Providence; a member of the Judiciary Committee; and chair of its important Antitrust, Commercial, and Administrative Law subcommittee, which was leading the investigation into Facebook and the other social media giants, David was a fast-talking, fast-thinking legal workhorse and would be the first openly gay impeachment manager in American history.

Madeleine Dean (PA): If one person on the team spoke to proverbial "middle America" more than anyone else, it was my charming friend Mad, a mother, grandmother, former trial lawyer, former professor of English and rhetoric, former state legislator, and the eloquent, silky voice of the all-important Philadelphia suburbs. Mad had come to Congress in 2018 after winning a feisty campaign in newly redrawn Pennsylvania districts that emancipated the true-blue Democratic majority in her state. She had just recently published a powerful book, *Under Our Roof*, with her son Henry Cunnane about their family struggle with his addiction.

Diana DeGette (CO): The dean of the Colorado House delegation, a tried-and-true trial attorney, and a serious Washington insider, Diana was always ready with political tips, gossip, and insight. She was as attuned to the internal dynamics of our team as she was steeped in the practical trial questions we would encounter.

On the floor, even before this meeting, she offered to help me coach members on their eventual presentations to the Senate, and I had gladly accepted.

Ted Lieu (CA): My seatmate on the Judiciary Committee, a naturalized American citizen born in Taiwan, a decorated air force veteran, and a member of the Air Force Reserve, Ted was a savagely methodical lawyer who loved to demolish right-wing fallacies in detail during committee hearings and in compressed form on Twitter, where he was a demon. When the egregious Devin Nunes's lawyer sent Ted a threatening letter regarding his comments about Nunes's relationship with a figure in the Ukraine scandal, Ted wrote back, "I welcome any lawsuit from your client and look forward to taking discovery of Congressman Nunes. Or, you can take your letter and shove it." Although Ted had been dealing over the last year with a recently diagnosed heart condition, he told both the Speaker and me that he was up to the task, and that was all we needed to hear. He was a fighter through and through.

Joe Neguse (CO): My constant collaborator and co-provocateur on the Judiciary Committee and on the Democratic leadership team, where we served together in the 117th Congress, Joe was a powerful legal researcher, extremely creative thinker, and superb oral advocate deeply invested in the justice system and social progress. At thirty-seven, he was not only the youngest member on our team but also the youngest impeachment manager in American history—which is just one of many firsts attached to his name, including first Eritrean American in Congress.

Stacey Plaskett (Virgin Islands): The first nonvoting delegate ever to serve as an impeachment manager; a former federal prosecutor in New York, Washington, DC, and the Virgin Islands; and a Republican (who still spoke fluent Republican) until 2008, Stacey was my law student at American University Washington College of Law in the 1990s. Her amazing energy for cross-examination in argument and her magnificent rhetorical skills led me to tell her once, in front of the whole class, "You need to be a prosecutor one day." Stacey was a person of great warmth and levity, and she was profoundly upset and offended by the storming of the Capitol and filled with righteous anger and passion over it.

Eric Swalwell (CA): A member of both the Intelligence and Judiciary Committees. Eric had been deeply involved in the Impeachment 1.0 investigation and, as a former prosecutor and the son and brother of police officers, brought great trial skills and rhetorical talent to the table. He had a savvy social media presence, engaged constantly with the media, and had actually run for president of the United States in 2020, on a gun-safety platform and new-generation pitch.

As I took them all in and began to compile in my mind where each member's gifts could best be deployed, Speaker Pelosi was thanking everyone for the tremendous devotion of time and energy they were making to this overwhelming endeavor. "You have the chance to bring accountability back to government, to rescue Congress, the Constitution, and the country from this insurrection and this president.

"The Confederate flag did not wave in these halls during the Civil War, but these people brought it in and waved it and desecrated the John Lewis exhibit. Remember Abe Lincoln," she said, gesturing behind her to the portrait. "'A house divided cannot stand.' So let's be unified. Let's be focused. And please coordinate media strategy with Jamie, and he can coordinate with Drew and Henry from my office," she said. "It's very important for all of us to stay on message."

She then called on me to speak.

"I join the Speaker in thanking you for your service, and I thank you, Madam Speaker, for your confidence in all of us," I said. "This is, above all, a team of great lawyers and prosecutors, and as impressive as each of you is individually, together we are far more powerful. This is going to be the most low-ego, high-performance team in the history of impeachment trials.

"We are not going to get up and make a series of disconnected speeches of great oratory that vanish into thin air. We are going to tell America and the Senate *one single, unforgettable story* with short, vivid chapters brought to life by you and by video, one spellbinding and appalling narrative about how Donald Trump became obsessed with denying the reality of his defeat and Joe Biden's victory and then set about using every means in the world, first lawful and nonviolent, then unlawful and criminal, and then, finally, violent and criminal, to overthrow the lawful presidential result and replace it with what he called his 'continuation' in office."

I explained that we would meet on Zoom every single day and that we could do our work remotely because of COVID-19. I then invited the members to say a few words. What followed was a moving symphony of gratitude and determination to defend democracy. There were so many firsts around the table—first Asian American impeachment manager, first openly gay impeachment manager, first nonvoting delegate impeachment manager—that you could feel history coming alive all around us.

"Finally," I said, following through on the Speaker's last point and warming to the task, "please cancel all media interviews and, if you think you have one you need to do, please tell me and Henry and Drew, so we can discuss it. Right now, each of us is speaking for the whole team, and we have not yet worked out our trial strategy and tactics or the complete theory of our case or all the constitutional answers. The things they want to ask you about are things we either don't know the answers to or we can't talk about yet, like: Will we have witnesses? Will there be police video at trial? When will the trial start? And how long will it go? So we'll just look like idiots if we speak now. Don't worry—you all are going to be more famous than you ever dreamed of when we get to the trial, but the chances of us messing up by doing an interview now are huge, so let's not take any chances. I hope everyone's cool with that."

A few people, like Mad and Diana, seemed relieved, and said so. "I just want to focus on the case," Madeleine said. Stacey said she welcomed the chance "to be a prosecutor again and just say, 'No comment, no comment.'" Others may not have been disgruntled, "but they weren't exactly gruntled," either—a great and fitting line I once read in P. G. Wodehouse.

And then the Speaker was whisked away by staff and security and, suddenly, as I looked around the room at my new band of brothers and sisters in this final battle for justice and the rule of law in the Trump administration, I realized that we were almost completely, totally, and irrevocably on our own.

To start off with, we needed a team of attorneys and a professional staff. The House Judiciary Committee and the House Committee on Oversight had remarkably good lawyers and professional staff who would become the infrastructure of the operation, but we needed an experienced chief counsel who would coordinate everything, one person I could caucus with and who could make everything happen: motions, briefs, video compilation and editing, exhibits, trial scripts, coaching sessions.

We needed Barry Berke.

Barry, a partner with Kramer Levin in New York, had been special counsel to the Judiciary Committee in the year leading up to Impeachment 1.0 and had impressed us with his calm, cheerful, and relaxed demeanor; his authoritative command over voluminous amounts of law and evidence; and his sincere and winning techniques of argumentation. While the key cross-examiner in the earlier impeachment proceedings had been the eloquent and effective Daniel Goldman, Adam Schiff's trusty right-hand man from Intelligence, Barry was the main lawyer on the Judiciary Committee and a universal favorite among our members as a team builder. I spoke to Barry about joining us, and he was delighted and ready to get started on Friday, January 15. He said in his cheerful voice that he could be down from New York in two days.

I also made a point of talking to him about Joshua Matz, another impressive veteran of the last round. We would be drafting a long case brief for submission to the Senate, a reply brief, and some pretrial motions; I knew we would need a proven constitutional lawyer with a golden pen who could work with me and Barry full time on various written pleadings. That would be Joshua, who joined us as impeachment counsel.

As we took a couple days to nail down our team, we drew on the sensational young lawyers eager to sign up from the Judiciary Committee and the Committee on Oversight and Reform. Like impeachment managers themselves, these lawyers, too, were the face of democracy's defenders in twenty-first-century America. Sarah Istel was counsel to the House Judiciary Committee and joined us as a key lawyer on the staff assisting Barry and Joshua. Aaron Hiller, deputy director of the Judiciary Committee, would do a lot of interfacing with the Senate staffers, and Arya Hariharan, its chief Oversight counsel, would come to command a small army of researchers and clerks poring over mountains and mountains of videotape, audiotape, and photographs. A shrewd political strategist, Arya would not only help ferret out the explosive video and audio evidence from the Metropolitan Police Department, the Capitol Police, and all over official Washington, but would also help make sure the still-raw experiences and emotions of members and staff never got lost in the shuffle. Finally, ever there for me in a pinch was Amy Rutkin, a level-headed political savant who always saw around any corner to the coming collisions.

Meantime, the Committee on Oversight and Reform, another committee I served on and loved like a family (along with Judiciary and Rules), would play the central role in assembling all the evidence that would become the heart of our case: the official government videos from indoor and outdoor fixed cameras and other stationary sources; the videos from body cameras; TV footage; unofficial private videos from reporters, cameramen and women, tourists (*real* tourists), protesters, rioters, and insurrectionists; social media posts; livestreaming tape; old news clips; presidential press conference segments; tweets and Facebook posts; and police records.

Finally, because Trump's pernicious scheme for insurrection followed multiple lines of attack on the Capitol and different kinds of chaos, Barry took the lead in organizing the video editing team with the tried-and-true trial consultants at the DOAR Group. DOAR had the job of reducing thousands of hours of videos and audio clips, countless headlines and arresting photographic images, into thirty- or sixty-second illustrative cuts and more substantial presentations, like the longer, thirteen-minute video I would end up using in my opening statement on Tuesday of the trial. We had a meeting with a rapidly forming video production team on January 14 to explain our preliminary sense of what we were looking for, what we were finding, and how we wanted the vast database of video, audio, and still images to be catalogued, edited, labeled, and presented.

The first raw clips I saw were shocking and almost unwatchable. I saw rioters lifting gas masks off the faces of cops and spraying them directly with mace and other contaminants. I caught images of fascist rioters gouging the eyes of officers who were heroically trying to seal off the Capitol. And I watched, spellbound, as gangs of insurrectionists marched and fought in military-style formation against badly outnumbered officers, some of them knocked unconscious, their faces covered in blood.

Upon seeing this bewildering collage of intimate assault, terror, and hatred, I realized that, for Trump, the red-hot center of the action may have been his inside political coup and all this wretched violence just a strategic weapon leveraged toward that political goal. But for the enraged masses he had corralled into coming to the federal city to "stop the steal," for the violent domestic extremist groups and all the wildcat aggressors drawn to this "wild" fight, the violence *itself* was the draw. What made Trump such an effective and diabolical leader in the right-wing authoritarian mode was that he understood that he could deploy the violence instrumentally for his own political purposes, while his followers enjoyed it intrinsically for its own sadistic delights. In the meantime, our video editors were able to capture both the tactical and the sadistic dimensions of this bloodthirsty clamor and free-for-all.

Here, the problem was not how to find the right video team—because we blasted off with a first-rate and experienced squad—but how to turn off the gushing stream of film directors, documentarians, and extremely big-time and big-name movie producers who were calling me to insist that we needed their help because they knew how to "produce" the precise film we needed. It was natural for (and authentically good of) them to call because (1) filmmakers were as horrified and outraged by the attempted coup and violent insurrection as tens of millions of other Americans and wanted to do something to help, and (2) this unprecedented insurrection offered a spellbinding and kaleidoscopic festival of images—of sickening fascist violence, terrifying political danger, and authentic civic and physical courage by police officers—all of which these filmmakers considered cinematically irresistible and instantly iconic. The towering creative volunteers who called from New York, Hollywood, and across the country wanted no compensation or recognition other than the chance to contribute to our team and the defense of democracy. I was moved by their patriotism and their conviction that they had the crucial ideas we needed to win a two-thirds majority in the Senate in the impeachment vote.

Now, I love movies as much as the next American, but I was convinced that if we brought on a big-time movie producer from Hollywood, an Academy Award–winning director, or a Sundance-winning documentarian to edit our clips—and we did have representatives of all these groups in the profusion of overtures I was fielding—it would almost certainly destroy the entire case. No matter how popular or mainstream the figure, the moment the GOP discovered that we had Hollywood blockbuster film director Alfred Hitchcock making our videos, they would spend fifty million dollars describing our case to the public as a "made-for-TV Alfred Hitchcock fantasy flick by the liberal Hollywood elite—what's wrong? Michael Moore not free? Oliver Stone too busy?"

It would have been game over for our case in the Senate. We were looking to build some high bridges and underground passageways between the two partisan sides of the Senate, not reignite and spread the right-wing culture war.

So I had to turn down these totally well-meaning feature film and documentarian Hollywood luminaries, who all struck a disarming and completely confusing tone in our conversations, as if we were old friends just suddenly getting back in touch after a few years, despite the fact that we were, to my knowledge anyway, complete strangers.

The point was that a Hollywood role in producing our film would have been a "huge creative blessing and a total political curse," as I kept politely insisting. Some protested that it could all be kept secret and confidential—which in Washington, I responded, means you tell *only* your chief of staff and your spouse and children and their best friends, and which in Hollywood means you tell *only* your agent, your life coach, and your massage therapist. No, I was emphatic with all the would-be volunteers: it can't happen. I felt good about just saying no (a word that doesn't come easily to liberals and middle children like me), because the truth is: We didn't want slick videos. We needed it to be raw and jagged like January 6 itself—like reality.

Something else did need to be set up quickly: security. The flip-side of having the most diverse team of impeachment managers in history is having to defend them against all the incoming racist, anti-Asian, anti-Latino, homophobic, misogynistic, and anti-Semitic death threats and online harassment. What all of us had experienced as a steady trickle of abuse at the start of the Trump period had turned into an overflowing septic tank around January 6 and a gushing sewer of threats during impeachment. Rank-and-file members of the House ordinarily have no security detail, but in the wake of the insurrection, we met with the Sergeant at Arms and the Capitol Police to devise a plan for the team.

My security detail was excellent in every respect: professional, discreet, reassuring. Whether I was at a meeting on the Hill, crossing from the House side to the Senate side, or taking a hike in Rock Creek Park or Northwest Branch with family and friends, they had scoped it out beforehand and walked quietly at a distance, watching for anything untoward. I was meeting them, of course, at a tough time for the Capitol Police force. More than 140 officers had just been injured on January 6. Officer Brian Sicknick had died a day after being brutally assaulted with bear mace and the nasty chemical concoctions that turned the west front of the Capitol into what the cops that day were calling "the gas chamber." The whole force was physically exhausted, mentally drained, and thoroughly traumatized. Two officers were so demolished by the hatred and evil they experienced that day that they took their own lives within a few days of the riots, a fact that haunted me. (Sadly, two more would come to take their lives and hundreds of officers and their families continued to be crushed by the trauma of these events.)

One officer in my own detail was racked with remorse because he had been off duty on January 6. Several officers I spoke to in the immediate aftermath told me that they or their friends were thinking about leaving the Capitol force or police work altogether because of the searing mental and physical wounds they carried and the seeming indifference of so many politicians eager to sweep the whole vile plot under the rug. And when they went back to work after the insurrection that shook the Capitol, these young officers passed by camped-out National Guard units, boarded-up broken windows, doors taped together, shattered glass, bloodstains, and— far worse—Republican politicians doing sinister TV interviews in which they experimented with different Orwellian twists on the violent insurrection that we had all experienced, in order to diminish and disguise the horror and criminality of these events.

I would come to learn a lot about the attack on the Capitol from these young officers, who were not much older than Hannah, Tommy, or Tabitha, and who redeemed a lot of my hope for our country. They were young people right out of college or a master's program who wanted to succeed, prosper in their careers, but, above all, serve their country. They worked hard, and they worked long hours. I saw patriotism as their overriding professional motivation. These officers and their colleagues on the Capitol Police force and in the Metropolitan Police Department almost certainly saved our lives on that bleak day. I hope our experience on January 6 showed progressive America that we need the police to defend the republic against organized fascist violence. I repeat: *We need the police to defend the republic against organized fascist violence.* I hope January 6 showed conservative America that all the extreme right-wing rhetoric about "backing the Blue" was nothing but propaganda and lies; the only handful of officers they truly advocate for are rogue racists in uniform who abuse the rights of our African American and Latino citizens. We need to continue to diversify, train, and invest in our police forces to protect democracy and our people against criminal and terroristic violence.

By Friday, January 15, our lean, nimble team of members, lawyers, and investigators was in place and assembling for action; we had only about three weeks to develop and refine our "theory of the case," as the lawyers call it, and compose trial strategy for the strange courtroom that now doubled as a crucial part of the crime scene: the United States Senate.

In order to define our arguments, it would be critical for everyone first to understand the nature of this "constitutional trial" and to grasp how it differed completely from criminal trials in standard-issue courtrooms across the land. In the Senate, the jurors are the one hundred senators from the fifty states. They have not been chosen at random or screened for their impartiality on the facts of the case. On the contrary, in the perfectly split fifty–fifty Senate, all the jurors stand for particular political values and platforms. Moreover, while they swear to, and sign an oath of, impartiality before an impeachment trial begins, they cannot be dismissed as jurors for harboring, displaying, or even proudly declaring bias, prejudice, or unfitness.

Furthermore, in a real courtroom, the judge almost always rules on questions of law while juries decide questions of fact. But in a Senate trial, the Senate jury itself decides on questions of both law and fact, ruling, for example, on questions of jurisdiction to hear the case or on the question of whether to allow witness testimony. The senators write the rules; they interpret the law; they approve or disapprove procedural, constitutional, and legal motions; and then they vote on the ultimate culpability of the defendant. Given all their assorted roles and functions, it is better to think of the senators as operating not solely as jurors but as jurors and judges combined.

Formally speaking, there *is* a judge in the impeachment context, but his or her role has been defined to be something ministerial and administrative in nature. The judge isn't there to interpret, articulate, or decide contested issues of law. In a presidential impeachment (as opposed to impeachment of a judge or a Cabinet official, for example), the chief justice of the United States is supposed to preside, a constitutional rule that would come into curious behind-the-scenes play in our case.

From these essential and fundamental differences in how a Senate impeachment trial differs from normal trials, countless other differences flow. While real criminal juries meet and deliberate and vote in seclusion, delivering their verdicts only afterward, in open court, Senate jurors vote in public and do not deliberate together and analyze the facts and charges. Jurors in the real world operate under a complete gag order with respect to media interviews until the trial is over, while Senate jurors often do TV interviews during breaks in the daily action of the trial itself.

You now have a sense of why people say an impeachment trial is far more "political" than it is "legal." Three weeks before trial, the media and political worlds were filled with predictions that the 50 Democrats in the Senate would vote to convict Trump and the 50 Republicans would vote to acquit him. At most, it was said, the GOP would suffer the "defections" of Utah's Mitt Romney, the 2012 Republican GOP nominee for president who voted to convict Trump on one count in the first Trump impeachment trial, and the independent-minded Susan Collins of Maine, creating a 48–52 split and leaving "the Democrats" far short of the 17 Republican votes "they need to reach 67 and convict."

I did speak, weeks before the trial, with a few senators and former senators—not about the specifics of our case, but about how to break through the Republican ideological sound barrier in such a sharply polarized tribunal, just so our case could be heard fairly. The media was saying that our presentation, no matter what it consisted of, would simply fall on deaf ears in half the chamber.

The advice they gave was influenced heavily by the fact that the Senate had been through Trump's first impeachment trial just a year before. These senators did not consider that trial a constitutional, political, or moral success. Although they had nice things to say personally about the preparation and argumentation of the members of our first team, starting with the able chairs of the two committees involved, Adam Schiff and Jerry Nadler, they recalled the proceeding itself as protracted and polarizing, "totally endless," as one said, filled with "political science lectures and moral sermons," according to another, oftentimes confusing, tedious, and exhausting.

I heard from Maryland's two fine senators, my friends Ben Cardin and Chris Van Hollen, at the start of January, right after we lost Tommy. More than two decades before, Ben and his wife, Myrna, had lost their thirty-year-old son, Michael, to suicide, and they had offered Sarah and me a lot of consolation and loving advice. Chris Van Hollen was similarly compassionate. When I asked the two of them for any thoughts on how to make the impeachment trial a real proceeding rather than a partisan rumble, they were both sober and fatalistic about the prospects for any real breakthrough on the facts with their colleagues, who were already going on television and saying that Trump was neither culpable nor triable in the Senate.

"They're afraid of Donald Trump putting someone up against them in their primaries," Ben said, cutting to the chase.

"They are going to say immediately it is partisan and boring," Chris told me. "They will try to get everyone to just look away. The numbers will be around the same as last time."

I had spoken to Sen. Elizabeth Warren, whom I had known and admired a long time. She had also phoned Sarah and me in the New Year to send her love and sympathy. Tommy had been in Elizabeth's husband Bruce Mann's first-year Property course at Harvard Law School, and he had regaled us with numerous stories about Professor Mann, tracking his development from a forbidding Socratic professor who sternly upbraided a tardy student in the first meeting of the semester to a popular favorite who joked with students and always remembered their best points from prior classes. When I asked Elizabeth about impeachment, she said, "Your main audience here is the American people and history. If you do your job right, America will never forget the facts. But most of the Republicans in that room are a lost cause. Think about history instead."

That was the sum total of conversations I had with senators prior to trial, and I was careful never to approach any senator about trial strategy or tactics, witnesses, or arguments before or during the proceeding. On one of our daily manager calls before trial, I warned the managers to avoid all substantive conversation with senators about evidence in the case during the trial. It would be easier to have a categorical rule like that, the same kind we had regarding media contact. Although there was no Senate or House rule against managers talking to the Senate jurors, the rule against ex parte contacts in judicial trials (individual parties talking to the judge without the other side present) was etched in my mind from my first days as a state assistant attorney general. While there were some impressive lawyers in the Senate whom I would have loved to consult in advance, such as Rhode Island's Sheldon Whitehouse and Minnesota's Amy Klobuchar, I thought it would be bizarre for us to engage in substantive dialogue with the senators on trial issues and tactics.

All this came flooding back to me during the trial when it came out that a group of GOP senators—Lindsey Graham (SC), Ted

Cruz (TX), and Mike Lee (UT)—were meeting with and advising the Trump defense team during the trial, to help them craft arguments and coordinate messages after their thoroughgoing debacle of an opening day. Cruz defended their Thursday night meeting as an opportunity for "sharing our thoughts" about legal strategy, and Trump attorney David Schoen waved off questions about the propriety of soliciting advice from the Senate jurors, stating in the most revealing way, "There's nothing about this thing that has any semblance of due process whatsoever." This all struck me as cynical contempt for what an American trial is.

As I thought about the trial in the Senate, I kept hearing the question from pundits demanding to know whether the Democrats could "get to 67" votes in the Senate. I realized my honest answer to this was: no. No, we couldn't get to 67, but we had a good chance, with McConnell's help, to get to 76 and even a decent shot at convicting Trump with all 100 votes.

Let me explain my political math. To achieve 67 votes to convict meant getting 17 out of 50 Republican votes, which seemed extremely unlikely to me if Mitch McConnell decided he wanted to block us. He had too many ways to cajole and control his members to prevent our getting to 67. Similarly, even if he was appalled and repelled by Trump, as a lot of evidence suggested, there was still no way McConnell would end up inside a group of 17 dissident Republicans voting to convict, or in *any number* constituting a minority in his caucus. This is because his paramount objective had nothing to do with the trial and everything to do with staying majority leader and reclaiming the Senate majority, and that dream could never be achieved if he found himself on the wrong side of a pro-Trump majority in his own caucus. We managers might have been counting to 67 to convict, but McConnell—even if, in his heart of hearts, he wanted to convict—would be counting to 26, a majority within the caucus and the bare minimum he needed in order to hang on and avoid being overthrown by a more Trumpified loyalist in his own caucus, like Josh Hawley or Ted Cruz. In theory, then, McConnell could side with 25

Republicans and become the 26th GOP vote to convict. Thus, 76 was actually the magic number, the threshold at which McConnell could lead a group of Republicans to break from Trump while remaining the leader of a majority within the GOP Senate Caucus.

Thus, 76 was, paradoxically, far more likely than 67.

I honestly thought 100 was a possibility, too. If we delivered a devastating, overwhelming, and irrefutable case for Donald Trump's culpability, and if McConnell had had enough of Trump's fascist mayhem and wanted to be done with it all, perhaps he could convince *all* the Republicans to hold hands and jump together. I was convinced that Trump spelled not only a clear and present danger to the survival of American democracy but also certain doom for the Republican Party. Surely a shrewd operator like Mitch McConnell understood the peril they were all in with Trump. Surely most GOP senators could see that Trump brought ruin to everyone in his entourage and chaos to anyone who chose to come under his influence. If we created a case so overwhelming that the vast majority of Americans turned in anger and disgust against Trump and his Proud Boy shock troops and anyone complicit with them, McConnell might just use it as the opportunity to go to the mattresses and force the GOP to be done with Donald Trump once and for all.

But this also meant that Trump's current crop of diehards— Marco Rubio, Josh Hawley, Ted Cruz, Rand Paul, Mike Lee—would have to break from their craven and cultish allegiance to him. I say "allegiance" rather than "loyalty" because I didn't believe that any of them was truly loyal to him. These were not true friendships, or even true political friendships; they were temporary shotgun marriages of political convenience. All the right-wing senators crowded around Trump in the hope of inheriting his fund-raising lists and hypnotic control over the fanatical nationwide insurrectionary mob that had dramatically expanded the ranks of traditional Republicans. That was it. These GOP senators could turn on Trump on a dime, and I thought they really might do that, given the chance to strike a collective blow that might be the be-all-and-end-all of Trump's tyranny. It was just a shot, nothing more than that, but it was a real shot, and we had to take it.

The bottom line was that regardless of the math, I had to believe we could win, so the others could believe it, too. And I did believe it, with all my might. To take our shot, we needed to make a devastating and airtight case based on a mountain of uncontradicted facts, an overwhelming and irrefutable stack of case precedents and legal authorities and, not least, irresistible moral claims about the meaning of democracy that would shock and stir the conscience of America and even the world.

The facts were out there. Barry and the gang were hot on the trail of evidence of the president's shocking actions and shocking inaction, and the direct consequences thereof, that would stun the country and leave Trump's lawyers flummoxed and flabbergasted. And it would not just be *facts* and *statements* and *proof*, the kinds of things that Trump and Fox News were well practiced in dismissing and heaping scorn upon. It would all be backed up by irrefutable *video images*, the official language of the internet and pop culture, not so easily demolished by ultra-right talking heads.

Meantime, the law was overwhelmingly and incontrovertibly on our side. Like the nation, the Senate understood instinctively what "incitement to insurrection" meant. Like its first cousin "incitement to riot," the concept produced strong mental images. What I liked about incitement to insurrection as a charge was that it fit Trump's conduct like a glove. It was precisely what he wanted to do and, indeed, had succeeded in doing: inciting people to stage a violent uprising in the Capitol to interfere with the counting of the electors to force Vice President Pence to do his bidding. Would a single senator be able to argue that it was *not* an impeachable constitutional crime for the president to incite violent rebellion and insurrection against the union? Would any senator or lawyer really assert that a president had a constitutional right to incite violence against our government without being impeached and convicted for it? We would have to come out swinging on this point right away and, essentially, dare the president's lawyers to claim just that.

Trump's lawyers would have to argue that the First Amendment protected his speech and that impeaching him for his speech would violate his First Amendment rights. According to this mind-bending argument, the mastermind inciter of a violent insurrection to defeat the popular will and thwart our national presidential election would magically become an innocent victim of liberal political correctness and the "cancel culture" mob. This sleight of hand, which I predicted would become Donald Trump's Fox News go-to defense, always made me laugh, because everything about it was such an absurd inversion of the law.

To begin with, "incitement to imminent lawless action"—which is exactly what Donald Trump's conduct consisted of—is not protected by the First Amendment anywhere, as found by the Supreme Court in *Brandenburg v. Ohio* in 1969. *Brandenburg* is the standard that applies to government prosecution of a guy on the street who deliberately incites violence in a way that is actually likely to produce it. If it's mere advocacy, it's protected by the First Amendment; if it crosses over to "inciting imminent lawless action," as it did here, it loses its protection and is punishable by government. So even if *Brandenburg* were the right standard to apply to presidential conduct in an impeachment trial, Trump would lose outright. All the factual evidence in the case demonstrated that his speeches and actions were not geared to mere public education or policy advocacy but, rather, to *incitement of imminent lawless action*, to the end of whipping his crowd into a mob frenzy that would unleash sordid violence on the Capitol. This was the unusual case of incitement to violent insurrection that fit the strict constitutional standard to a T—and the whole world saw it. Trump was not being impeached for his *ideas* or his *views* about incitement to violent insurrection; he was being impeached for *inciting* violent insurrection.

Ultimately, though, the First Amendment framework is an irrelevant distraction from the real issue here. The actual question is whether inciting insurrection against the government is an impeach-able offense for a president who has sworn an oath to "preserve, protect and defend the Constitution of the United States." And here the answer is obviously yes, because there are lots of things that any private citizen might say or do that are protected by the First Amendment that would nonetheless constitute a blatant violation of a president's oath to "preserve, protect and defend" the Constitution. Nobody made Donald Trump run for president, but his voluntary de-cision to swear the presidential oath imposed upon him a unique and weighty duty not to discard, disparage, or violate the Constitution. The First Amendment had nothing to do with it.

Of course, a lot of legal argument, especially in spoken word, turns on metaphor and analogy, and this is especially true in the First Amendment field, where the image of fire is replete and central to First Amendment jurisprudence. What was gnawing at me was perhaps the single most dominant metaphor in free speech rheto-ric and jurisprudence, Justice Oliver Wendell Holmes's famous statement, in the 1917 *Schenck v. United States* case, that even "the most stringent protection of free speech would not protect a man in falsely shouting fire in a theater and causing a panic," which has been translated and reduced over a century of absorption into the bloodstream of popular vernacular to the (now rather misleading) cliché that "you can't shout fire in a crowded theater."

This overworked modern fire-related metaphor was putatively helpful to us. I noticed that, as the Trumpified right began to float its burgeoning First Amendment defense of Trump's incitement to insurrection, a lot of the liberal respondents on the talk shows were insisting upon Trump's culpability by automatically asserting that "you can't shout fire in a crowded theater."

This was okay, I supposed, but it left me vaguely uncomfortable for two reasons. One was that the compressed popular form of the phrase—"You can't shout fire in a crowded theater"—left out the word *falsely*, which was in the Holmesian original. That word was necessary for Holmes to safely make his point, because you can obviously shout "fire" in a crowded theater if it is true that there *is* a fire and you need to get people out of the theater. But repetition of the modern version of the cliché without the word *falsely* invited the Trumpian response that Trump was simply calling for his supporters to "stop the steal" because there was a real election steal going on. Already Rudy Giuliani, in his inept and fumbling way, was stumbling toward this answer when he said, "If it's true, it's not incitement." This amusing formulation conflated libel law ("if it's true, it's not libel") with incitement law, but it was slouching toward the argument that Trump's lawyers might have arrived at: that Trump was not *falsely* shouting fire in a crowded theater. This would have been wrong and answerable, because his statements about the election were indeed demonstrably false and had been rejected in more than sixty federal and state courts across the land, but it also completely changed the subject and bogged us down in talking about the election, something off topic and unnecessary, rather than the violent insurrection itself.

This was my second problem with promiscuous use of the "shouting fire" trope. As a metaphor, it implicitly and dramatically *understated* what Donald Trump had done in the events leading to January 6 and even *sanitized* his role in those events. He was not an innocent, or even a malicious, bystander who falsely reported a fire to cause a mass exodus; he deliberately created the crisis himself and then did nothing at all to address it but keep other people from intervening effectively to end it.

We needed to rework the metaphor to make lightbulbs (or at least fireflies) go off in people's heads. So I called my favorite constitutional law professor, Laurence Tribe, again. He had been thinking about the "crowded theater" problem, too, and suggested the perfect amendment to Holmes's trope to fit this context:

"Trump is not like a citizen who falsely shouts fire in a crowded theater," Larry said. "He's like a fire chief who sends the mob to burn the theater down."

"Yes!" I cried. "That's it. And then, when the people at the theater call the firehouse to report the blaze," I said, continuing the metaphor, "the chief does nothing for several hours but sit back and enjoy the whole thing on television. He incites them to set the fire and does nothing to douse the flames. He *wishes* he were just like some guy who falsely calls out fire in a crowded theater." I felt like I was back in law school myself.

After the anemic First Amendment defense, there was really only one constitutional argument that we had to address, the one that Trump's followers truly wanted to hang their hat on: namely, the jurisdictional claim that the Senate could not conduct a trial of Trump because he had already left office. This, we were all convinced, would become the catch-all final rallying point for Trump supporters in the Senate. It required no messy encounter with the brutal facts of the case, and it had enough surface plausibility not to sound ridiculous, as the "free speech" argument did. But the jurisdictional argument had no basis in the text, structure, or history of the Constitution, much less the precedents and understandings of the U.S. Senate.

My rationale went like this: According to Article I, Section 3, the Senate has the "sole Power to try all Impeachments." Donald Trump was indisputably impeached by the House while he was president (on January 13, 2021) for acts he committed while he was president (on January 6, 2021, and before), so this *was* an impeachment. Therefore, the Senate has the power to try the case.

Now, that textually based argument is backed up by more than two centuries of relevant and controlling history, both the Founders' clear understanding that trials *could be* and *would be* conducted of former officials and the Senate's unbroken pattern and practice of actually *conducting* trials of former officials. Trump's jurisdictional argument had been tried before repeatedly and had always failed. There was the famous *Belknap* decision after the Civil War, when a corrupt secretary of war got caught taking kickbacks and immediately tendered his resignation to President Andrew Johnson in the hope of avoiding the disgrace of impeachment in the House and trial in the Senate. Yet, the House proceeded to impeach Belknap, and the Senate proceeded to try him, on the powerful theory that a high-level governmental criminal should not be able to escape the constitutional consequences of his actions simply by quickly resigning and then running for office or taking appointive office later on. Instead, he should be tried, convicted, and removed and then disqualified from further service (if the Senate sees it that way), because impeachment exists in the Constitution as a protective and proactive measure to guard the union against future harm from known official miscreants and violators. The logic of this argument was overwhelmingly satisfying to me, but we still needed to dig up, master, and mobilize several centuries of, first, British and, then, favorable U.S. Senate precedents to support it. Joe Neguse and David Cicilline were hard at work, and we would do it together.

We had all taken a break from press interviews after the impeachment vote, but the media was ravenous to hear from me about how the managers were preparing and where we were going with it all. They also wanted to know how I was handling my new assignment while grieving Tommy's death. My communications director, Samantha Brown, had logged hundreds of interview requests, and the Speaker and her team thought I should go on one of the Sunday talk shows and recommended sitting down with Jake Tapper of CNN. They were willing to do it over Zoom but there were just too many things, I knew well by now, that could go wrong on Zoom, including dogs barking in the background, someone calling me during the interview, the cell phone or computer slipping from its position, or even a total interruption of the connection. Rather than worry about all that, I decided to get dressed in a full suit (no sweatpants and socks) and drive myself down to the CNN studio and do it in person.

The interview caused a sensation because Jake Tapper effectively unified his questioning about the insurrection and our plans for trial with his questions about who Tommy was, the deepening crisis of mental and emotional health among the young, and what I would say directly to depressed young people today. The reaction to our extended and profoundly emotional conversation was astounding as tens of thousands of messages immediately began pouring in from across America. Jake gave me the chance not only to talk in depth about Tommy, his poetry, his compassion and brilliance, and his love for the world, but to connect the struggles of millions of families like ours to the miserable, depressing, and ever-declining state of our politics.

When Jake asked me how I could deal with "trauma on top of trauma" and go forward to lead the managers' team, I froze for a moment and then simply said, "These events are personal to me. I'm not going to lose my son at the end of 2020 and lose my country, my Republic, in 2021. It's not going to happen. The vast majority of the American people—Democrats, Republicans, Independents—reject armed violence and insurrection as a new way of doing business in America."

When he asked me what I would say to young people in crisis, with one in four Americans telling pollsters they had seriously considered suicide in the last thirty days, I said that they must seek help from a family member or a doctor, or by calling 911 if they are alone. I told them that "there is a lot of love out there," and the "outpouring of love for Tommy has been astounding," but we do not need to wait for people to die to show them love and we are all here for them right now.

Most interviews leave most politicians feeling pretty disappointed, but I will always remember and treasure this interview because I learned so much doing it. The power of the conversation was that Jake Tapper gave me the space and the prompt to connect the private trauma of our losing Tommy to the public trauma of the violent insurrection. In speaking this way, I felt Tommy strongly in the core of my being, right where he was on January 6 itself, right next to my heart. And I felt the strong bond between parenthood and patriotism. For just as parents must take care of their children because they are our children, citizens must take care of our country because it is our country. The word *patriot* comes from the ancient Greek word for fatherland, *patris*, which captures two relationships at once: the love we have for our country and the sense of passionate common cause and solidarity we have with our fellow citizens.

Several weeks into our mourning, I was a bit more relaxed now in talking about Tommy and was able to have sustained conversations about him with colleagues and friends without breaking down and dissolving completely. I felt his spiritual and emotional presence best when I was with lots of family, but as family gradually peeled away and returned to their hometowns, I experienced our son as a spur and call to action driving me ahead.

I knew lots of people were saying of me that I was dealing with my pain and grief by "throwing myself into my work," but that phrase did not quite capture my sense of this period. It was very hard for me as a father to disentangle my thoughts and feelings from Tommy's thoughts and feelings, and I felt that, in this case, the braiding of our intellectual and emotional identities permitted me to do serious hard work with political meaning and moral integrity. While some people seemed to be humoring me when I talked like this, all my talk about Tommy's being "with" me, in my heart and in my chest, felt true, and still it does.

But I was about to be put to a real test: solitude. After we made a brief and entirely restful and healing trip to Malvern, Pennsylvania, to be with Tabitha and Ryan at the home of Ryan's loving parents, Susan and Gary Vogel, Sarah was going out west to be with Hannah and Hank in Lake Tahoe, and I would be home by myself working on the case. It was scary to see her getting ready to go—I would be alone in the house with my thoughts, my memories of that day, and new habits of self-cross-examination over every decision I had ever made as a father and a person that I could lapse into in moments of despair. But it did make me smile to think of Sarah's being with Hannah and Hank, who had been expressing a sense of loneliness and isolation so far from home. I had never been crazy about the idea of the two of them living out there, precisely because you never knew what might befall people in your family, and I wanted Hannah closer, both to *have* our support and to *be* our support. I missed her sharply, and I was planning to make the trek out west to see her and Hank's beautiful new home in Nevada as soon as the trial was over.

In the meantime, I would work at home. If I made enough progress, I would fly to Florida with Tabitha to spend a few days with my brother Noah and his wife, Heather, and other family who were vacationing down there.

I knew I could find a lot of mental clarity while traveling away from all other demands and from the giant, ubiquitous, sprawling stacks of mail, which my wonderful constituent caseworker and aide, Christa Burton, had so kindly offered to sort into more manageable categories, which we then built on at home—"Tommy's Friends from College," "Tommy's Friends from Law School," "Lost a Child to Suicide," "Supporting Your Work on Impeachment; Condolences," "Thanking You for Work on January 6," "Hannah and Tabitha's Friends," "Supportive, Loving Constituents," and so on.

Christa's generosity and foresight saved me, because by the time the trial was over, we would have a backlog of more than *fifteen thousand* letters and messages and emails. While friends seeing the stacks would say that none of these sympathy writers was expecting an answer, these friends obviously did not know my close-to-desperate compulsion not only to answer every single letter, email, and call I received, but to answer them *personally* and *promptly*. I experienced this impossible mound of unanswered mail, a lot of it from close friends and family members and a lot of it from hurting young people, as a terrible, constant reproach and a staggering emotional burden. But it wasn't the sheer volume that weighed on me; it was the astonishing beauty and poignancy of the individual messages coming in, all of them fitting together, like a jigsaw puzzle of Tommy's life and his beliefs, all of them teaching me something important about America and about what a good life is. Whenever I took an hour to go through them, I could read and answer only three or four letters, usually in tears, before becoming mentally overwhelmed and emotionally drained. So I resolved, based on the urgings of my district director, Kathleen Connor, and of Julie, to write one general letter to everyone sending us condolences, and one general letter to everyone who wrote about our work on impeachment, and then one general letter to everyone who offered both condolences and political solidarity. There were a lot of this last variety—one person actually wrote me to say, "I was looking for a condolence card for the loss of your son which also said, 'and thanks for saving our country too,' but Hallmark apparently doesn't make those." When I sent these form letters back, it gave me a bit of relief just to know that my correspondents had at least heard something from me.

But when I contemplated actually trying to respond to each person individually, it stressed me out, and that is when I decided to write this book. It would be my own attempt at a personal answer, a labor of love and a way to respond to all those people who told me, in such fine-grained detail, about the love and the crises in their own families, about their grievous personal losses and their incremental triumphs, and about the desperate fears they have for our nation's future and the most cherished hopes they have for what America may still become in a world of so many frightful dangers.

On Monday, January 25, I got to see the impeachment managers, lawyers, and staff in person. For more than a week, everyone had been working around the clock, together from afar, on this project. Our Zoom meetings now bore no resemblance to most Hill meetings, which involve a lot of posing, preening, and showing off. All that status-seeking behavior was gone. These were *work* meetings, full of brainstorming, conceptual breakthroughs, and steady substantive progress. We were getting stuff done, and it was pleasurable for that reason.

When I walked into a session that day and saw Stacey, Madeleine, Joaquin, and Diana deep in conversation about our factual presentation, I felt overwhelmed with gratitude. Based on what I had been reading about the predictable chaos unfolding in the Trump camp, I knew in my gut we were making huge strides forward on the track before they had even found the course.

For Trump, everything in a case is about knowing, and working, the *judge*—not the *law*, much less the overarching political *morality* of the moment. His is the right strategy for a corrupt society, but not for a healthy, self-respecting democracy. I was not following the continuing upheaval on Trump's legal team, but I felt certain that his ultimate defense team and strategy would reflect his own slovenly and expedient approach to the rule of law. We would be speaking only to the highest values and impulses of American political culture.

At 6:45 p.m. on Monday, January 25, we gathered in the Rayburn Room to line up for the four-minute procession over to the Senate side. The Clerk of the House lined us up, with me in front, followed by seniority pairs: DeGette and Cicilline, Castro and Swalwell, Lieu and Plaskett, Dean and Neguse. Someone handed me a copy of the Article of Impeachment that I would be reading over on the Senate side, and I realized that I had not practiced it, so I read it through quietly several times while we waited for the bell to ring and to start walking.

When we began to process, the enormity of the moment seized me. Walking through Statuary Hall and the Rotunda, both of which had been trashed and defiled in unspeakable ways by Trump's insurrectionist army, I looked behind me at the managers in their four rows of two. Everyone had been joking before about how Ted Lieu was such a great walker, with his military gait, but they all grew solemn and quiet as we began to process. All you could hear was the staccato fall of our shoes as we paced through the Rotunda, where the loud clicks of reporters' cameras punctuated our footfalls.

When we crossed into the Senate section, a huge throng of reporters on either side of us began to snap photos and record our approach. When we got up to the main doorway to the Senate chamber, we waited outside as the Clerk of the Senate received us and relayed word of our arrival. A few moments later, we were allowed in, and we walked to the front.

I was shown my way to the main dais, where I proceeded to read the House resolution and the Article of Impeachment. I was conscious of slowing down and speaking the words "Donald . . . John . . . Trump" with prosecutorial gravity, emphasizing "high Crimes and Misdemeanors," and particularly accenting Section 3 of the Fourteenth Amendment's bar on future officeholding for anyone who has "engaged in insurrection or rebellion against" the United States.

When I reached the final sentence, I looked up and took in the room in one quick sweep. All the Democrats seemed to be there, on my right-hand side, just a handful of Republicans on my left, including Mitt Romney, in the back, and Roger Marshall of Kansas, a former House colleague of mine whose presence I erroneously took as a promising sign of his intention to keep an open juror's mind. I saw Ed Markey, Kirsten Gillibrand, and the genial Tim Kaine, who I heard was trying to shop around a substitute censure resolution that apparently had found little traction on either side. When I saw Corey Booker, a vegan and big-time animal rights person, I smiled and thought about Tommy.

And I read the several pages of the House indictment of Trump and this closing paragraph:

> *Wherefore, Donald John Trump, by such conduct, has demonstrated that he will remain a threat to national security, democracy, and the Constitution if allowed to remain in office, and has acted in a manner grossly incompatible with self-governance and the rule of law. Donald John Trump thus warrants impeachment and trial, removal from office, and disqualification to hold and enjoy any office of honor, trust, or profit under the United States.*

And then we walked out of the Senate—solemnly again, breathing a sigh of relief—and into another clicking and clattering clutch of photographers and reporters shouting out their questions. We continued on. We would let the Article of Impeachment speak for itself.

That night, before I flew out to Florida, I was visited by our friend Dar Williams, who had driven down from New York to make a memorable vegan Indian dinner for me and to hang out before I left for Florida the next day. We cried over Tommy, our dear boy whom all our friends, like Dar, had loved and cherished, and she told me her thinking about death and life, meditations that gave me great solace. A religion major at Wesleyan as an undergraduate, Dar shared with me that evening different religious approaches to comprehending death, and I found her descriptions immensely soothing. She said it was important not to censor the emotions as they came—grief, agony, sadness, anger, love, nostalgia, loneliness, guilt, longing.

"You take these things and these intense feelings as they come to you and encounter them without taking them personally," she said. "There doesn't have to be some destiny or karma attached to them. You don't say there was some God that wanted them to happen to you. Feelings are like the weather, and you let the weather happen without taking it personally." She was urging me—I think in a kind of Buddhist way, although I wasn't sure—to observe my own feelings and not to become dissolved in my sadness and pain. I appreciated her thoughts a lot.

I told Dar about the Amish-quilt theory someone had shared with me, which is that a family is not best seen as a sequence of generations that just leaves the oldest ones behind but, rather, as an Amish quilt in which each of us is represented by a square, whether we happen to be alive or not, and no one ever loses his or her place in the quilt, but the whole quilt just keeps getting bigger and more beautiful over time.

I told Dar about our close friend from Texas Jennifer Lord (a conservative Republican, no less), who lost her sister Rebecca, a brilliant woman and my first political director, to a horrid case of thyroid cancer. When Jennifer came to stay with us in the wake of Tommy's death, she had offered to make us a quilt out of shirts from Tommy's closet. We agreed right away. One day, perhaps, I would like my shirts to become part of that quilt, too. And one day, some kid (a great-grandchild or a great-grand-niece or -nephew) will curl up under that family quilt on a sofa, on a Sunday afternoon on a freezing winter day, and fall asleep with it, maybe with a dog at their feet, near a crackling fire, and our neighboring patches in that quilt, mine and Tommy's, added long ago by someone's loving hand (someone whom they'll likely never know), will help keep that kid warm through a nap in a snow-storm.

As I grew increasingly tired from the day and was thinking these soothing melancholy thoughts, I asked Dar, who had brought her guitar down with her, to play me Tommy's favorite Dar Williams song, the melodic and enthralling "The Ocean." As Dar played, I tried to hang on, but I fell into a vast and deep sleep that would carry me to the morning.

CHAPTER 9

REVERSE UNO

Humor is not a mood but a way of looking at the world.

—LUDWIG WITTGENSTEIN

When Tabitha and I arrived in Florida, I kept thinking, *Why am I in Florida? Why is a grieving father preparing madly for an impeachment trial in all this sunshine?*

It seemed fundamentally discordant for me to be there, in the land of fresh squeezed orange juice, Early Bird Specials at golf courses, and blinding sunshine at the beach, although my friend Rep. Darren Soto (from Kissimmee) assured me that people in Florida know how to mourn and grieve, too. In fact, he told me that there had been a new trend developing, even before COVID-19—I think he called it "grief tourism"—where people arranged travel packages to the beach, Disney World, or the Everglades in connection with their coming to Florida for a funeral.

Only in Florida.

I would continue to get a lot of work done down there, especially on our trial brief and my opening statements, but the psychological truth, I realized, was this: I had come down to Florida to see Noah and his family because I wanted to find out whether I could laugh again and whether it would be all right to do that.

Nobody on earth ever made me laugh more or laugh harder than Tommy Raskin, but it was always a close race between him and my brother, Noah, who had been making me laugh his entire life. His and Tommy's sense of humor were similar: zany, ridiculous, utterly transgressive, slightly more on the profane side for Noah, slightly more on the philosophical side for Tommy. So when Noah and Heather urged me to join them in Florida in January for a few days, it made me think a whole lot about what it means to mourn and grieve.

The truth is: I had been pretty miserable. I cried at different points throughout each day and had little control over it. Sometimes, I would cry while opening a letter just because the name of the person who sent it to me was from my past or from Tommy's inner circle, and I would cry all over again when I read the letter, and then cry once again when I answered it. I'd been a pretty good sleeper my whole life, but my sleep was suddenly awful. In the first few weeks after Tommy left us, Sarah and I would both "wake up" crying from "sleeping." I dreamt one night that I was carrying Tommy around with me everywhere, slung over my shoulder. I dreamt another night that there was a sly thief in our neighborhood who broke into our house and stole everything. Our beloved dogs, Toby and Potter, saw me suffering and grieving one night, and they did everything they could to comfort me. But they soon turned sheepish and sullen like me. They began grieving, too.

The grief therapy I'd begun immediately after Tommy's death continued to help me to take the full, terrifying measure of the different dimensions of our loss and our grief: about losing Tommy; about losing our only son; about losing a best friend and intellectual soul mate; about losing Hannah and Tabitha's beloved brother; about losing the person who most embodied the spirit of my father; about losing the person who best articulated the kind of America I wanted to live in; about the agony and appalling bottomless mystery of suicide; about not ever again knowing what life would be like; about Tommy's resplendent, dazzling future remaining unwritten and unlived and slipping away from us; about all the lost promise of his gifts (private and public); about missing Tommy's presence so painfully each and every day.

Weekly therapy dramatically improved my ability at least to articulate the meaning of my grief and my capacity to connect with my own thoughts and feelings and the thoughts and feelings of other people. It restored powers of critical insight to my emotional life. It gave me the urge to write. It also helped me to see that I was baffled by the problem of finding joy and laughter in the wake of the death of someone I loved more than life itself.

When I see Tommy Raskin, I see him laughing uproariously, smiling at the center of attention, darting around and making everyone laugh, telling jokes and intricate stories, spreading joy and soaking it up.

I identify Tommy with humor and laughter.

When I see us together, I see us laughing about this or that story; about a scene in *Borat* or *Wedding Crashers*; about something ridiculous a politician said during the day; about some form of tone-deaf, authoritarian political correctness expressed by someone anywhere along the ideological spectrum; about something typically outrageous and lovable that Hannah or Tabitha has done.

But my experience of *losing* Tommy is dominated by misery and suffering, suffocating darkness, a free fall into despair.

And yet, I asked myself, did I best honor Tommy's memory by drowning in negative emotions indefinitely? This assumed, of course, that I had a choice in the matter. (Did I have a choice?)

How will we do justice to his exuberance and playfulness, his unfinished projects and longings for change, his irreducible essence as a friend in *good humor* and *good fun*; this beautiful boy who was the life of the party, a prankster, a blood donor, and a do-gooder who wanted nothing but *happiness* for all other people and all sentient beings? Do we owe it to Tommy and to ourselves to find happiness and laughter again? In addition to all the grief, this philosophical conundrum confounded me, and I faced it without my most trusted philosopher, Tommy, there to help me think it through.

I had accepted the invitation to go down to Florida because Noah and Heather promised that I would have lots of time to work there and ample opportunity to be with Mariah, Boman, and Daisy, Tommy's beloved cousins, my enchanting nieces and nephew. Perhaps with my brother and Tommy's cousins, and the other family and friends who would inevitably materialize, I could recover my laughter, or at least my will to laugh.

There were several lingering elements of the case I wanted to think about in the final week before the trial in the Senate, and I had a lot of writing to do. I had this weird feeling that being in the same state as Donald Trump, being closer to him physically—we were forty-five minutes away from Mar-a-Lago, his new legal domicile— would enable me to see things from his disturbed vantage point. I would view Washington and the trial at a clarifying distance, from a key swing state, a place where (despite COVID-19) people were waterskiing, getting drunk in tightly packed bars, and throwing large weddings, barbecues, and business conventions. Down here, I could think and write, Zoom all day with the managers and lawyers, and come up with big-picture themes and aggressive tactical ideas. Having watched Trump for four years, I knew that he thrived on the element of aggressive surprise. Perhaps we could generate a surprise or two to spring on *him*.

While down in Florida, I viewed at length for the first time some of the footage that the video production team had been collecting and inventorying. It was, of course, gory and sickening to watch organized groups of hoodlums, racists, and neofascists pummeling, spearing, and torturing Capitol Police and MPD officers. But in order to illuminate the real meaning of this madness, it was necessary to broaden and extend the time frame, to track Trump's cultivation of such violent political extremism during his time in office.

The enormous violence that shook the Capitol on January 6 was *political* in nature, unlike, for example, a barroom brawl, a sexual assault, or a Mob hit against a rival criminal gang. It was political in a double sense: First, it had the political quality of dividing the social world between "friend and enemy," which has been the organizing principle of reactionary political thought for centuries, as best expressed in the writings of political theorist and jurist Carl Schmitt. It drew upon and activated deep racial enmity and hostility, as was experienced by my fearless constituent in Silver Spring, Harry Dunn, an African American officer for thirteen years on the Capitol Police force, whom I interviewed and who told me he had been subjected to what he called a "torrent" of racist abuse and invective during the January 6 riot. Dunn had never been called the N-word before when he was in uniform, but he heard the word uttered dozens of different times in taunts and provocations as he tried to protect members of Congress from the mob that day. Similarly, Sgt. Aquilino Gonell, a U.S. Army veteran and naturalized citizen whose family came to the United States from the Dominican Republic, was not only punched, pummeled, kicked, and sprayed in the face with bear mace, but also taunted as a "traitor," a "foreigner," and "not even a real American." From the standpoint of democracy's defenders on the ground and in the Capitol that day, the ideological rage of the rioters and insurrectionists transformed the violence into terror.

The political ideology of the mob went even beyond the deep pools of racial hatred, xenophobia, anti-Semitism, and immigrant bashing from which participants drank on that day. They imagined that they had converged on Washington to destroy corrupt politicians of both parties, traitorous police officers, lying media, agents of George Soros, defenders of the Clintons and the Obamas, and other sinister, shadowy forces identified by conspiracy theorists in QAnon and by Trump's authoritarian polemicists like Steve Bannon. Their foes were anyone who dared get in the way of the perpetual reign of Donald Trump, and their friends were anyone who joined in the festival of merciless violence called to restore Trump to his American throne.

But, as we looked farther back in time and to a broader panorama, we could see that the violence was political in a second sense: it was a core part of Trump's long-running *strategic* plan to maintain power, and on January 6, it became a specific tactical maneuver. I knew a lot about the "Unite the Right" festival of racist violence in Charlottesville, Virginia, in 2017, of course, but I was especially struck and dumbfounded by the footage gathered on the storming of the Michigan State Capitol in Lansing on April 30, 2020. Rioters and militia groups armed with AR-15s showed up in response to Trump's call to "LIBERATE MICHIGAN," with plans of disrupting legislative renewal of Governor Gretchen Whitmer's COVID-19 stay-at-home policy order. That video was eerily foreshadowing. The participants looked just like the January 6 insurrectionists: they were heavily armed and masked, wearing combat fatigues, and striking the poses of gangsters or terrorists, with no respect for the rule of law. The menacing of legislators by protesters in the gallery armed with AR-15s, the terroristic threats of the militia groups, and the attempt to physically charge the Michigan House led to adjournment and closure of legislative proceedings there.

The events in Michigan were Trump's early opportunity to test how far his most violent followers would go in response to his incendiary and thinly coded rhetoric. The siege of the Michigan statehouse laid down a gauntlet with his movement that legislative institutions and buildings would not in any way be off-limits in his plans for upending the rule of law and unleashing the ambitious designs of his most avid followers.

I also reviewed scenes from the brutal violence that erupted at the December 12, 2020, "MAGA II" rally, which drew thousands of Trump followers to DC to denounce Biden's victory as fraudulent. Hundreds of members of the extremist fringe cultivated by Trump—especially the Proud Boys, who were pumped up with Trump's September message to them from the debate stage to "stand back and stand by"—roamed the streets of Washington, from the Capitol to the Mall to Black Lives Matter Plaza to the downtown bars, to initiate bloody brawls and rumbles against their real and imagined opponents. When night fell, the violence turned especially vicious and deadly. They jumped and casually stabbed numerous counter-protesters whom they considered to be "Antifa." Dozens of arrests were made.

Trump, of course, had nothing negative to say about the perfor-mance of his violent enthusiasts on that day; much less did he do anything to shore up the forces of order in the capital city. In fact, he tweeted his enthusiastic embrace of the rally, feigning ignorance that such a mass uprising was even taking place: "Wow! Thousands of people forming in Washington (D.C.) for Stop the Steal. Didn't know about this but I'll be seeing them! #MAGA."

Like the one in Michigan, this event also displayed all the basic elements present on January 6: propaganda and organizing around electoral lies; rejection of the constitutional order; the unleashing of violent, primitive, and racist impulses in the streets to advance Trump's political agenda; and Trump's enthusiastic embrace of the violent ardor of his followers and his acquiescence to even their most criminal actions. Trump's long-running fascination with political violence and his cultivation of extremist elements would become part of our story at trial as we showed how mob violence had created the context for the president's efforts to coerce and subdue Mike Pence, destroy Biden's Electoral College majority, and prevail in a contingent election.

From the moment Speaker Pelosi asked me to lead the team, I was convinced that we could convict Trump in the Senate if he came and testified, almost regardless of what he said. If he told the truth, an admittedly unlikely hypothetical prospect, everyone would see that he had not only *incited* the insurrection but also *worked to create it*.

If he came to testify and lied, which was a near certainty, he would not only expose himself to perjury prosecution for all his false statements, but he would cause the whole defense strategy (of avoiding direct factual claims and railing instead about imagined constitutional injuries suffered by *him*) to come tumbling down on the GOP.

At the same time, if he categorically refused a subpoena to testify—and surely, if there were any sane lawyers left in his entourage, they would fight a witness subpoena at all costs—he would destroy the defense team's newly emerging, whiny argument about his not being afforded enough "due process." After all, the heart of due process is the opportunity to be heard, and if he rejected an opportunity to testify at length before the Senate, this argument would be revealed as a sham. Trump would also be exposed as a craven and incorrigible turncoat even to the brainwashed malcontents who joined him in betraying the country and attacking the Congress.

But, then again—and this was where I was getting hung up—Trump's lawyers would be certain to tie a subpoena up in court forever and to invent a new "former president's executive privilege" doctrine that conservative judges in the judiciary would be inclined to embrace. And we would be none the better for it. Everyone on the Hill to whom I discreetly mentioned it couldn't stand the idea of our calling Trump as a witness. They had waited years for White House counsel Don McGahn to testify, had watched Trump and his cronies give us the finger for years and thumb their noses at both Congress and the courts. Members were sick of subpoena fights; they had been there, done that, and they hated it.

So how to proceed? It was gnawing at me. I needed a serious litigator's counsel on this.

I called my best friend, Michael Anderson. We were roommates through college and law school. (He was also in Larry Tribe's Constitutional Law class, but his attendance was, charitably speaking, spotty, as the class met before high noon, when most gentleman anarchists arise.) A creative union lawyer, Michael is one of the finest practicing attorneys I know, in a class with people like

Michael Tigar, Angela J. Davis, Jimmy Klein, and Abbe Lowell. You talk to Michael, and you know you're talking to a *lawyer*, but he also speaks with absolute authority and a little bit of an attitude about the things he knows, as if he's reproaching you for not doing with *your* life precisely what he has done with *his* life.

"Of course you have to call Trump," he said, when I asked him if it was worth doing.

"But he'll fight the subpoena," I said. "It's three years later, and we're still struggling to get Don McGahn to comply with a subpoena. How does it help us? Everyone on the Hill is skeptical."

"Well, you don't need to subpoena him. Just invite him to sit down for a voluntary interview."

"How does that help us? We can't comment on his refusal to testify, so where does it get us?"

"No, bro, that's where you're wrong," he said. "Under the Fifth Amendment, a criminal defendant or suspect's refusal to testify cannot be used against him in a *criminal* trial."

"Oh, right," I said.

"But there's nothing stopping the government or anyone else from discussing his refusal to testify in a *civil* proceeding," he said.

Michael was right. An impeachment trial was totally civil, of course. Trump wouldn't spend five minutes in jail if he were convicted.

"Wow," I reflected. "You're saying we can just *rail* about his hiding under his bed."

"But it's much better than that," Michael said. "There's a well-developed body of precedent saying that a tribunal in a civil proceeding can not only *consider* the refusal to testify but can draw *an adverse inference* against someone who exercises the right not to testify. If a person would naturally be expected to testify in their own defense in a civil case and refuses to do it, the jury can be instructed that it has the right to draw an adverse inference from the simple refusal to show up. And guess who's the strongest justice on the government's right to use your no-show decision against you in a civil proceeding?"

"Justice Scalia," I rightly guessed, knowing how Michael thinks. Michael is famous for using the doctrines and arguments of right-wing justices on behalf of his union and progressive clients. It was a classic Andersonian maneuver. The right wing forgets that legal doctrines can apply in lots of different directions if—and this is admittedly a big if—the judges are not being totally partisan and partial about it.

"So are you going to call him to come testify?" Michael asked. "You gotta do it."

And with that, I had a plan. After we filed our brief, I would wait for Trump's answer to come in. And if he went crazy, as I thought he would, and started denying that he'd done certain things, instead of just sticking to the argument that the Senate has no jurisdiction, I'd write him to say that we now had a factual dispute and that I'd just love to come interview him at Mar-a-Lago or in Washington to clear up the controversy. And then, when he decided he'd rather not talk about it, we'd insist that the jurors of the Senate draw an "adverse inference" (i.e., a negative conclusion) from his unwillingness to discuss what he did on January 6, using Supreme Court precedent to support us.

The Senate is not governed by normal rules of evidence, but these constitutional doctrines and Supreme Court authorities would have a lot of rhetorical power in shaping discourse on the Senate floor and public understanding in the impeachment trial. It would be a tactical breakthrough in dealing with Donald Trump, omnipresent talker, commentator, and heckler but also elusive defendant and narcissist snowflake. Michael's "adverse inference" strategy could help change the dynamics of the trial before it even started.

With the help of some other Trump-era freedom fighters, I had another breakthrough on the case while in Florida—this one semantic, not tactical. Our constitutional motions team (Neguse, Cicilline, and I) had prepared truly devastating arguments against the Republican rallying cry that the trial was unconstitutional because Trump's term in office was over. This argument collided with the clear text of the Constitution, the British impeachment antecedents that informed our Founders' handiwork, and more than two centuries of crystal-clear Senate impeachment precedent (although none of it presidential).

The problem was that, while we had a vastly superior legal argument, Trump's argument had the virtue of alluring simplicity: How can you conduct an impeachment trial of a *former* president? Of all the rhetorical spaghetti and lasagna that Trump's lawyers were throwing at the wall to see what might adhere, this one was, in my judgment, by far the stickiest.

I was convinced that we needed to show people that letting impeached presidents off the hook simply because they had committed their high crimes against America on the day of January 6, two weeks before the new president was inaugurated, would mean that this "sacred constitutional day" for counting the electors would become a new free-fire zone, a moment of open season on constitutional democracy for incumbent presidents who lost and chose not to accept defeat. Under a "new January 6 exception to the Constitution," I kept insisting, any frustrated loser in a presidential campaign could try to mount an insurrection, rebellion, or coup against Congress, or simply go out and round up opposing members of Congress or the Supreme Court and jail them all. This could not be right, because the president's Oath of Office to "preserve, protect and defend the Constitution of the United States" begins on the first day of office and lasts through the final day in office. The Senate, we must recall, has the power to "try all impeachments."

But even this rather compelling argument, in a frenetic world of vast social media and scarce public attention, seemed too bulky. I had asked Larry Tribe to give the matter some thought, and he, too, was frustrated in trying to boil our argument down to something more memorable and vivid. He called me back to say that he had explained the problem to Professor Timothy Snyder, who was now on the phone, too.

This was great to hear. I greeted Tim warmly. He had appeared, just a few months before, as a witness in the subcommittee I chair, Oversight's Subcommittee on Civil Rights and Civil Liberties, in a hearing on white supremacist infiltration of law enforcement, and had been typically perceptive and eloquent that day.

A professor of European history at Yale, Tim Synder is a compelling author whose tiny little book *On Tyranny: Twenty Lessons from the Twentieth Century* was a runaway best seller at the start of the Trump administration. You could see people devouring it at the pro-immigrant airport protests, the Climate March, and other mass demonstrations. All of Tim's works breathe the struggle for freedom, justice, and democracy against fascism, communism, and the new kleptocratic authoritarianism that the Putin-Trump axis ushered onto the world stage.

Larry had asked Tim to think about a good conceptual counter to the false but seductive claim that the Senate could not try a former president. Tim suggested that we frame Trump's argument as a runaway executive's special pleading for a "January exception" to the Constitution, which had a properly authoritarian and ominous ring to it and improved dramatically upon my focus on just January 6— as Trump's argument would, in truth, effectively immunize *any* mischief or corruption that took place in a president's final month in office. Indeed, it's really more like a December–January exception, but that doesn't produce quite the same cadence. The rhetorical simplicity of the January exception was ominously pleasing to the ear. I was extremely grateful to both Tim and Larry.

As these strategic plans came together, I told the managers and staff that I was thinking carefully about calling Trump as a witness. I talked everyone through the potential steps that might get us there and the "adverse inference" line of civil cases that would invite the Senate to judge Trump severely for not testifying. Barry and Joshua endorsed it, and the reaction among members was greatly enthusiastic. Plaskett, Swalwell, and Lieu—our high-octane prosecutors, who routinely got each other fired up—loved it because they were eager to go on the offense against Trump and his legal squad, which newspaper reports suggested was already in chaos and flux. I think all three of these swashbuckling prosecutor Democrats were dreaming about being deputized to go depose Trump at Mar-a-Lago.

In the meantime, the one and only self-inflicted lawyering wound on our side happened: somebody leaked. Barry called me late at night to tell me that he had received two different calls from reporters with the tip that Raskin was going to call Trump as a witness. Barry played it off nicely, telling one reporter that it was an overwhelmingly academic and hypothetical conversation (*You know those law professors*) and the other that he would be the first to know it was going to happen if he just laid off the story now so as not to tip off Trump's team. But if a couple of reporters had heard the leaked news, there must have been a lot more out there circling the story, too. Someone had leaked, and I was outraged about it.

The next morning's managers' meeting was the first and last time I expressed disappointment to the managers and the small group of lawyers and staff who joined our meetings.

"Somebody at the meeting *must* have leaked," I said, "because nobody else knew about it. But how can I discuss these essential matters of trial strategy and tactics if someone is going to run off to the media like a freshman city councilman trying to impress a reporter with his inside knowledge? This is just profoundly disappointing. If it happens again, whoever did it is off the team, no appeals and no questions asked and no matter how much I love you—and I love you all, but it won't make any difference. You'll just be gone. And if we don't find out who it was, I am going to have to start keeping more and more of the case to myself, and that helps none of us."

That was a pretty solemn and forlorn Zoom session. But we never again had a problem with leaking. Different people called to offer theories about who it was or how it had happened, but that was beside the point for me. I just wanted everyone to know that it was a major breach of trust, one we could not afford to repeat. We were—we will forever be—a band of brothers and sisters who came together and threw everything we had into this journey for democracy, but it threatened and cheapened our labor of love to traffic in political gossip with reporters about confidential and unfolding trial tactics. They would learn of our plans in due course.

I also had a chance in Florida to reflect on a single troubling word that would have huge implications for our trial strategy, for public understanding of what had happened on January 6, and for our chances of winning conviction votes from GOP senators.

That word was *coup*.

To understand its meaning in this context, consider where it fits into January 6. By my lights, the events of the day had three essential rings of activity.

The outermost ring was, by far, the largest. It was the "riot ring," containing multitudes of protesters turned rioters. They brought down massive poundage and ferocity on the defending officers and chanted "Stop the steal" and "We want Trump" for the masses of Trump supporters who had internalized "Big Lie" propaganda and disinformation. These people wanted to voice their displeasure with what most of them honestly believed was a stolen election, but they did not have a specific plan in mind for how to keep Trump in power.

The middle ring was the militarized "insurrection ring," which included the Proud Boys, the Three Percenters, the Oath Keepers, Texas Freedom Force, diverse militia groups, Christian white nationalists, and assorted Ku Klux Klan and neo-Nazi factions. These paramilitary elements trained before coming, arrived heavily armed, carrying or secreting everything from baseball bats and hammers to Confederate battle flags, bear mace, and heavy firearms and explosives. They were dead-set on smashing windows and crashing into and entering the Capitol. They wanted to shut down the congressional vote-counting process and beat and subdue any officer, staffer, or member who got in their way. It was not clear to me what their precise plans were beyond that, but obviously a lot of them had been talking about "race war," "white revolution," and "1776," and they all felt that their hero Donald Trump had a God-given right to rule, despite the 2020 election. Some of them wanted to hang Pence and all the "corrupt politicians."

The innermost, central ring was what I called the "ring of the coup." In my mind, it was here, in the bull's-eye center of the action, that Trump operated, likely along with Chief of Staff Mark Meadows, Rudy Giuliani, House Minority Leader Kevin McCarthy, Michael Flynn, Sen. Josh Hawley, Rep. Jim Jordan, and the most extreme elements of the GOP House and Senate conferences. This is where the actual strategy for Trump to stay in power was being executed. The basic idea turned on getting Vice President Mike Pence to declare and announce a hitherto-unknown and unilateral power in the vice president to repudiate electoral votes from specific states—in this case, Arizona, Georgia, and Pennsylvania. If and when he did that, by vaguely citing allegations of fraud in those states and "returning" their votes to the state legislatures, Pence would succeed in lowering Biden's electoral vote total to below 270, which would "immediately" trigger, under the Twelfth Amendment, a contingent election. The House contingent election was the one place where Trump could still "win" the election.

Later revelations at the end of July, that Trump had been constantly pressuring the Department of Justice to declare the election "corrupt," provided decisive support for this working theory. When Trump met with then-Acting Attorney General Jeffrey Rosen and then-Acting Deputy Attorney General Richard Donoghue on December 27, they tried to convince him that the information he had about election fraud was false and that the department could not "snap its fingers" to nullify the election. Trump responded that he wanted DOJ, according to Donoghue's notes from the phone call, to "just say the election was corrupt + leave the rest to me and the R. congressmen." It is telling that Trump referred to the Republican *congressmen* and not to the Republican *senators*. The key difference in this context, of course, is that senators do not vote in a contingent presidential election in the House.

What would happen immediately after a contingent presidential election victory for Trump would be anyone's guess, but Trump's militarized core of national security advisors, led by Michael Flynn, had been urging him to declare "martial law" and a state of siege. It is easy to imagine Trump invoking the Insurrection Act after a contingent election victory and imposing martial law to put down all the chaos and insurrection he had unleashed against us in Congress. He would declare simply that Speaker Pelosi had proven herself unable to keep and maintain order in the House and that the newly reelected president would have to do it instead. A lot of people think that the DC National Guard, which is under federal control, was being held back on January 6 because Trump wanted to deploy it later in conjunction with a martial law order.

In any event, assuming that Pence had been successfully persuaded to "return" the electors to the state houses and that this sequence of events or something similar to it had actually taken place, would we consider such a return to power a "coup"?

To my mind, yes, absolutely. This would definitely have been the first American coup. Specifically, it would have been what the political scientists call a "self-coup," where a sitting executive in power overthrows the electoral process and constitutional order to prolong and continue his reign. Trump had orchestrated the violent dimension of the January 6 assault on the government partly to demonstrate his rabid popular support to Mike Pence, Mitch McConnell, and others who were appalled by his plan and who stood in his way. He wanted to scare them both politically and physically.

But, of course, there were a lot of Republicans who supported, either explicitly or implicitly, the inside strategy related to the Electoral College and the contingent election. Some on my team and staff expressed the fear that, if I pressed this point too hard, that

Trump was actually engineering a coup, we would lose many of the Republican senators who did not support the outer rings of violent action and whom we otherwise might be able to get to vote to convict on the narrow grounds of "incitement to insurrection."

Barry and some others thought I should shy away from using the word *coup*, even though I felt certain the nomenclature was correct for the sequence of moves that the third, innermost, central ring had planned for us. I decided I would use it sparingly at trial but would nonetheless explain it, so that people understood precisely *why* Trump had continued to incite insurrection even when leaders of his own party were crying out for help: he wanted to keep turning up the insurrectionary violent heat on Mike Pence to get him to "reject" enough electors to throw the whole contest into a contingent election.

I had no doubt—and I have no doubt—that both the nonviolent and the violent parts of this plan constituted an attempted coup. But we had to focus at trial on proving "incitement to insurrection," the specific article approved by the House. Yet, I felt confident that we managers would also have to do our part to show America what had happened each step of the way in this pernicious plan. We would lay the groundwork for a future outside commission or legislative committee to investigate thoroughly and report back to America both on what did happen to us and on what almost happened to us, both in the insurrection and in the closely coordinated attempted coup, on January 6, 2021.

On our last day in Florida, January 29, Tabitha and I spent the morning splashing around and playing Ping-Pong and Uno with Mariah, Boman, and Daisy, as well as Noah and Heather and our close family friends Michelle Carhart and Zina and Shammy Khalid and Shammy's daughter, Myrna. I had been working non-stop on my oral arguments, my refutations, my moral framing of the constitutional issues, and I promised everyone I would set my computer down and just hang out and play. There was a whole lot of laughter that day—and a river of tears, too, a lot of hugging. Noah and Shammy kept trying to predict what Trump's nickname for me was going to be. Given that Trump had called Adam Schiff "Shifty Schiff," Noah said the "obvious choice" for me would be "Ratskin," a go-to playground insult for all of us since elementary school. But he thought "Pointy Head," "Giganto Head," "Tom Paine-ful to listen to," or "the Professor" would be "more on target." Shammy put his money down on "Tom Paine-in-the-ass" or "Thomas Jerkeson." Boman said he would just call me "Harvard" or "little Mr. Sunshine" and leave it at that. More candidates rolled in throughout the day, prompting great hilarity. When there was joy and laughter in the air, I kept feeling Tommy in the interstices of all the banter and play. That was a warm feeling. At other times, I felt that, yes, life was going to be a party again one day, a party like this with the people I loved, but the guest of honor would have vanished without a trace, and we would search from pillar to post, but the person whom everyone most wanted to see, the person we had all come for, would be nowhere to be found, and no one would know what to do about it. How could there be a party without Tommy?

In this melancholy space, we talked a lot about Tommy's memorial service, which had been postponed because of the impeachment trial coming up in early March (although the date was still being kicked around in endless House-Senate negotiations) and COVID-19 and all the strictures and prohibitions in place, and also because Tabitha and Hannah were adamant that their brother's memorial service would not be a Zoom event. With everyone still so shaky and me being diverted to the trial, Ryan had stepped forward to volunteer to take over the service and organize it for us. It was now set for Saturday, April 3, to be held on the makeshift stage of the parking lot at RFK Stadium. I could not wait for proper honor and love to be bestowed on Thomas Bloom Raskin, but I also dreaded this date, because it would somehow mark a final good-bye, which I simply could not accept in any way.

I had come down to Florida to be with my beloved brother Noah, who made me laugh uncontrollably like Tommy did. But, in the total narcissism of my parental grief, I had been blocking out an emotional reality which suddenly hit me down there: the loss of Tommy was nearly as brutal and traumatizing for his uncles and aunts—and his cousins—as it was for his mom and dad and sisters. Noah and Heather, Erika and Keith, Kenneth and Abby, Eden and Brandon, Tammy and Gary, Emily, Zacky, Maggie, Mariah, Boman, Daisy, Phoebe, Lily, Gray, Tessa, Maddox, Kai, Emmet—all of them have suffered a terrible grievous loss and all are working their way back to equilibrium themselves.

I think I expected Noah to be his madcap self and to cheer me up and make me laugh at all moments, as he used to do, but he has been wrecked like me. We spent a lot of time talking, a lot of time commiserating and philosophizing. He is still an absolute riot, you might say, but we will have to recover our laughter together.

And my nephew Boman has Tommy's spirit.

Before Tabitha and I left, we all played Uno once more, the addictive and silly card game of reflexes and outbursts. While we played, Boman asked me about what Trump was going to argue at trial.

Trump will say he never explicitly called for violence, I told him; he just wanted them to fight "like hell," and people say that all the time, don't they? And then he'll say that his speech was protected by the First Amendment.

"Isn't it?" Boman asked, dealing the cards. Boman was in his junior year of college at Virginia Tech, an extremely intelligent and logical young man.

"How can the president claim that free speech gives him the right to destroy our constitutional democracy, steal our election, and make a mockery out of everyone else's free speech?" I asked.

"Maybe a bum on the street could try to claim that he's just spouting off, but not a president who has sworn an oath to preserve, protect, and defend our Constitution."

"I like your answer," Boman said, nodding with an appreciative frown. "It's like Reverse Uno. You take their words, and you just send them right back to them."

Boman urged me to turn all of the Trump arguments back against them and shout out, "Reverse Uno!" each time.

Mariah and Daisy joined in and began screaming "Reverse Uno!" at us.

This made me smile. I did not completely recover my sense of humor while in Florida, as I had hoped I might, but I could see that, even with all the tragedy and evil around us, there was still going to be humor in the world, and Tommy and I would both be part of it.

CHAPTER 10

WRITING TRUMP

Be bold and mighty forces will come to your aid.

—JOHANN WOLFGANG VON GOETHE

W e returned to Maryland in time for two birthdays: Tommy's, on January 30, when he would have turned twenty-six, and Tabitha's, on January 31, when she turned twenty-four.

Sarah and Hannah would be back home, too. Our close friend Judy Minor came to spend the weekend also, and we had decided to be all together at home and to read letters, but there was also a charitable outing on the morning of January 30. It was being conducted in Tommy's name by Wendy Kent and the Takoma Park Middle School Cluster Food Support program, a distribution of food and toiletries from the driveway of the Takoma Park Adventist Community Services building. For an hour, we loaded bags of groceries, soaps, and baby shampoos into cars that had lined up for blocks to receive the offering of one bag per car. This local project, now permanently named Tommy's Pantry, was one of more than a dozen new projects launched in our community and around the world by people who were moved by what they had learned of Tommy's life and his moral passion.

We spent the rest of that emotional day talking about Tommy and reflecting on the clarity of his ethical convictions. We kept getting notes from people going vegan or vegetarian in his honor. We got a beautiful note from Alexandria Ocasio-Cortez, who had gone on a long hike with Tommy, Tabitha, Sarah, and me one day in the fall, along with Alex's boyfriend, Riley. AOC was planning to go vegan for Lent, in Tommy's honor, which made us happy. We read through many dozens of letters and registered astonishment at the way Tommy's words and just the few pages we had written about him had struck a major chord with people all over the world whom we had never met, from Bethesda and Wheaton to Thurmont and Baltimore and Boulder and New Orleans and Soweto and Berlin and Milan.

Monday, February 1, marked one week until the trial and the tempo intensified. The first order of business was to meet with Mark Patterson, Senator Schumer's general counsel and right-hand man for impeachment. We had been in telephone and Zoom contact for the past few weeks, and Mark and Barry were in close touch, but my impression was that everything always seemed to get pushed over to the next meeting. I also noticed that the most straightforward matters of scheduling and procedure between the House and Senate—like when the trial should begin and how long it should go—seemed to elicit drawn-out and weirdly elliptical and inconclusive conversations.

Still haunted by the last trial, the Senate wanted to keep the proceeding as short as possible. The last one had "dragged on," I kept hearing, for "two weeks." I assured Mark and our friends in the Senate that this would be a one-week enterprise. In truth, we never had any interest in a prolonged trial. On the contrary, I was aiming for a tight, riveting, compelling presentation of the facts of Donald Trump's long-running, desperate attack on the 2020 election culminating in deadly violence on January 6, interspersed with punchy and irrefutable constitutional arguments and propulsive video images of that horrific day. We wanted our case to hit them like lightning, and I didn't want any of the senators to be bored for a moment or rolling their eyes at anything we said. I quoted my colleague Missouri congressman Emanuel Cleaver, who cracked me up one day in caucus when a visiting speaker, who had already gone on way too long, promised he'd be able to finish it up in just another ten minutes. An ordained minister who reveals an apt witticism for pretty much every occasion, Cleaver leaned over to me and whispered, "There just aren't that many sinners who get saved in the *second hour* of the sermon."

A more serious issue involved who would be presiding over the trial. I felt strongly, along with the other managers, especially Neguse, Cicilline, DeGette, and Lieu, who were outspoken on the subject, that Chief Justice John Roberts had to preside. The Constitution specified that this was the case when the president was the defendant, and it stood to reason: the Framers thought there was no way that the president of the Senate—that is, the vice president of the United States—should be forced to preside over the trial of the president, whose conviction and removal would immediately pave the way for the vice president's ascent to the presidency. Having the vice president preside over the president's trial would embed a serious conflict of interest in the Constitution, or, at the very least, create a serious *appearance* of a conflict.

But I was being told from the beginning, by lots of staffers and senators, that Chief Justice Roberts was refusing to participate in our trial. He had apparently hated presiding over the first Trump trial, had experienced the indignity of various senators angling their questions during the Q&A period to insult the Supreme Court's jurisprudence or its impartiality, and had been deeply offended by the refusal of the Democrats to join a ceremonial resolution at the end of the trial, which Senator McConnell had drafted, thanking him for his service. All these things were aggressively contested by the Democratic leadership in the Senate and by numerous senators and staffers who spent hours with me refuting every detail of Roberts's bill of particulars. But it was all for naught: Roberts had made up his mind.

Leaving aside the substance of his complaints, they were all irrelevant, we thought, to the chief justice's crystal-clear *constitutional obligation* to cross the street and enter the Senate chamber for the duration of a trial involving a president.

But Mark Patterson kind of changed my mind by advancing the plausible position that, whatever the merits of Trump's dubious claim that he could not be tried in the Senate, the analytically distinct command that the chief justice preside over the trial of the president did not apply strictly to a *former* president, as there was no danger, when he had already left office, that the vice president would be induced to prejudice the proceedings against him in order to take his place in the White House. Thus, the Senate president here, Vice President Kamala Harris, could reasonably preside over Trump's trial without a real or apparent conflict of interest, because the conviction of former President Trump would not lead to *her* elevation to the presidency and his acquittal would not block her elevation. But if she chose not to preside (which Senate staffers were telling me was likely, since her vice presidential persona was just forming and she would rather be known by a policy commitment than presiding over the trial of the former president), then that role would fall to the president pro tempore, which was Sen. Patrick Leahy from Vermont.

Although this was a fairly persuasive interpretation of the constitutional language—and it would of course be a "political question" for the Senate to decide—I thought that the Senate's treating the presiding officer issue this way could become a political headache for us. Leahy is a great senator and a man of old-fashioned integrity and conviction, but he is also a well-known liberal and had already voted to convict Trump in the Ukraine impeachment round; we predicted Republicans would rail against his perceived lack of neutrality and use his presiding to disparage the proceedings, which definitely came to pass.

More seriously, if a Democratic senator were seated at the dais, everyone would conflate the substantive jurisdictional question of the Senate's *power* to try a former president with the sharply different *personnel* question of whether the chief justice must be the one to preside over the trial. The glaring absence of the chief justice in the Senate would be falsely read by many—and no doubt, aggressively interpreted for the public by Trump's advocates—as a statement by the highest Court in the land that there was something constitutionally suspect about trying a former officer.

But ultimately, there was just no room to maneuver. Roberts had made himself clear to McConnell, who made himself clear to Schumer, who thought he had no real alternative other than to accept this as a fait accompli. By the time it got to my level in the constitutional food chain, all I could do was quietly accept Senator Leahy's being in the chair and praise such a studiously fair-minded and excellent senator and juridical thinker—all of which was fortunately true. The less said about the missing chief justice or any of this, the better.

Besides, we had other things on our minds. For one thing, I had a couple of motions I had been contemplating. One was to request that the Senate jurors vote by secret ballot on the ultimate question of the guilt of the former president. Everyone knew that tremendous pressure was being brought to bear on Republican senators—lobbying by Trump's acolytes, political intimidation by his avid supporters, and even violent threats by the hornet's nest of extremists that had been shaken loose by the insurrection. One Republican senator confided in me his private guess that a secret ballot could result in upward of 80 votes to convict, but otherwise, the members were "far too frightened of the repercussions to vote their conscience." Both sides described a push for a secret ballot as a "nonstarter" and a radical break from Senate practice, one that could only inflame Republicans against us. As we talked about it, it began to dawn on me that a rule change along these lines wouldn't work anyway, since it would be relatively easy to figure out which

Republican had voted which way, as the Trump supporters would come out and publicly declare their votes "for Trump." I dropped the idea.

We went through a similar thought exercise with another motion I wanted to bring: to change the partisan seating arrangement in the Senate into an alphabetical one. As matters stood, all the Republican senators would be seated to the right when facing the rostrum and all the Democratic senators to its left (following parliamentary history going back to the French Estates General, which gave us the terms "left" and "right" as applied to the political spectrum when the church and nobility sat to the king's right and the Third Estate sat to his left). This configuration obviously established and reinforced a partisan mentality, which is why party leaders all over the world favor it for legislative deliberations and voting. Sitting together has a profound cognitive and social peer pressure effect. As I learned in Gov. 30 in my freshman year of college, in politics, "where you sit is where you stand."

But the event of an impeachment proceeding was not—or, at any rate, it *should not have been*—a partisan exercise calling for party discipline. This was a *trial*. Who ever heard of jurors for a bank robbery case sitting according to political party registration? But, again, we were told by everyone on the Senate side that making this change would constitute a radical departure from Senate practice and impeachment trial practice, one that its members would reject. The request to relocate themselves and their belongings would annoy the senators, and as one of them told me, it "might not be the best way to introduce Professor Raskin to the Senate." Once again, questioning these time-honored party-line seating arrangements, I was clearly barking up the wrong tree, a towering oak of Senate tradition—and I quietly relented.

The main thing that we wanted and—because of Mark Patterson's focus and the nimbleness of Barry Berke and Aaron Hiller—we got from all these preliminary negotiations was an open-evidence rule. Here, Senate tradition worked in our favor, based on the last several impeachment resolutions, which permitted any verified and authentic evidence to be considered as long as it was properly submitted forty-eight hours in advance. The danger was that we would be confined to evidence taken directly in the House Judiciary Committee, which of course had not had the time to conduct any hearings at all, much less assemble a comprehensive and meticulous factual record. We were afraid that Trump's lawyers were going to push for something like that and embroil us in a nasty political fight over the rules of evidence, which we wanted to be open enough to allow in the steady stream of startling facts demonstrating Trump's culpability. I am not quite sure how Barry and the gang got exactly what we wanted on this, but it seemed to me that sometimes the genteel traditions of the Senate, perhaps combined with the tumultuousness of the Trump legal team, worked out for the best.

We turned in our seventy-seven-page "Trial Memorandum of the United States House of Representatives in the Impeachment Trial of Donald J. Trump" on Tuesday, February 2, which set forth the factual arguments and legal conclusions we had worked on for a month. That memo was sizzling hot, with every line flying off the page in patriotic fury over what had happened to us, starting with the introduction:

> By the day of the rally, President Trump had spent months using his bully pulpit to insist that the Joint Session of Congress was the final act of a vast plot to destroy America. As a result—and as had been widely reported—the crowd was armed, angry, and dangerous. Before President Trump took the stage, his lawyer called for "trial by combat." His son warned Republican legislators against finalizing the election results. "We're coming for you." Finally, President Trump appeared behind a podium bearing the presidential seal. Surveying the tense crowd before him, President Trump whipped it into a frenzy, exhorting followers to "fight like

hell [or] you're not going to have a country anymore." Then he aimed them straight at the Capitol, declaring: "You'll never take back our country with weakness. You have to show strength, and you have to be strong."

Incited by President Trump, his mob attacked the Capitol. This assault unfolded live on television before a horrified nation. But President Trump did not take swift action to stop the violence. Instead, while Vice President Pence and Congress fled, and while Capitol Police officers battled insurrectionists, President Trump was reportedly "delighted" by the mayhem he had unleashed, because it was preventing Congress from affirming his election loss. This dereliction of duty—this failure to take charge of a decisive security response and to quell the riotous mob—persisted late in the day. In fact when Congressional leaders begged President Trump to send help, or to urge his supporters to stand down, he instead renewed his attacks on the Vice President and focused on lobbying Senators to challenge the election results.

We told the story of how Trump had many times urged his followers to come to DC for a "wild" protest—what president speaks like that? We explained how Vice President Pence had resisted Trump's coercive entreaty to reject dozens of swing-state electors that had gone to Biden and to overturn the election, kicking it all into the House for a contingent election. We explained how Trump's jacked-up supporters had attacked the Capitol and beaten up hundreds of police officers, leaving several people dead and hundreds wounded. We explained that this incitement to violent insurrection against Congress in joint session was a high crime and misdemeanor, a cardinal impeachable offense that threatened the stability of American constitutional democracy. We showed that this was an egregious violation of his Oath of Office and that there was literally no defense for his conduct. We dealt with the jurisdictional question in fine detail, showing how Senate history and constitutional text and structure all blocked the claim that the former president could not be tried for his crimes.

I signed the brief on behalf of the whole team and proudly included all the managers' names on it, and all the lawyers' names, too, leading with Barry Berke and Joshua Matz, who had been indispensable, working day and night on it along with Sarah Istel, and of course our House counsel, Douglas Letter. I was exultant that we had written a vivid and engaging brief that people were talking about and analyzing all day, and I aggressively recommended it to reporters who were calling me looking for gossip about what kind of pizza and snacks the managers were eating while we practiced for oral argument. The trial brief dominated the news cycle that day and prepared the Senate for what was coming.

The day got better when we received the president's formal answer. It went way beyond his asking for dismissal for lack of jurisdiction because of his claim that the Senate could not try a former president. That would have been the smart way for him to go. Instead, Trump's lawyers, apparently reacting to their client's wounded pride, explicitly denied a number of factual allegations, thus jumping right into the jaws of the "adverse inference" trap.

After I read President Trump's answer, I set about composing a letter to him revealing the mistake he and his lawyers had made: he had contested key facts of the case, putting them in issue. We thus invited him to testify, at the pains of the Senate's drawing every possible negative conclusion against him if he refused. We sent it, printed on my letterhead, directly to President Donald J. Trump on Thursday, February 4, via his attorneys Bruce L. Castor Jr. and David Schoen. It read, in part:

Dear President Trump,

As you are aware, the United States House of Representatives has approved an article of impeachment against you for incitement of insurrection. See H. Res. 24. The Senate trial for this article of impeachment will begin on Tuesday, February 9, 2021.

Two days ago, you filed an Answer in which you denied many factual allegations set forth in the article of impeachment. You have thus attempted to put critical facts at issue notwithstanding the clear and overwhelming evidence of your constitutional offense. I write to invite you to provide testimony under oath, either before or during the Senate impeachment trial, concerning your conduct on January 6, 2021, and not later than Thursday, February 11, 2021. We would be pleased to arrange such testimony at a mutually convenient time and place. . . .

If you decline this invitation, we reserve any and all rights, including the right to establish at trial that your refusal to testify supports a strong adverse inference regarding your actions (and inaction) on January 6, 2021.

. . .

Very truly yours,
Jamie Raskin
Lead Impeachment Manager

We got a letter back from Trump's lawyers in far less time than it took Trump to respond to the violence at the Capitol. It was barely literate and did not bode well for the lawyers' performance at trial. Calling my letter a "public relations stunt" (which I thought was rich coming from lawyers for a man who had originally run for president partly *as* a public relations stunt) and missing the point completely, the lawyers offered this classic non sequitur and jumble of nonsense: "The use of our Constitution to bring a purported impeachment proceeding is much too serious to try to play these games."

With a lot of work to do for trial, and having already established that Trump had put factual issues into play that he then quickly refused to testify about, we left it at that. We would return to make the point at trial itself.

After weeks of coaching the managers and kicking around arguments on Zoom, we had the final weekend to moot the managers (including me) in person. We went back to our familiar haunts in the Rayburn House Judiciary Committee Hearing Room to run everyone through their presentations. It was pretty brutal. Barry, the team, and

I had dozens of comments, revisions, and ideas about wording, pacing, pauses, body language, and emphasis. I was at my professorial pickiest—telling people to use "who" instead of "which" and to eliminate too many adjectives; going ballistic over any use of "impact" as a verb (which Julie had been warning people was my explosive pet peeve); urging them not to read their presentations to the senators but to speak them directly to America; and telling the managers to look up high, as the cameras would be staring down at the tops of their heads. (Aaron and Arya had been involved in a negotiation with the Senate staff to improve the cinematography, the lectern, and the microphone setting, but there was still an overall top-down aerial view effect we would need to keep in mind.) The key thing to remember, I urged them at the end, was that we were there to tell the Senate and the country the story of this atrocious attack on America and that each part of the story fit together like a jigsaw puzzle and led to one inexorable conclusion about the central culpability of Donald Trump. If any of us decided to give a moving speech for the ages that lost the overall narrative thread, that portion would be no success and could throw the whole train off course. *Stay focused, and keep it tight* was my message.

The managers may have felt slightly flustered and frustrated by the volume of notes they got, but I was thrilled with their performance and knew they were going to be sensational, and I told them this. Still, I thought it was advantageous that they were going to keep working up until the end, incorporating all the feedback, using all that surplus anxiety to their advantage.

What a team they were. Stacey was spellbinding and had me in tears. Neguse was brilliantly careful and lucid in his delivery. Swalwell's rap about his family of cops was stinging now. Castro spoke with an authority that was towering and commanding. Dean spoke directly to the heart of America about the sick violence trained against our institutions. DeGette was personal and relentless, like the effective trial lawyer still dwelling in her soul. Cicilline was razor-sharp, unflappable, and never missed a detail. Lieu was thrilling to watch, filled with moral passion and dangerous logical rigor. The whole thing was going to be mesmerizing for America. I let everyone go at 7 p.m., so the managers and staff could get some rest and so our friends on the Capitol Police who were there could take over the Judiciary Room and watch the Super Bowl. I didn't even know who was playing.

On Monday, February 8, we did a walk-through of the ornate and intimate U.S. Senate chamber with Mark Patterson. Everyone individually tested the microphone and lectern and started reciting stuff on command. When I got up for my turn, I offered the favorite Tom Paine passage with which I had driven everyone crazy in the 2020 campaign season and which I had revised and improved—at the urging of Speaker Pelosi, who assured me that the forward-thinking radical democrat and Enlightenment liberal Tom Paine wouldn't mind so much—so as not to offend modern sensibilities:

> These are the times that try men and women's souls. The summer soldier and the sunshine patriot will shrink at this moment from the service of their cause and their country. But everyone who stands with us now will win the love and affection of every man and every woman for all time. Tyranny, like hell, is not easily conquered. But we have this saving consolation: the more difficult the struggle, the more glorious in the end is our victory.

As the Senate tech crew adjusted the sound, they asked me to test more. So I continued, reciting some favorite passages from W. H. Auden's "Ode to W. B. Yeats," which I recited for my mom's memorial service two decades before and which the loss of Tommy had brought surging back to me. "Earth, receive an honored guest," it opens, "William Yeats is laid to rest." When I finished, I got choked up, and I saw some managers look at one another with a glance that said, *Is Jamie actually going to be able to do this?* I had detected undercurrents of anxiety from my friends in Congress from the start. Some of them thought I was throwing myself into trial as a form of emotional avoidance and that, at any moment, the dam of emotion might burst all over everything and everyone.

Stepping away from the lectern, I thought back to the time when Tommy, who was a seriously mesmerizing poet, recited his master-piece, "Where War Begins," at the DC Veg Fest in 2017. He and I drove down there together on a beautiful Saturday afternoon. I was over the moon just to be with him. I spoke first to the crowd—all prose, clunky theory, and rhetoric—a politician's speech. And then Tommy got up and demonstrated how verse, idea, energy, brilliance, and physical grace and movement could combine in one spellbind-ing young man to touch everyone in his presence. "Let us discuss the meaning of war / not just to the children, disabled, and poor," he started, and proceeded to unleash a dazzling profusion of images and rhymes that included the passage "When it comes to the right to live free from the blight / of aggression, oppression, from tyrannous might / how smart you are, friends, shouldn't matter at all / trauma is still trauma for the creatures that crawl."

I missed my dear boy. I missed my son.

Madeleine came over and gave me a big hug—well, a big hug for COVID days—and told me all the managers were with me.

Neguse, Cicilline, and I spent the rest of the afternoon and eve-ning practicing for our constitutional motion establishing the Senate's jurisdiction, which we would argue the next day, the first day of the trial. The Trump team thought they might be able to seize the early advantage by bringing their best-sounding argument forward at the start and distracting everyone from the facts of the case with their basically empty constitutional theory. But we had a big February sur-prise in store for them, and they would receive it in words and pictures when things got rolling Tuesday.

PART III

CHAPTER 11

VIOLENCE V. DEMOCRACY: THE JANUARY EXCEPTION

Sovereign is he who decides on the exception.

—CARL SCHMITT

P eople kept asking me how I was doing this.

The truth is that I thought it might be the last significant thing I would ever do with my life. Right after we lost Tommy, I did not know if I could continue working productively in a world without him. I did not know what it would mean even to live in such a world. I was grateful beyond measure to Speaker Pelosi for pushing me to lead this project; she threw me a lifeline and I thought that at least I could end my career with a powerful statement about democracy and violence.

Back when we first lost Tommy, I was tempted, in my darkest moments, to say, *Let them have it*. And by *them*, I meant all the forces aggressively ruining our world, the people denying COVID-19, climate change, the Holocaust—the people clearly wanting to bring back the plagues of fascism and chaos.

But then I would think of Hannah and Hank. I would think of Tabitha and Ryan. I would think of Sarah. I would think of all of Tommy's amazing cousins, of my brother and my sisters. I would think of my friends and their children; my cousins; my students; the people I worked with; my staff; the people I represented; the thousands of people who wrote us, whom I can never properly thank. I would think of people's grandchildren whom I know and those I don't know, the grandparents I know and I knew, the people whom I loved.

So much goodness in this hard world. Tommy used to quote Camus's statement: "The world I live in is loathsome to me, but I feel one with the men who suffer in it."

I would think of Tommy and the last part of his final message to us, which was not different in substance from the message of his life:

Look after each other, the animals, and the global poor for me. All my love, Tommy

He did not tell us to give up, to despair, to surrender, to become selfish or despondent.

He gave us, and he gave the world, *all his love.*

Did I not have more love to give? I was not suffocating in illness like my dear son was at the end. What else could I do but carry on?

People asked me how I understood the confluence of these shocking, traumatic events in our lives, a puzzle I continue to wrestle with to this day. This much I knew, as we drove down to the Capitol on Tuesday, February 9, for the first day of the trial: the life of Tommy Raskin and the violent insurrection on January 6 were polar-opposite events in the universe.

This precious young man of boundless talent had given all his energy for the idea of the dignity and worth of all human beings— and of all sentient beings. For Tommy, violence was the enemy of humanity and of all living things. The purpose of democracy and its operating system, the law, was to control, even to end, political violence, state violence, criminal violence, racial violence, gender violence, mob violence, the deliberate and needless infliction of pain and suffering on others. The purpose of democracy is to dignify and uplift each person on his or her path in life, to address misfortune as best we can, to make this life a gentler proposition.

January 6—that stomach-churning, violent insurrection; that desecration of American democracy; that demoralization of all our values; that explosion of seething hatred that caught his sister and his brother-in-law in its tentacles—would have *wrecked* Tommy Raskin.

So as a congressman and a father of a lost son and two living daughters, I would take a stand, with everything I had left, against that violent catastrophe in the memory and spirit of Tommy Raskin, a person I have, alas, not even begun to properly render in words.

At noon on Tuesday, February 9, an hour before we kicked off arguments over whether the Senate could even hear the trial, managers and staff assembled in S-219 in the Senate, which everyone called "the war room." I noticed upon entering that it bore the same number as Steny's hideaway office in the House, which on January 6 became a sanctuary and safe room, the place where Tabitha, Hank, and Julie hid out: H-219.

From 219 to 219.

From *sanctuary* to *war*. I didn't really want to call it a "war room," so I suggested, perhaps, "headquarters," but no one paid that suggestion any mind. It had been called the war room for a long time. That's what our friends Jason Crow, Val Demings, Sylvia Garcia, Hakeem Jeffries, Zoe Lofgren, Jerry Nadler, and Adam Schiff had called it during Impeachment 1.0. Maybe that's what Henry Hyde and the GOP team called it back in 1998, when they impeached Bill Clinton for not being perfectly forthcoming about his sex life under oath.

When I got to S-219, everyone was talking, debating, and furiously reviewing their notes. It quieted down ever so slightly as I said good morning to all, took a seat, and started to go over mine. And the quieter I got, the quieter everyone else got, too.

I remembered something my dad once told me when I was in high school, about his experience working in the Kennedy White House. There was always a state of high chaotic energy there, he said, but the closer you got to the Oval Office and the president, where the most important decisions were being made, the quieter and calmer it got. When I asked why this was, he said that everyone on staff was scrambling to organize or respond to the events of the day, but the president was communing with history itself. That is why a president may seem to have time for extracurricular distractions like poetry, tweeting, squabbles with celebrities, romantic affairs, or golf: the actual work of government is being done elsewhere, and the president is just steering a very big ship, making minor adjustments to the ship's wheel, adjustments that have vast implications for our collective destinations and destiny. I imagined President Kennedy—whom my father liked very much but never idolized—communing with Jefferson and Madison, with John Quincy Adams, with Lincoln and Grant and Frederick Douglass, with FDR and Eleanor Roosevelt, and with other leaders of the misty past as well as hypothetical presidents of the future.

With Trump, of course, I imagined the ambient noise level in the White House operating in reverse from the way my dad described it: it was probably eerily quiet in the staff quarters of the Trump White House, where people worked in abject terror of attracting Trump's attention, and louder and more frenetic as you got physically closer to Donald Trump himself—at which point, in the inner sanctum, you finally entered a world of careening, chaotic mental derangement and screeching rage.

As the lead impeachment manager, I wanted to commune for a few moments in quiet with my predecessors in impeachment history. Until this moment, America had seen only three Senate presidential impeachment trials—of Andrew Johnson, Bill Clinton, and just the prior year, Donald Trump—and the three lead managers had been Benjamin Butler of Massachusetts, Henry Hyde of Illinois, and Adam Schiff of California. I thought about each of them.

The only one I knew was Adam Schiff. Although way senior to me—he was in his eleventh term, and I was just starting my third—and far more important in the House hierarchy as chairman of the Intelligence Committee, Adam was my friend; owned a house in my district with his lovely wife, Eve; and had offered me good advice when I started out as lead manager. He told me to make sure our managers had no ex parte contact with senators during trial that could come back to haunt us and did not go off on their own tangents in argument or in the media. Still, through no fault of Adam's or of his partner Jerry Nadler, the New York liberal warhorse and distinguished chair of the Judiciary Committee, the House presentation for the first Trump impeachment was powerful but fell short in the Senate, with votes of 48–52 and 47–53 on the two articles, leaving us well below the two-thirds constitutional threshold for conviction. But during his role as lead manager Adam was always soberly focused on confronting the lawlessness of Trump, especially in his dealings with foreign countries, and I admired him for that. One strange twist of history is that Adam and I both are vegan, meaning that 50 percent of the lead presidential impeachment managers in American history have been vegans (although I have definitely been known to cheat on goat cheese).

The author of the infamous anti-choice Hyde Amendment, Henry Hyde was a right-wing Republican culture warrior and chairman of the House Judiciary Committee caught up in the anti-Clinton mania of the 1990s. He led the absurd drive in the House to impeach President Bill Clinton for lying about whether he had had sex with White House intern Monica Lewinsky and, embarrassingly, whether oral sex constituted sex. The Senate acquitted Clinton of perjury on a vote of 45–55 and of obstructing justice on a vote of 50–50. This was a disgraceful use of the impeachment power, which was designed by the Framers to protect us against presidents who had engaged, as George Mason put it, in "great and dangerous offenses" of a political nature against the integrity of the republic. The only saving grace was this: If lying about a private romantic affair met the threshold of a "high Crime or Misdemeanor" for GOP senators, then surely inciting a violent insurrection against the government to overthrow an election would have to qualify, too, no?

Benjamin Butler was a fascinating and irascible figure, a Massachusetts Radical Republican in Congress who chaired the House Committee on Reconstruction and played the leading role in passing both the Ku Klux Klan Act of 1871 and the Civil Rights Act of 1875. A clever and wily politician and, by most counts, inept and bumbling military officer who was forced out of the Union Army after some humiliating disappointments in the First Battle of Fort Fisher, Butler clashed with President Andrew Johnson, who was doing everything he could to thwart congressional Reconstruction and rebuild the system of political white supremacy and economic exploitation of the African American population in the South. Butler led the impeachment team against Johnson, but the eleven-article indictment was filled with makeweight charges, and its central claim was flawed. Johnson escaped conviction in the Senate by only a single vote, and it's a pity that the articles were not stronger, because he was richly deserving of impeachment and conviction for his unconstitutional assaults on Reconstruction and the rights of African Americans. His acquittal helped gut Reconstruction and the Reconstruction amendments, replacing the cause of freedom, justice, and equality with the Jim Crow system that lasted nearly nine decades before *Brown v. Board of Education* (1954) and the civil rights movement emerged. Benjamin Butler was as imperfect as the rest of Congress, but he earned his salary and fought hard, if not always effectively, on the side of social and constitutional progress.

I toasted these lead managers in my mind and drew whatever strength I could from their journeys and hard work.

But now it was time to rally the troops. I called everyone together in the center of our room before we would leave to meet in the Rayburn Room back over on the House side to process back over formally to the Senate. I had only three things I wanted to say. The first was that I had been on a lot of teams in my life, with a lot of talent in different sports, but I had never been on a team like this one, whose every member placed the work and success of the group itself above his or her own interests and desires. I had seen Diana spending hours helping to edit other people's parts of our presentation; I had seen Joaquin and Mad exchanging tips and feedback; David had filled in for me at the Rules Committee—every day, I had seen this group becoming far more than just the sum of its parts. This group cohesion and idealism, even more than the palpable brilliance and eloquence of the managers, are what I would always remember about the experience.

The second thing I told them was that while our argument was built on a bedrock of reason, cold facts, and constitutional law, I did not want them to be afraid to show their passion now that they'd mastered their arguments. It was their passion that had brought us here, passion built on righteous anger over what Trump and the mob had done to us, and that anger, in turn, was built on a tremendous love of country, an appreciation for what America has done for us and for our families, a fierce patriotism in their hearts that America needed to see.

Finally, I ended by quoting a great Marylander, Frederick Douglass, who was born into slavery forty-five minutes from where we stood in the Capitol and who escaped to become our great freedom fighter: "If there is no struggle," said Douglass, "there is no progress. The struggle may be moral or physical, or moral and physical, but there must be struggle. Power concedes nothing without a demand. It never has and it never will." I reminded them that the Capitol itself had been built with slave labor and we have had to struggle against the constant undertow of violent racism in America in order to become a more perfect union, and we were now part of that essential historical project.

We made our way quickly over to the Rayburn Room, where we had fifteen minutes before we would process back over to the Senate. I had only one other issue to conquer before we left: my mouth was suddenly extremely dry, and I was having a serious problem speaking.

It was bizarre. I was with Arya and asked if she could find me some water, and she did. I drank an entire bottle, but it had no apparent effect. I drank another, in vain. I reflected that speakers with dry, sticky mouths make audiences cringe and look away.

"I think you're just having stage fright," Julie said.

"I don't get stage fright," I said, which was technically true for my time in politics. I did get it in high school once, when I played piano for my friend Nick Viorst's rock group, the Last Chance Band, and simply forgot how to play all the songs. And it happened again in 2017, in the House, when I had the chance to play against Garry Kasparov in chess—he had come to talk to me about Russian human rights violations—and, seated before the greatest chess player who ever lived, I forgot everything I knew about chess. Fortunately, both times I recovered midway into the experience. Kasparov crushed me, needless to say, but he told the chess reporters who had come to watch the game, "He's not bad for a politician."

"Well, you sound fine," Julie said.

"No, my mouth, it's completely dry. It's not working. This is a nightmare."

"It does sound fine," said Arya. "It might be a little bit of stage fright. You've never spoken to an audience of like twenty-five million people before."

"It doesn't sound fine," I said. "I'm calling Dr. Monahan."

I dialed him and, as always, he picked up right away. God bless Dr. Brian Monahan, the House doctor—how did he tend to 535 patients, all of whom had his cell number?

"Hello, sir," he said, ever calm and reassuring. "What can I do for you? I know you have a big presentation in a few moments."

"Dr. Monahan, my mouth is extremely dry, and if it stays this dry, there's no way I can do this."

"Well, it's probably just stage fright, sir," he said. "Why don't you send someone to pick up a cough drop, and it should clear it right up."

Indeed, that cough drop proved to be the miracle medicine I needed. I thought about how this wonderful unassuming man, a great doctor, had been vilified by the unvaxxed, unmasked, and unhinged amongst us. I called back to thank him for saving the day.

Everyone was telling us that the vast majority of GOP senators had made up their minds to vote to dismiss the trial on the grounds that it was illegitimate because there was no Senate trial jurisdiction over a former federal official. This "exit ramp," as they kept calling it, would give them a way to acquit while never addressing the damning and overwhelming facts of the case.

The point of our opening was to shock them into rethinking their position on this phony technicality. Our constitutional arguments were a complete home run, but how would we get them to listen? If they had cemented their minds shut, how could they assimilate our arguments over the text and structure of the Constitution, the brick wall of Senate precedents favoring trials of former officials, and the sheer illogic of saying that a corrupt or criminal president could escape conviction and disqualification from holding future office simply by resigning?

As the floor leader in the Maryland Senate for marriage equality, I'd discovered that not a single senator had changed his or her mind because of the compelling and rigorously logical equal protection and due process arguments I set forth, but many of them changed their minds because their favorite cousin was gay or their next-door neighbors were a long-term lesbian couple or they really thought our gay (and proudly out) colleague Sen. Rich Madaleno was a brilliant budget analyst and a good guy. The arguments over *equal rights* failed to enter their minds until they took the *humanity* of LGBTQ people into their hearts.

Likewise at the impeachment: no level of brilliant legal persuasion and no thick body of precedent would move a single Republican senator unless and until we shocked their consciences and woke up their human feelings to the meaning of the violence that had entered our House. This was why I told Speaker Pelosi in our very first conversation over this trial that we would set about to create a meticulous video presentation of the facts that would be riveting and unforgettable.

We needed to explain that, if adopted by the Senate, the totally concocted new "January exception" would make January 6 a lethally dangerous new norm in our politics, inviting outgoing presidents and their mobs to attack the presidential election process, storm the Capitol, threaten and intimidate government officials, and smash people over the head with Confederate battle flags and baseball bats to overturn and destroy elections. We would show them the bloody chaos of January 6 as a preview of the new, dystopian American future awaiting us if Donald Trump got away with yet another crime against the republic.

As I began to speak, I took on this idea right away—that, per the Trump team's argument, if you commit an impeachable offense in your last few weeks, you do so without constitutional accountability. I definitely saw a number of faces register interest and comprehension at this point. A live audience with critical intelligence is what we needed, and it seemed to be attentive:

> This would create a brand-new January exception to the Constitution of the United States of America—a January exception. And everyone can see immediately why this is so dangerous. It is an invitation to the president to take his best shot at anything he may want to do on his way out the door, including using violent means to lock that door, to hang on to the Oval Office at all costs, and to block the peaceful transfer of power.
>
> In other words, the January exception is an invitation to our Founders' worst nightmare. And if we buy this radical argument that President Trump's lawyers advance, we risk allowing what happened on January 6 to become our future.

And then I *showed* them what a January exception would mean for America's future, hitting Play on the stunning thirteen-minute video we had compiled. The video told in excruciatingly graphic detail the story of Trump's incitement and mobilization of mass violence against the police to halt Congress in its work and over-throw the results of the 2020 election.

The video began with Trump's repeated exhortations and incite-ments to the increasingly agitated mob: "We will stop the steal . . . When you catch somebody in a fraud you're allowed to go by very different rules . . . We won the election and won it by landslide, and now we're going to walk down to the Capitol, and I'll be with you . . . I hope Mike has the courage to do what he needs to do . . . If you don't fight like hell you're not going to have a country anymore."

Then it switched over to heart-pounding scenes of savage vio-lence and mayhem unfolding at the Capitol: rioters overrunning the physical barriers, helmeted insurrectionists assaulting officers with pipes and flagpoles, protesters working to erect a gallows, all while people are screaming "Fight for Trump!" and "Fuck *you*, police! Fuck *you*!" and "Take the building! Take the fucking building!" and "No Trump, no peace! No Trump, no peace!"

I was glad we kept in the scene of Sen. Mitch McConnell him-self, now seated fifteen feet in front of me, speaking on the floor on January 6, saying, "We're debating a step that has never been taken before in American history," and challenging the weak argu-ments and "sweeping conspiracy theories" being advanced by GOP colleagues to cancel out the election. "Nothing before us proves illegality near the massive scale that would have tipped the entire election," he said.

The video shifted to the battle inside, where insurrectionists were taunting cops with face-to-face provocations like "We outnumber you a million to one, dude," and "Are you gonna beat us all?" and "There are a fucking million of us out there, and we're listening to Trump," and "Traitor Pence! Traitor Pence!" and then, ominously,

"Break it down! Break it down!" as rioters pounded on the doors of the House.

Other scenes were etched in my mind: An insurrectionist yelling, "Is this the Senate? Where the fuck are they?" Another yelling, "That's what we need, thirty thousand fucking guns in here." Another shouting back into the mob, "We need fresh patriots to the front!" There was the heroic Capitol Police officer Daniel Hodges stuck in the doorway screaming and writhing in agony as a rioter lifted his gas mask and sprayed him in the face with bear mace. The mob pressing forward and screaming in unison, "Heave-ho!" Rioters stealing police shields and using them as weapons.

The scene then turned to Trump's pathetic tweeted video to his followers, two hours after the chaos began, clearly not disapproving in any way and still whining about his stolen election: "Never has there been a time like this where such a thing happened where they could take it away from all of us, from me, from you, from the whole country." He then made a glancing effort to counsel peace, a day late and a dollar short: "Go home, we love you, you're very special . . . I know how you feel, but go home and go home in peace."

Before it was all over, the senators watched U.S. flags being used as weapons to spear officers and a visionary insurrectionist urging his comrades to look to the future: "Go home and storm your county buildings and take down every one of these corrupt motherfuckers."

I watched the senators watching the footage. The Democrats were uniformly horrified and aghast. Because members of Congress were inside when these events happened, all of us had missed the full magnitude of the "American carnage" Donald Trump had unleashed outside on the perimeter of the Capitol and against our police officers downstairs and en route to the chambers. The senators were sheltered from the bloodshed and had had no idea about the *military* character of a lot of the violence: the extremists marching in formation, their arrival with dangerous chemicals, the coordinated poundings and beatings of officers, the use of walkie-talkies to plan ambushes of our officers.

Later that day, I must have heard a dozen times from senators of all stripes some variation on the statement "I had no idea the violence was this terrible," or "I never actually saw Trump's speech before, and it's amazing how he ordered the whole thing," or "I had no sense of how close we were to being apprehended or killed." The senators' shock over the magnitude of the violence was inevitable given that we all had extremely limited perspectives on that chaotic day. But the video created a 360-degree panorama of the battle and a coherent timeline of key events. The managers were floored when we saw the first cut, and now it seemed that a lot of senators were devastated by the video, too, awed by the scale of the violence. It was quickly dawning on them that we had all narrowly missed being killed.

I observed divergent reactions on the GOP side of the aisle. Many Republicans looked spellbound and transfixed. I noted on my legal pad that Lisa Murkowski looked "shell-shocked," Mitt Romney "shaken" and "disapproving," Richard Burr "ashen-faced, pissed." Mitch McConnell, who sat maybe fifteen feet in front of me, looked "pained, confused—crying?" The woman next to him, who I believe was his chief of staff, seemed "completely miserable and disgusted." When, in the video, Officer Hodges screamed out in pain while being crushed in the doorway, I thought McConnell was "tearing up, could lose it" and seemed physically uncomfortable.

But not all Republicans.

Senators Ted Cruz, James Lankford, and Marco Rubio were watching intermittently and in a distracted, paranoid way, alternately whispering to their neighbors like naughty fifth-graders and squinting hard at the screen as if they might detect some big liberal conspiracy of Antifa spies hiding in the bushes in the background. Rand Paul seemed not to be watching at all for long periods, but was "scribbling away smugly," as if noting down gaping logical fallacies in our presentation.

When the video stopped playing, the silence in the room felt like a moment etched in history. I gave the senators an extra beat to catch their breaths and let the enormity of the crime sink in. Then I spoke:

> *Senators, the president was impeached by the U.S. House of Representatives on January 13 for doing that. You ask what a "high crime and misdemeanor" is under our Constitution. That is a high crime and misdemeanor. If that is not an impeachable offense, then there is no such thing. And if the president's arguments for a January exception are upheld, then even if everyone agrees that he is culpable for these events, even if the evidence proves, as we think it definitively does, that the president incited a violent insurrection on the day Congress met to finalize the presidential election, he would have you believe there is absolutely* nothing *the Senate can do about it—no trial, no facts. He wants you to decide that the Senate is* powerless *at that point. That can't be right.*

With that statement, I sought to preempt any effort by the defense to shift to the question that had tied the Senate in knots in the three prior presidential impeachment trials: What exactly is a "high Crime or Misdemeanor"? I wanted them to equate what they had just seen with the very concept of an impeachable offense so that Trump's astounding conduct would become not just a possible case, but the *paradigm case* for impeachment, conviction, removal, and disqualification. This strategy was, in fact, daring Trump's lawyers to *deny* that incitement to insurrection was an impeachable offense or, in the alternative, to concede that it was an impeachable offense but then refute our factual proof that Trump's conduct *was* incitement. The Senate needed to see that Trump's "January exception" jurisdictional argument would prevent even the most blatant and undeniable cases of violent impeachable conduct from ever being heard in the Senate.

What I aimed to reinforce, repeatedly, was the fact that the Framers understood just how crucial a peaceful transfer of power is in a democracy. There is a reason that the State Department monitors elections in other democracies as it does; there is a reason that having a peaceful transfer of power for over two centuries had been a source of immense pride for our country until January 6. The Framers and democratic theorists understood that the transition of power is always the most dangerous moment for democracies. While every historian will tell you this, we'd just learned it ourselves the hard way.

I explained to the senators that the Framers wrote an oath right into the Constitution to bind the president from his very *first* day in office until his very *last* day in office and every day in between. One specific reason the impeachment, conviction, and disqualification powers *exist* is to protect us against presidents who try to overrun the power of the people in their elections and replace *the rule of law* with the *rule of mobs*. By definition, these powers had to apply even if the president committed his offenses in his final weeks in office, because that's when elections get attacked, and that is precisely when we need them *the most*. I told them that we would show that everything that we knew about the language of the Constitution and the Framers' original understanding and intent, along with prior Senate practice and common sense, confirmed this rule.

This historical context framed a crucial piece of our argument. Neguse and I were about to detail in compelling fashion the precedents for trying a former government official after he has left office, but I wanted to set forth the rationale for these decisions first, to grab them by their common sense. There can plainly be no exception, for January or any other month, to the rule of complete presidential fidelity to the Constitution.

We had put together a comprehensive historical argument demonstrating the unbroken understanding that former officials could be impeached, tried, and convicted. British history provided the model for the American practice of impeachment, so I began there. Notably, every single impeachment of a government official that occurred during the Framers' lifetime concerned a *former* official. In fact, the most famous of these occurred while the Framers gathered in Philadelphia to write the Constitution: the impeachment of Warren Hastings, the former governor-general of the British colony of Bengal and a corrupt guy. The Framers knew all about his case, and they strongly favored the impeachment, actually invoking it by name at the Constitutional Convention, which made it the only specific case of impeachment they discussed at the convention. And even though everyone there surely knew that Hastings had left office two years before his impeachment trial began, not a single Framer—not one—raised an objection when George Mason held up the Hastings impeachment as a model for our Constitution.

The early state constitutions also supported this understanding—so much so that every single state constitution in the 1780s either specifically said that former officials could be impeached or were entirely consistent with the idea. There was an overwhelming presumption in favor of allowing legislatures to hold former officials accountable in this way. Some states, including Delaware, even *confined* impeachment to officials who had already left office. These precedents confirmed that removal from office was never seen as the sole reason for impeachment in America; the overall goal was always about accountability, defense of society, and deterring corruption.

The debates at the Constitutional Convention further bolstered this interpretation. As they discussed impeachment, the Framers repeatedly returned to the threat of presidential corruption aimed at corrupting elections, the machinery of self-government. As William Davie of North Carolina explained in 1787, impeachment was for any president who spared "no effort or means whatever to get himself reelected." Meanwhile, Hamilton, in Federalist No. 1, said the greatest danger to republics and the liberties of the people comes from political opportunists who begin as demagogues and end as tyrants and from the people who are encouraged to follow them.

"President Trump may not know a lot about the Framers," I told the Senate, "but they certainly knew a lot about him."

Given the Framers' intense focus on danger to elections and the peaceful transfer of power, it is inconceivable that they'd designed impeachment *not* to govern the president's final days in office. There are simply too many opportunities to interfere with the peaceful transfer of power in those precarious weeks, a time when abuse of power would be most tempting and most dangerous. As a matter of history and original understanding, there was no merit to President Trump's claim that the Senate lacked the power to gather evidence and conduct a trial.

After laying out these points, I could tell that the vast majority of the Senate was paying attention and was moved. The video evidence of Trump's aggressive incitement to dangerous mob violence and insurrection seemed to have come as an epiphany to them. Now they got the fundamental danger of inventing a January exception and had assimilated the full weight and logic of constitutional text and history.

But Joe Neguse and David Cicilline sealed the deal with knockout arguments about controlling Senate case precedents and about the embarrassing fallacies and errors in Trump's answer to our arguments. Joe showed them that, from the beginnings of the republic, the Senate itself had considered challenges contesting its jurisdiction to try former officers who had been impeached by the House, but it had always rejected these challenges and landed squarely on the side of there being proper constitutional power to try them. He demonstrated that, although we made this elaborate argument rooted in text, structure, and precedent in our trial brief, the president's lawyers had completely ignored it in their response: "The trial brief offers no rebuttal on this point—none," Joe said, and I saw several Democratic senators shake their heads at that moment. Trump's lawyers had simply dodged the central and overriding issue in the case.

Joe also showed that the star legal scholar of the Trump defense team, George Washington University Law professor Jonathan Turley, who had addressed the GOP Caucus to convince everyone that an impeached former president can never be tried in the Senate, had actually been, in his prior scholarship, a steadfast *champion* of our position that "the Constitution authorizes the impeachment of former officials." Joe showed how Turley's body of scholarship over many years radically contrasted with his recent political acrobatics and intellectual contortions as he tied himself into knots for the Trump defense. The juxtaposition of Turley-the-scholar with Turley-the-pro-Trump-political-crusader was devastating.

David was similarly relentless and exhaustive in preemptively refuting all the flotsam-and-jetsam Trump legal arguments we had been reading about in the press: that Donald Trump had not been given due process to be heard (that, of course, is what the Senate trial is for, not the grand jury indictment process, and Trump didn't want to speak at the trial, much less in the House for the impeachment phase), that there was some suggestion by virtue of the chief justice's not being present that the Court lacked jurisdiction (a complete non sequitur, as David calmly explained), and that Trump should be free from any incitement charge because Democrats have also used the rhetorical phrase "fight like hell" (but never, of course, as part of a sequence of actions and words deliberately designed to incite a violent insurrection against the government).

I wanted to close our presentation back on a personal note, humanizing the jurisdictional decision for everyone so that America could see what inventing a "January exception" would mean for our way of life in America's increasingly polarized politics, government, and society. Personalizing the issue would remind the senators of the stake that they and their families had in preventing future presidents from rolling over Congress and treating us like roadkill. It would also show the wounded and stressed-out Capitol officers and their families, our hardworking staff and their families, and everyone who worked on Capitol Hill—from journalists to engineers to janitors—that we shared their anguish and would always have their backs. It was vital to remind America how personal democracy is for all of us and how personal the *loss* of democracy is, too.

Humanizing the jurisdictional issue succeeded in bringing the proceedings on Tuesday to an unplanned emotional crescendo when I got choked up talking about Tabitha and Hank's experience and what Tabitha had said to me about never coming back to the Capitol:

> *The reason they came with me that Wednesday, January 6, was because they wanted to be together in the middle of a devastating week for our family, and I told them I had to go back to work because we were counting electoral votes that day on January 6. It was our constitutional duty. And I invited them instead to come with me to witness this historic event, the peaceful transfer of power in America. And they said they heard that President Trump was calling on his followers to come to Washington to protest, and they asked me directly: Would it be safe? Would it be safe? And I told them: Of course it should be safe. This is the Capitol.*

Walking the senators, the media, and millions of people watching at home through Tabitha and Hank's experience of that day, I told the story of our being in Steny Hoyer's hideaway before the counting began, as people came to greet us and offer condolences. I told the story of how they came with me to the floor and watched from the gallery during the powerful show of bipartisan support for our family during our week of agony. I told of how they watched me deliver a speech that included the words of Abraham Lincoln's famous 1838 Lyceum speech, in which he said that if division and destruction ever came to America, it wouldn't come from abroad but from within. I observed that Lincoln deplored mob violence and mob rule and had said that the uncontrolled use of mobs for political ends would lead to tyranny and despotism over the people of America.

I recalled how, when my speech was over, they'd gone back to Steny's office off of the House floor, but that they didn't know—none of us did—that the House had already been breached, that an insurrection, or a riot, or a coup, had come to Congress, and that by the time we learned about what was going on, it was too late. I couldn't get out there to be with them in that office. They were stuck there alone with Julie, my chief of staff.

Sharing the story from my perspective, I detailed how, all around me, people were calling their wives and their husbands and their loved ones to say good-bye, how members of Congress in the House removed their congressional pins so they wouldn't be identified by the mob as they tried to escape the violence. The sounds I heard—the voice of our new chaplain saying a prayer for us, someone telling us to put our gas masks on—and then the sound I said I would never forget: the sound of pounding on the door like a battering ram, the most haunting sound I had ever heard.

I told the world about how Tabitha and Hank had locked and barricaded themselves in Steny's office, hiding under a desk with Julie and sending their final texts and whispered phone calls. They feared an extremist with an AR-15 and thought they were going to die.

Throughout all this, I'd managed to maintain my composure. But it was as I remembered my final words with Tabitha as she prepared to go home that my emotions got the better of me:

> I told her how sorry I was, and I promised her that it would not be like this again the next time she came back to the Capitol with me. And do you know what she said? She said, "Dad, I don't want to come back to the Capitol."
>
> Of all the terrible, cruel things I saw and I heard on that day and since then, that one hit me the hardest, that and watching someone use an American flagpole, with the flag still on it, to spear and pummel one of our police officers, ruthlessly, mercilessly tortured by a pole with a flag on it that he was defending with his very life.

People died that day. Officers ended up with head damage and brain damage.

People's eyes were gouged. An officer had a heart attack. An officer lost three fingers that day. Two officers have taken their own lives.

Senators, this cannot be our future. This cannot be the future of America. We cannot have presidents inciting and mobilizing mob violence against our government and our institutions because they refuse to accept the will of the people under the Constitution of the United States. Much less can we create a new January exception in our precious, beloved Constitution that prior generations have died for and fought for, so that corrupt presidents have several weeks to get away with whatever it is they want to do. History does not support a January exception in any way, so why would we invent one for the future?

When things go well in court, you know it, just like you know it when they go south. We were soaring when we closed. Barry passed me a note in which he'd written "fantastic job," and a number of senators were looking directly at me and smiling, including the wonderful new senator from California, Alex Padilla, Wisconsin senator Tammy Baldwin, and Sen. Richard Blumenthal from Connecticut. I remember New Hampshire senator Maggie Hassan smiling kindly at Joe, David, and me. Barry passed me another note saying that we had ended with thirty-three minutes left for a possible rebuttal after the defense's remarks, which meant we had essentially answered every possible question and resolved every possible logical doubt over Senate jurisdiction in less than ninety minutes and had done it in a way that previewed major facets of our presentation about the facts of January 6 to the country.

Senator Schumer moved for a ten-minute break and then came over to our table and said, "Thank you. That was excellent."

I was elated by the crisp and punchy effectiveness of our performance, which meant we had avoided the worst outcome of boring and annoying the Senate. At the same time, I felt awkward that I had started to choke up when I quoted Tabitha. Luckily, I was able to catch myself and make it through, but I was afraid it looked sloppy, and I felt certain that the right-wing blogosphere would be nominating me for an Academy Award for Best Dramatic Performance; after all, any emotion shown by liberals is presumed to be insincere and counterfeit. I was actually surprised by what happened, because I had not broken down at that point in my argument when we practiced it on Sunday, but the intensity of the moment in front of the Senate and the country was just dumbfounding. I was thinking about Tommy, I was thinking about Tabitha, I was thinking about Hannah, I was thinking about January 6 and America and what would come next for all of us. I was thinking about my parents and what they had wanted America to be like for us and our family. I was realizing that I was going to be able to express my love for the country over this week, but I felt that I had no way to know where our country was actually headed. In the wake of the insurrectionary riots and rampant right-wing violence, American democratic "exceptionalism" felt like a fragile and endangered dream.

When we came back from our break, the defense team took the stage. It was their turn, and we were prepared for the worst. We expected them to set forth detailed arguments seeking to explode our case. They would try to debunk our textual analysis, refute our constitutional history and study of impeachment antecedents in England, expose distortions in our canvass of Senate impeachment trials of former officers throughout American history, and undermine our predictions about the corrosion of American democracy that would follow from adoption of the "January exception."

But they did none of this. They never laid a glove on us. They did not contradict, much less refute, any of our legal arguments.

What did they do? I'm not quite sure.

It is hard to say precisely what they did except leave senators, House members, lawyers, law professors, law students, and tens of millions of Americans across the country baffled, befuddled, and bemused.

Some people in the room experienced the Trump lawyers' arguments as comic relief from the crushing gravity and solemnity of the proceedings. But I experienced them more as tragedy than comedy. People had died because of Trump's callous recklessness and hunger for power, yet he and the people around him were apparently so convinced of his political invulnerability in the divided Senate that he had sent in a couple of lawyers who apparently never felt required to develop a coherent theory of their case. In fairness, they had been dealt a terrible hand, because Trump was guilty as sin on the facts, which were disturbing and irrefutable, and he had no serious legal arguments to acquit him. His defense was a purely political one (jury nullification, you might say) based on his accelerating stranglehold over the Republican Party.

We were surprised when Bruce Castor got up to speak first, given that David Schoen was supposed to open. We had seen the two gesturing furiously at each other while our arguments were going on, so it seemed like Castor had called some kind of last-minute audible reverse play. Madeleine Dean and our colleague Mary Gay Scanlon knew Bruce Castor from the tight-knit Philly legal community and had told us that he was famous for having refused, as the Philly district attorney, to prosecute Bill Cosby for sexual assault. Castor had briefly been Pennsylvania's acting attorney general, but both Dean and Scanlon said that he loved to talk and had the "gift of gab."

Once he'd arrived at the dais, Castor showed how understated our colleagues had been in this assessment, but his opening was completely different from what we were expecting. For one thing, rather than going for the jugular, he complimented us on an "outstanding presentation." I nodded gratefully to Castor, who stood perhaps five feet away from me. I had been striving mightily to be kind to the Trump lawyers, but I was still surprised by this high praise and gentlemanly remark. I thought he was going to turn now to the legal arguments and try to eviscerate us, knowing that the best lawyers put you on a pedestal to make sure you fall from an exceedingly high height. But he seemed to have been genuinely moved by our "outstanding presentation" and proceeded to strengthen our position by condemning the violence and the "rioters" of that day in terms that seemed to reinforce our position rather than strengthen their own. These were comments I quickly recorded, because I felt certain that it would be just a matter of weeks, if not days, before Trump started praising and whitewashing the work of the insurrectionists.

I was now bracing for an onslaught because I thought Castor had shrewdly distanced the Trump *team* from the Trump *mob*, which would make sense if he had brought a legal argument to maintain that distance. But had he? How would he disentangle Trump from the events he, Castor, was denouncing as "horrific" given Trump's key role in organizing and inciting those events? This was the point at which Mr. Castor should have launched into the argument for the Senate's jurisdiction, but instead, he began to wander and free-associate, and the only argument I could identify tucked into this soliloquy was that, as I wrote down on my notepad, we were "looking for someone to blame"—which was amazing and, again, had nothing to do with the jurisdictional argument. Essentially, he was claiming that Trump was being unfairly targeted because the public was out for blood and just looking for someone to hold responsible for this terrible misfortune of a riot, which had really been more akin to a natural disaster than a violent crime. We saw Republican senators on both sides glancing at one another and furrowing their brows, shrugging to confess complete confusion.

And then, Castor's remarks went from a rudderless, unfocused presentation to one that was self-indulgent, weird, bizarre, impro-vised, and completely self-defeating. It was a rambling diatribe of folksy anecdotes, attempted flattery, and endless digressions. On and on it went, relentlessly, fearlessly, aimlessly.

Then Castor's partner David Schoen got up and took glancing and conclusory shots at our argument, never seriously engaging with the systematic analysis we had offered. Schoen's presentation, too, was filled with rhetorical fluff, but he adopted a far more biting and stinging edge, as if he needed to offer Fox News and the right-wing blogosphere some meaty chunks to feed their audience. Schoen's angry tone intimated no hints of reconciliation or healing, but he repeatedly called *us* to the carpet for dividing America. It was not the insurrectionists and rioters who had damaged the country, as Castor at least had been honest enough to acknowledge, but the *impeachment managers*. Schoen said, "And to help the nation heal, we now learn that the House managers, in their wisdom, have hired a movie company and a large law firm to create, manufacture, and splice for you a package designed by experts to chill and horrify you and our fellow Americans."

Hiring a "movie company" was precisely what we had *not* done. The DOAR Group was a firm that worked to compile video clips for trial. And every single clip from our video was authentic and documented.

But Schoen continued: "They don't need to show you movies to show you that the riot happened here. We will stipulate that it happened, and you know all about it."

Actually, none of the senators "knew all about it"; indeed, virtually *no one* had had any real sense of the complete event before we got up to provide an overview of it for them. Of people who had served in office, Donald Trump would have known the most about the causes and main events of the insurrection, but of course, their client had refused to testify. Most senators were shocked by what they had seen, and they had barely seen anything yet.

After long meditations on due process, Schoen ended by reciting a Longfellow poem, and then he sat down, to the relief of a clearly exhausted and benumbed Senate.

At that point, I had a huge decision to make. I had thirty-three minutes left to respond to a potpourri of trivial arguments that were maddening to me. I longed to refute and knock down each and every one of them, crushing them into fine dust. And I knew a lot of bad-ass lawyers who would have eviscerated the defense's presentation.

Yet, every lawyer also dreams of yielding back rebuttal time as a gesture of supreme confidence in the arguments offered, as a subtle dig at the weakness of the opponent's arguments (*you didn't lay a glove on us*) and as a form of courtesy and magnanimity to the tribunal. It is undeniable that thirty-three minutes would have been a deliciously huge amount of time for me to offer my own take on the role partisanship plays in an impeachment trial, an issue hammered by Trump's lawyers and something I was eager to discuss. But I remember thinking, *The side that whines the most about partisanship is always the side that ends up sounding the most partisan.*

I was torn. What to do? And then I returned to Tommy. In his own political struggles, he never went for the jugular. He always tried to uplift even his greatest adversary. This was the young man who had shown remarkable grace and gentleness with even Dinesh D'Souza, a rhetorical knife fighter. Tommy was the boy who advised his classmates to "make friends with someone you don't get along with." Like my father and his grandfather—indeed, like Sarah's father, too—Tommy never gave up on anyone, and he always tried to model decency and good manners. *The smart thing to do here*, I thought, *would be to do the Tommy Raskin thing.*

I got up and looked at Senator Leahy, who had been in the hospital just the day before, and I looked at all the weary and still video-shocked senators. After telling them that we strove to be bipartisan and nothing was more bipartisan than the desire to recess, I waived the remainder of our time—our work was done for the day. They clapped heartily.

A revived Senator Leahy proceeded to call the question: "The question is whether Donald John Trump is subject to the jurisdiction of a Court of Impeachment for acts committed while President of the United States, notwithstanding the expiration of his term in that office."

We won on a vote of 56–44, with 6 Republicans—Cassidy of Louisiana, Collins of Maine, Murkowski of Alaska, Romney of Utah, Sasse of Nebraska, and Toomey of Pennsylvania—joining all the Democrats to declare that Trump was clearly subject to the jurisdiction of the Senate.

It was exhilarating to have won the first and most important procedural skirmish and exciting to know that Senator Cassidy of Louisiana—who had not been counted before as an "open mind"—actually was in motion. It was not the landslide defection from Trump we were hoping for, but it was at least the swing in a positive direction we needed. It had shown that people's "minds are open," as I told reporters afterward.

As we left for the hallways, we could hear the commentators praising the managers and pillorying Trump's lawyers for their meandering and unfocused speeches. I heard one commentator say it was one of the "greatest mismatches in the history of legal arguments."

For the briefest of moments, I let the satisfaction of those words sink in, but I wasn't about to dwell on them. After all, the actual trial hadn't even begun.

CHAPTER 12

IS THIS AMERICA? TRUMP ON TRIAL

It is certain, in any case, that ignorance, allied with power, is the most ferocious enemy justice can have.

—JAMES BALDWIN

We were on our way down North Capitol Street with the security detail on Wednesday, February 10, with Julie seated to my left in the backseat, running through notes to prepare for the trial, a million details swirling in my mind. Someone was phoning me.

"Congressman Raskin, please hold for the president," a velvety, professional voice said.

And five seconds later, I heard: "Jamie, it's Joe Biden."

This was no prank call. I recognized his voice and accent immediately. I always loved the upbeat, Mid-Atlantic, Irish American pitch to his voice, which reminded me of the accent of the conductor on Amtrak's Northeast Regional traveling from Union Station "Warshington" to Thirtieth Street Station Philly. Biden spoke in the accents of my childhood. It was the way I heard English spoken at Funland, at the Rehoboth Beach Boardwalk or a pancake and bingo night at a fire station in Chestertown, Maryland, where they brought us kids from Echo Hill camp. I had heard that local brogue coming through loud and clear in the voice of Metropolitan Police Department officer Michael Fanone, the DC undercover drug detective who, on January 6, threw his uniform on and rushed down to the Capitol with his partner to join the battle after hearing distress calls over the radio. I loved this accent, not north like New York or New England, not south like North Carolina or Georgia, not official Washington, but the Mid-Atlantic states, the Chesapeake Bay, Virginia through Pennsylvania and New Jersey, the verbal soundtrack of my childhood.

"Mr. President," I said. "What an honor."

"Jamie, I'm calling to tell you that you're a helluva lawyer but you're an even better father. God love ya."

I was stunned. The sweetness of the remark, the kindness of his calling, bowled me over.

"Mr. President, I . . ." I could find no thought to complete, and I choked up.

"No, no," he said. "I know what that was like. About a month after Beau died, I had to make a major speech at the FBI after some agents were killed, and so I know what that takes. It's presumptuous of me to say this but I'm incredibly proud of you. I saw you up there and I saw myself in it and I'm so proud of you."

I could find no words, and I clutched the phone to my mouth and swallowed hard.

"We are forever linked by that," I said. "Thank you for the great job you're doing and for giving us all so much hope. I'm honored to be part of your party and to be one of your citizens."

"I want you to know Jill and I are here for you," he said, "and we're sending you and Sarah and your family our love."

And then we said good-bye.

I remembered that President-elect Biden had called Sarah and me just three days after we lost Tommy, too. I was in such a blur of shock then that I could hardly even register it. I could only remember him promising that the day would come when Tommy's name would bring a smile to my lip before tears to my eyes, and that the agony of the end would dissipate as time restored the fullness of our sense of Tommy's wonderful life. (His kind call was followed by another beautiful call, this from Vice President–elect Kamala Harris, but I was again too steeped in agony and grief to absorb the particulars.) The president's kindness now floored me.

Julie and our security guys were quiet. They always were when I began to cry over Tommy, but I read into the silence something else: amazement, and spreading relief that Joe Biden was our president. It was dazzling appreciation for him that I felt in that car.

Then there was too much silence, like five minutes of it.

No one was going to speak, so I said, "He has the greatest emotional intelligence of any president since Abraham Lincoln. That was like a year's worth of grief therapy."

I think everyone was kind of choked up, so I just wanted to make them laugh.

"He is our president of pathos," I tried. "Our chief emotional officer."

Nothing.

"Our commander in grief," I said, and everyone laughed finally. "I *love* that guy."

And I do, and always will.

On the ride down, we got to hear a lot of the reaction from the arguments of the first day. A number of Republicans praised our performance. "If I'm an impartial juror trying to make a decision based on the facts as presented on this issue, then the House managers did a much better job," Senator Cassidy said on CNN, and I took note of his intellectual seriousness, reminding myself to establish eye contact with this senator I had never met. "The issue at hand is: is it constitutional to impeach a president who has left office, and the House managers made a compelling, cogent case— and the president's team did not."

Mississippi senator Roger Wicker told CNN that "Democrats sent a better team" with "very eloquent" managers, but he remarked that nothing could change his mind on the constitutional question, apparently not even the arguments. Many Republicans, like Ron Johnson of Wisconsin, were bending over backward to say they were not going to change their votes. A number were on message, circling the wagons to prevent any further erosion in Trump's position.

CNN's Kaitlan Collins reported that "Former President Trump was unhappy with Bruce Castor's opening argument," and that "Trump was almost screaming as Castor made meandering arguments that struggled to get at the heart of his defense team's argument." Reports kept flowing in of Donald Trump's tirades against his lawyers, who actually had done not too badly with the terrible hand they had been dealt. I started to feel a little sorry for them.

Professor Alan Dershowitz, who had been my Criminal Law professor at Harvard Law School and one of Trump's lawyers in Impeachment 1.0, was all over the news railing about the new team's performance. "I have no idea what he's doing," Dershowitz had said directly of Castor.

There had been rumors that Dershowitz was going to lead Trump's team, and I was delighted about that prospect. Just as getting to kick around ideas on a daily basis with my beloved Larry Tribe gave me a sense of completing a journey from my Constitutional Law class, getting to kick around the right-driving ideas of the Trumpified Alan Dershowitz promised a similar kind of arrival and return from my days in Criminal Law. Dershowitz had given a stunning, improbable speech in Trump's defense in the first trial, where he asserted that Trump could not have abused the powers of his office if his motivation was to get himself reelected. This astounded me because motive (robbing a bank to give the money to Amnesty International) does not negate the existence of either criminal conduct (sticking up the teller and taking the money) or criminal intent (choosing to rob the bank), and in any event, the Framers believed that presidents driven to do anything out of political self-advancement or self-preservation were not the *least* dangerous kind of politician but the *most* dangerous kind. Yet, like most acrobatic Trump apologists, Dershowitz was now uttering, with a tone of solemn authority, what he once would have recognized as an illogical, even plainly absurd argument. Hearing my former Criminal Law professor proclaim that a president could not be guilty of a constitutional crime if he was motivated out of a concern for reelection (treason? bribery? inciting violence?) was when I most regretted not having been named to the first impeachment team.

The thought that I'd get a chance to oppose Dershowitz in Impeachment 2.0 had been invigorating. But then I heard from a former GOP congressman who seemed to know something that Dershowitz was pulled from consideration for the second trial team because he had represented a number of serious criminals who had obtained pardons from Donald Trump on his way out the door. Trump's team apparently did not want any more public focus on those pardons.

I certainly did not consider Dershowitz's blistering attacks on Trump's lawyers as a case of sour grapes, because there was plenty for trial lawyers all over America, especially in Philadelphia, to criticize in their performance. I had gotten messages from my friends Andy and Gwen Stern, two titans of the Philadelphia Bar, saying that lawyers in the City of Brotherly Love were scandalized by what they were seeing during the first day of the trial, not just the astonishing absence of legal analysis but the sheer amorality of the whole thing in the wake of such bloodshed. (Andy would die in a tragic accident at the Jersey Shore just a few months later.)

Still, I started to feel that all the criticism heaped on Trump's team missed the point. Yes, they talked about irrelevant and silly things, but only because Trump had no alibi, no factual defense, and a flimsy constitutional argument. Those guys basically had nothing to work with. If your best argument is that the outgoing president of the United States has a First Amendment right to incite violent insurrection against the United States without being impeached and convicted, you just don't have much of a case.

In the harsh world of real politics, everyone in the Senate knew that Trump's best argument was actually no argument at all, and his lawyers had nicely evoked that understanding. The case for Trump was simply everyone rallying around and guarding the raw power dynamics of the GOP. Trump was still king of his party, the dominant fund-raiser by far, the alpha guy who could destroy you in your primary by endorsing your opponent with a tweet or two, and the hegemonic angry voice of right-wing conservatism even *after* being banned from Twitter and Facebook for his compulsive and dangerous lying.

If you must choose between being feared and loved, Machiavelli says in *The Prince*, the ambitious tyrant must always choose to be feared. (Of course, as a ninth-grader, Tommy Raskin gave the best refutation of Machiavelli on this point I have ever seen: when the going gets tough, the politician embedded in love will have a lot more support to fall back on than the politician who is simply feared.) Fear was the controlling principle in the unhinged GOP of 2021, and no one was feared more than Donald Trump, because of his giant reach in the country and his famously explosive and vindictive temper. The Republicans I knew who had broken from him in some substantial way had either some kind of independent power base and famous family name to fall back on or some unusual set of moral and political convictions—or both. Mitt Romney and Liz Cheney were the paradigm examples.

The "impeachment Republicans," as they were called around here, inhabited a political planet galaxies away from run-of-the-mill GOP members of Congress who had more malleable principles and who depended on their public office and party position for everything in personal, professional, financial, and intellectual terms. Our challenge with Senate Republicans was staring us in the face: we needed somehow to break the iron political and financial discipline of Trump over the mainstream GOP conference by appealing directly to the senators' states and their constituents, going around them to shake public opinion and till the political earth under their feet.

Back in the war room, we prepared to move on to our factual proof of the president's offense against America. In the emotional odyssey of the trial, over the next sixteen hours of argument during the next two days, we would travel from the established *trauma* of the seething violence we had experienced to the specific arresting truth of the president's inciting and organizing *role* in that violence. We argued that this was a journey we had to take in order for any effective healing and reconciliation to take place. Human beings do not move straight from trauma to closure without first understanding how and why all their basic assumptions about the world were violently wrenched away or how they can reduce the chances of ever experiencing such a demolition of their expectations again. We made this a central theme of our case as Trump's GOP loyalists, now seized with a suspicious new "Kumbaya" spirit, worked to spread the message that "it's time to heal, let's all just move on and let bygones be bygones."

The managers understood that "Can't we all just move on?" was the go-to mantra of abusers and bullies *who got caught*. Whenever that sinister call to counterfeit healing surfaced, we pounced immediately to argue that we *could not* successfully heal without first confronting the reality of what had happened. As Stacey Plaskett reminded our team, there were women all over the country who recognized in appeals to a phony and unearned "unity" the same form of cynical gaslighting perpetrated against victims of sexual violence and harassment.

The actual truth may be a difficult path, our team all agreed after discussing it on Zoom one day, but it was the only viable path for redemption and renewal.

Over the next sixteen hours, we would make known the reality of the insurrectionary violence that shook the Capitol and of Donald Trump's long-running efforts to overthrow the 2020 election and replace it with a result more to his liking. We would track his efforts to reverse the election results, first, by some lawful if outlandish litigation; and then by multiple unlawful and coercive strategies that remained (barely) nonviolent; and then by the cryptic interplay of inciting brutal violence against other political leaders, including his own vice president, and exploiting vulnerabilities in the Electoral

College system to try to execute a political coup against the Constitution, the Congress, and the vice president.

As we had agreed at our very first managers' meeting and on pretty much every daily call, we would tell a single compelling story of these events leading up to the nightmare of January 6. Each chapter, assigned to a different manager or a small team of managers, would tease out a different dimension of these extraordinary events, pointing ineluctably to the culpability of Donald Trump for inciting and even organizing—although this was not necessary to the legal case—a violent insurrection against the vice president, the U.S. Congress, and our presidential election.

I opened our presentation by establishing multiple critical points with which we would frame our case over the next two days. I hoped my opening would serve as a preview, an overview, and a complete rhetorical preemption of the field of legal arguments.

The first point was that Trump's constitutional objection to the Senate's jurisdiction had been settled the day before, when the Senate voted 56–44 to reject it. That was now over and done with. The Senate jury had to leave this defeated technical objection behind and judge Trump's culpability based on the *facts*. I knew this was the essential starting point of the analysis because a number of GOP senators kept telling the media they still doubted that the Senate could try a former president. I wanted to emphasize to them that this was like a criminal prosecution in which the defendant alleged in pretrial motions that he could not be lawfully tried because his Miranda rights had been violated or the murder weapon had been found in violation of the Fourth Amendment. If the defendant succeeded on such a motion, the case would be over and the trial would end. But if the defendant lost those arguments, the jury had to forget them and proceed to hear and decide the case on the *facts*. There could be no backsliding to a *question of law* already cleanly disposed of by the court. I did not want the senators to confuse the matter simply because they were acting as both judges and jurors; they had to keep those two roles and functions distinct. Trump's

Senate supporters who wanted to use the jurisdictional objection as cover for an act of jury nullification had to drop it—although, of course, they never did.

I also wanted to pull out of the chaos of the Trump team's arguments from opening day a key proposition that we could contrast with our case. The House had voted to find that the ex-president had incited a violent insurrection against Congress, the Constitution, and the people. But the president's lawyers were arguing that his conduct was "totally appropriate," as Trump himself had put it, and that he was, rather, an *innocent victim of circumstances*, like the other innocent victims we would see getting caught up in all the violence as the evidence unfolded over the next several days. According to their argument, we were just rounding up the usual suspect: poor, misunderstood Donald Trump. I wanted to invite the senators to ask themselves, based on what they were about to see, whether Donald Trump was in fact an innocent bystander to these events, like the (vast majority of the) members of the Senate themselves, or the culpable central player who had incited the violence. Contrasting Trump's story with our story would be the best way for them to think about their fact-finding responsibility.

I wanted to introduce the phrase "inciter in chief" right away. I had come up with the locution during my preparations to define Trump's actual role in these events. I liked the way it captured both what he was spending his time doing and what he was spending his time not doing. It implied that Trump had traded the actual duties of his constitutional office for a new role of inciting a violent mass movement against our own government. It foreshadowed an argument we would hit very hard when we got to January 6 itself: that at a time when Trump should have been mobilizing the powers of the government to *defend* Congress and our constitutional order, he had continued to incite armed insurrectionists to deploy violence against us. It would be hard to imagine a more blatant violation of his oath to preserve, protect, and defend the Constitution. As Liz Cheney had put it when we were still over on the House side, this was "the greatest betrayal of the presidential oath in the history of the United States."

But I also wanted the senators to understand that there was a method to Trump's madness on that fateful day. It would be easy to be disoriented and paralyzed by the shock of all the violence. We did not want senators to be so overwhelmed by the viciousness of the attack that they forgot its underlying strategic dimensions. This was *political* violence. The insurrection did not take place on any old day in some random place. It was an organized attack on the counting of the Electoral College votes to stop Vice President Mike Pence and Congress from certifying Joe Biden's victory in the Electoral College. If all had gone well for Trump and his forces, he would have been declared president by the end of the day. It was not incitement just for the thrill of incitement—although, with Trump there was undoubtedly that sadistic, bloodthirsty element to it. This was incitement for the sake of staying in power.

It was also critical for us to establish early on that the incitement and provocation were not defined by a single moment in time or a single incendiary speech. They were the culmination of Trump's premeditated cultivation of violent actors and violent actions for a long period of time before all the manic violence was released on January 6. The crime needed to be understood in its full context and development, and that was the point of our presentation. We would show the long-running effort by Trump to delegitimize the election even before it took place. We would show the absurd and doomed efforts afterward in the courts to prove that there had been fraud or corruption in the electoral machinery. We would show the dangerous efforts to get the state legislatures to nullify their own elections and change the results. We would show the use of social media to cultivate an angry movement of extremists, fanatics, and supporters to come to Washington to "stop the steal." We would explore Trump's prior postelection MAGA gatherings that conditioned the crowds to street violence and focused their energy on the upcoming January 6 main event. We would examine the December 12 rally in Washington at which Trump said, "We have just begun to fight," and which ended in serious violent attacks and the burning of a church. We would look at the December 19 rally where he promised his followers that January 6 would be "wild." As the president of the United States of America put it, "Be there, will be wild!"

We would also show them how the attack was foreshadowed and mapped out online. The digital paper trail included social media posts celebrating and promising violence and credible reports from the FBI and Capitol Police that the thousands gathering for the president's "Save America March" were going to be violent, armed, and working to storm the Capitol Building. The insurrectionary masses being recruited for confrontation in Washington would arrive with a resolute conviction that they would later proudly scream in the faces of Capitol officers both outside and inside the building: *We were sent here by the president!* In their minds, this invitation and incitement completely justified their sacking of Congress, our workplace.

We would show why Trump's demonstrably passive and reportedly enthusiastic reaction to the storming of the Capitol was so incriminating. If violence had not been part of his plan and he had wanted his "Stop the Steal" protest to be like Dr. King's nonviolent March on Washington, why did he not leap into action when the violence began and call out the National Guard, mobilize federal law enforcement officials to join the fight, and reassemble the paramilitary strike force he had assembled to clear Lafayette Square of BLM protesters on June 1, 2020? In fact, he did the very opposite—and everyone knew it. According to those around him at the time, Trump reportedly responded to the attack he had incited with excitement, delight, and bafflement over why those around him weren't as gleeful as he was. He was glued to the TV as if watching a reality show. He justified and reveled in the attack—"These are the things and events that happen when a sacred landslide election victory is so unceremoniously and viciously stripped away," he tweeted after five hours of pitched violence, "Remember this day forever"—and he did nothing to help the people or the government in dire danger as our commander in chief. His lethal inaction told us everything we needed to know about his culpability in not just inciting but setting the stage for the attack.

In fact, we would show that the incendiary actions he took as the violence was going on exacerbated the destruction on the ground. His tweets, read aloud by insurrectionists to the rioters, incited them to further aggressive action. He lionized the insurrectionists, confirming the righteousness of their riot and further propagating the Big Lie at its heart. As the inciter in chief, he continued to betray his Oath of Office and to provoke the mob to more violence by reassuring them that he had won the election in a landslide and that everything had been stolen from them. At the end of the day, he began the process of glorifying and mythologizing the insurrection, which is why he told his troops to "Remember this day forever"— not as an episode of human disgrace, terror, and trauma, but as a moment of historic valor and heroism, a day for his supporters to celebrate. You could hear in Trump's instant nostalgia for the insurrection distinct echoes of the romantic Lost Cause mythology that still casts a spell over a lot of people who think the Confederates had a just cause to rebel against the Union that had nothing to do with the determination to enslave people.

I also wanted to put down Trump's First Amendment defense immediately, because the right-wing media had been working overtime to define impeachment as an expression of liberal "cancel culture." Of course, it was second nature for the American people that you can't falsely shout fire in a crowded theater, but Trump's case was much worse than that, I argued—with a hat tip to my beloved Larry Tribe: Trump was the town fire chief, who is paid to put out fires, but who instead sends a mob to set the theater on fire. Then, when the fire alarms go off and the calls for help start flooding the fire department, he does nothing but sit back, encourage the mob to continue its arson, and watch on TV, with unmitigated glee and delight, as the fire spreads.

This was the central metaphor we needed to lodge in their minds, and I was psyched to see a lot of the senators, at least the Democrats, nodding appreciatively at the point. We would hear a lot of irrelevant nonsense about free speech in the coming days, but none of it could overcome the devastating logic of that organizing image if it penetrated the consciousness of the Senate and the public.

I wanted to end my opening not with factual development, legal doctrine, or logical argumentation, but with an appeal to the deep moral meaning of these appalling events. By speaking to the common sense of the Senate and the entire nation, I hoped to stir the buffeted and battered conscience of the country.

Of all the stories I had heard during my preparation, one jumped out at me. While I was poring over everything I could about the violent interaction between the police and the mob, I came across a report about a thirteen-year veteran of the Capitol Police who had battled the mob for hours and been subjected to vile, racist abuse by the insurrectionists. The story was remarkable because it allowed us to see how the violence had destroyed the line between what was professional and what was personal for the police, forcing them to risk their lives to save democratic government and also to question their most basic understanding of who we are as a people.

This was how I closed:

I will tell you a final sad story in this kaleidoscope of sadness and terror and violence. One of our Capitol officers who defended us that day was a longtime veteran of our force, a brave and honorable public servant who spent several hours fending off the mob as part of one of those besieged and broken blue lines defending the Capitol and our democracy.

For several hours straight, as the marauders punched and kicked and mauled and spat upon and hit officers with baseball bats and fire extinguishers, cursed the cops and stormed our Capitol, he defended us, and he lived every minute of his oath of office.

*And afterward, overwhelmed by emotion, he broke down in the
Rotunda and he cried for fifteen minutes and he shouted out:*

"I got called an N-word fifteen times today."

And then he reported:

*"I sat down with one of my buddies, another Black guy, and
tears just started streaming down my face. [And] I said, 'What the
[f–], man? Is this America?'"*

*That is the question before all of you in this trial: Is this America?
Can our country and our democracy ever be the same if we don't
hold accountable the person responsible for inciting the violent
attack against our country, our Capitol, and our democracy and all
of those who serve us so faithfully and honorably? Is this America?*

That haunting question, raised by the then-anonymous officer,
Harry Dunn, who turned out to be my constituent, cut through
all the rhetorical sound and fury whipped up by Trump and his
apologists.

I doubted that most people were aware of it, but Officer Dunn's
question also had a profound echoing resonance in civil rights history. When Bob Moses and the Mississippi Freedom Democratic
Party's sixty-eight delegates brought their earth-shaking challenge
to the seating of the "regular" all-white Mississippi delegation at
the 1964 Democratic Convention in Atlantic City, New Jersey, the
figure who catalyzed the conscience (while alienating the bosses) of
the Democratic Party was voting and women's rights activist Fannie
Lou Hamer, who testified before the DNC's Platform Committee on
behalf of seating her interracial delegation, which included mostly
poor farmers and working-class people.

Declaring that "I'm sick and tired of being sick and tired," the
indomitable Hamer told millions of Americans about the political
apartheid and brutal violence she and other African Americans
endured in Mississippi. After reciting the savage assaults on democracy waged in Mississippi during this struggle to win voting rights,
she began to tear up and asked the question Officer Dunn would

later ask: "Is this America, the land of the free and the home of the brave, where we have to sleep with our telephones off the hooks because our lives be threatened daily, because we want to live as decent human beings, in America?"

"Mississippi: Is This America?" became the subtitle of episode 5 of the Ken Burns documentary series *Eyes on the Prize*, which I have assigned to my Constitutional Law students for decades. It is essential for understanding both the civil rights movement of the last century and the profound struggle we find ourselves in today.

When I saw that the eloquent and noble Officer Dunn had used the exact same words, I knew I had found the opening theme for the trial. It was the question we wanted the whole country asking when they considered the "American carnage" that was delivered to the Capitol on January 6.

For the rest of Wednesday and Thursday, we unspooled the spellbinding story of how the embittered president of the United States had come to deliberately incite a violent insurrection to steal the presidential election he had already lost. I thought the presentation, broken into nineteen chapters and assigned to managers in teams or as solo performances, was enthralling every step of the way. As captain of the team, I found that the most mind-blowing part of the trial was watching the managers bring to life in words and video the minutely diagrammed story we and the staff had pieced together over more than a month of round-the-clock labor.

But I also spent a lot of my time daydreaming about Tommy, whom I saw pretty much everywhere and in everything. Every time I looked at Sen. Bernie Sanders, I thought about Tommy, because Tommy loved telling the story of how he had left work one day at the Friends Committee on National Legislation and passed the senator, who was walking alone on the street. Tommy waved and greeted him with a compliment: "Hey, Bernie, great job. Stop the war in Yemen," and Bernie gruffly said, "Uh, yeah."

Tommy's Bernie imitation was priceless. I think he had a lot of affection for Bernie Sanders and the way his famously brusque demeanor one-on-one masked a true universal commitment to the welfare of humanity. Tommy told me that "Too many radicals love humanity in the abstract but don't like people concretely, while too many conservatives like the people in their group, but don't care about anybody else and can't stand humanity generally. We could take the best from both the radicals and conservatives and show love for everybody, or we could take the worst from both and just hate everyone, and that's Donald Trump for you."

Tommy would have been engrossed in the arguments at the trial and appalled by the evidence, especially given how attuned he was to the fascism, racism, and anti-Semitism bubbling up in cesspools in America and around the world. He always advanced the hope that political democracy, which depends on the quality of public sentiment and public education, would yield a love for human rights and justice in everyone's heart and mind. He had great physical and moral courage—especially when it came to confronting bullies.

Tommy was a commanding orator for a young man, winning prizes for speech, spontaneous composition, debate, and forensics. His debate partner Isaac always teased him because he was never content to score points on particular issues and then move on, as the coaches recommended. Instead, he wanted to convince the judges, while also coming up with beautiful turns of phrase that were worth zero in the competition scoring but that successfully captivated his listeners. Tommy's goal was to offer strong moral and political propositions that would engage his audiences, which were made up mostly (and humorously) of his competitors' parents, siblings, and coaches. His teammates may have wanted to win the competitions, but Tommy wanted to win the hearts and minds of the audience before him—it was the *audience* he went to debate for, not the trophies.

So this, too, would be my goal here, I realized. I wanted us to win the *audience*, to advance moral and political propositions about constitutional democracy that would lift America out of the mud and blood of the Trump nightmare. We would pin Trump's devious acts on him like medals of shame. But this trial ultimately would not be about him, for the whole world already knew who Donald Trump was and what he was capable of. This trial would be about *us* and who *we* were. That would be how I would try to make this experience worthy of Tommy Raskin. We would uplift the country the way Tommy tried to uplift everyone he met and all the lives he touched.

Thinking back to Florida, I also thought that maybe I would find a way to recover my sense of humor at trial. It would be hard—this was a pretty austere and stuffy environment—but I knew Tommy would have found a way to be funny in such a rarified context. He loved Sacha Baron Cohen, and it was a point of pride for him that his own grandfather, my dad, had been invited once on *Da Ali G Show* and had nearly gotten punked in a segment where Ali G interviews a group of unsuspecting Washington think tank talking heads about whether people's right to vote in federal elections should begin when they can prove they have entered puberty. The real joke was that all the think tank policy analysts and intellectuals started pontificating and debating the merits of the proposal. To his eternal credit, my dad exploded into laughter and could barely formulate a single sentence. While the other talking heads continued to evaluate the proposal and expound on their sudden theories, my dad just walked away from the interview—a response that was extremely funny in its own way and, according to Tommy, marked the only time anyone had ever done that on *Da Ali G Show* (although we were never able to verify that claim).

I kept thinking about a line Tommy invoked constantly from a hilarious scene in Sacha Baron Cohen's 2012 masterpiece, the dark comedy *The Dictator*. When Baron Cohen's character, the brutal dictator of a Middle Eastern autocracy, discovers that a body double has been switched in for him and is now appearing in public, Baron Cohen begins to yell in a thick Arabic accent, "He's not the legitimate leader, he's not the legitimate leader," a personal complaint that is quickly picked up as a political refrain and chanted by hundreds of human rights protesters. Whenever Tommy saw Donald Trump on TV, he would begin to chant, "He's not the legitimate leader, he's not the legitimate leader!" I had no doubt that, were he here today, he would be saying the exact same thing to me and daring me to say it to the Senate.

We launched our case on the merits with this: Donald Trump knew from the polls in the summer and fall of 2020 that he was going to lose the presidential contest to Biden, and he set about quickly to discredit the coming election as a "fraud" and a "total scam," propagandizing his followers with waves of bizarre disinformation and conspiracy theories before even a single ballot had been cast. Representatives Castro and Swalwell demonstrated how the slow burn of this incitement began months before the election, as Trump set the table for the Big Lie, telling everyone that the only way he could lose would be if the election were stolen, an outrageous statement for a candidate in a functioning democracy. Trump knew that if he could get his followers to accept this deranged premise, there would be very little work left for him to do.

When Trump did lose the election, he aggressively and indignantly denounced the results as a sham and a fraud, asserting that he had actually won *overwhelmingly*. There was no factual evidence to back up this rubbish, but Trump unleashed a multipronged attack on the election, using means both lawful and unlawful to have his way with the beleaguered American democracy and the exhausted and reeling American people.

On the lawful side, as Representative Dean explained, Trump and his partisans brought sixty-two absurd lawsuits, completely devoid of all legal or factual merit, in federal and state courts across the land, from the lowest county trial courts to the Supreme Court of the United States. All of them failed miserably—save one tiny partial victory in Pennsylvania that had nothing to do with fraud and that did nothing to change a single vote. Comically, the only examples I could find of a handful of people voting illegally were Trump supporters. One Colorado man who had murdered his wife actually cast her ballot for Trump days after her death because he said "all these other guys are cheating" and he "knew that's what she would have wanted." And, to be clear, no one should overlook how Trump's incessant phony claims about Democratic voter fraud planted the suggestion with his supporters not only that everyone was doing it, but that it would only be justified and fair to do it *back*.

With all his frivolous cases collapsing in court, before judges who upbraided many of his lawyers for their foolishness, Trump moved into far more illegitimate and criminal territory by trying to persuade election officials and state legislatures to nullify and topple Biden's statewide popular vote victories. Pouncing on leaders of GOP-run legislatures in states where Biden had defeated him, like Pennsylvania and Wisconsin, Trump tried to convince them to cite imaginary episodes of electoral fraud to pass laws completely voiding popular election results and then replacing them with appointed Electoral College slates pledged to Trump to send to Congress.

When this did not work, Trump applied coercive pressure directly on state election officials, most prominently Georgia secretary of state Brad Raffensperger, whom Trump tried to browbeat into "finding" him 11,781 votes to deliver him the election in Georgia. Representative Dean showed that this was a brazen attempt at election fraud, pure and simple—all recorded in crystal-clear tones on a long-distance telephone call. It was also, almost certainly—and definitely, in my view—an impeachable offense in itself. The audio evidence we brought for that phone call caused an epiphanic wave of shock to ripple through the Senate as it dawned on senators across the room that this really *was* election fraud in the offing and that Donald Trump was not the *victim* of it; he was the *perpetrator*.

A leading GOP activist and Trump supporter who had donated to Trump's presidential campaign, Raffensperger lamented that "my family voted for him, donated to him, and are now being thrown under the bus by him" for not agreeing to break the law to help him steal the election. Trump repeatedly savaged the secretary of state in public. In short order, the Raffensperger family received multiple death threats, including a call to Brad's wife telling her, "Your husband deserves to face a firing squad."

As victims of constant right-wing threats ourselves, the managers were emphatic about showing how refusing Trump's demands often produced a flood of death threats. This evidence would not only demonstrate the overall criminal character of Trump's inducements to election fraud, but also evoke the darker elements of the huge movement of fanatics and malcontents Donald Trump was assembling. In hindsight, this might have been a tactical blunder with the jury, which may have been afraid already about the blizzard of threats coming their way—as my best childhood friend, Jay Spievack, called to tell me in a typically profuse voicemail message on Wednesday night, "Stop reminding them that anybody who crosses Donald Trump gets death threats. You're doing his work for him—his best argument to Republicans is vote for me and nobody gets hurt. These guys are scared enough as it is."

In any event, after Trump's scandalous harassment of state officials and legislative leaders failed to shock them into overthrowing the election, he moved farther down two darkly illegitimate tracks, including the most fateful and dangerous one.

First, he made a central bull's-eye target of Vice President Mike Pence, using every tool and lever to force him to step far outside his role of neutrally overseeing congressional counting of the Electors. The goal here, as we'd understood for months now, was to try to force Biden under 270 electoral votes by rejecting, returning, or challenging the electors of specific states. This would trigger a contingent election in the House, which Trump would win, not on the normal legislative basis of one member, one vote, but on the Twelfth Amendment basis of one state, one vote, as the Republicans had 27 state delegations in the House—26 if you discounted Liz Cheney for Wyoming, which I felt pretty confident in doing—to our 22, with 1 delegation (Pennsylvania) tied. If they had succeeded in driving the race into a contingent election, it would have been game over for us, and they would have established the patina of legitimacy they were looking for.

Several Republicans told me that, at that point, Trump would probably have been ready to follow the urgings of Michael Flynn and the "MyPillow Guy," as everyone called him, Mike Lindell, to declare martial law to cement his victory and to put down the "chaos" he himself had rained down upon us. This, however, was not hard evidence and would need to be the subject of a broader investigation when the trial was over.

We don't know everything that went into the White House strategy to strong-arm Pence, but we know that the pressure was intense and that, on January 6 itself, Trump summoned Pence to the White House and read him the riot act (if you will), telling him that he would go down in history either as a "patriot or a pussy," depending on whether he participated in Trump's scheme. In the midst of the riot and insurrection, as Vice President Pence was being whisked out of lethal danger, Trump turned up the heat, tweeting out at 2:24 p.m.: "Mike Pence didn't have the courage to do what should have been done to protect our Country and our Constitution, giving States a chance to certify a corrected set of facts, not the fraudulent or inaccurate ones which they were previously asked to certify."

The usually obsequious Pence's admirably stubborn and still-unexplained refusal to propel Trump's plan meant Trump had to mount a final effort to coerce him into submission. He would have to galvanize and incite a private army of extreme right-wing insurrectionists who would shut down electoral vote counting and either intimidate Pence into complying with his wishes or create sufficiently violent and miserable conditions to be able to knock over the whole chess board and create the exceptional space for Trump to invoke martial law. Trump needed "some street muscle for the political tussle," as one Capitol cop bitterly told me, and that is why he went out to recruit thousands of the same kind of "very fine" people he had seen marching on the fascist side of the street in 2017 in Charlottesville, Virginia.

Then Trump proceeded to *incite* them. That was a matter of public record. Stacey Plaskett showed how Trump displayed a mad, long-running penchant for political violence and had repeatedly called on his followers to attack protesters at his rallies. "Knock the crap out of him, will you?" he declaimed from the rostrum at one rally. "I promise you I will pay for the legal fees. I promise, I promise." When called upon by moderator Chris Wallace at the September 20, 2020, presidential debate to denounce violent extremists and white supremacists, Trump answered, "Which ones?" When invited to denounce "the Proud Boys," he stated grandly, "Proud Boys, stand back and stand by"—which they promptly adopted as their official group motto and emblazoned on T-shirts. Why was the president of the United States telling a group of fascist hooligans to "stand by"? Stand by for what?

Stacey showed how Trump's recurrent efforts to activate aggressive and violent impulses in his followers produced serious dangers at the street level in the presidential campaign. The Senate got to watch on the video monitors a huge fleet of Trump trucks and cars driving at high speed on a Texas highway trying to run a Biden campaign bus off the road. This scary scene from the last few weeks of the election got a big thumb's-up on Twitter from Trump, who set the action to music, added the Lone Star State flag, and tweeted the video out to millions of his followers with the caption "I LOVE TEXAS." Trump was not once but repeatedly shown to be an avid cheerleader for violence.

After he lost the election, Trump's apparatus helped pump up not one but two "Million MAGA Marches" in Washington, on November 14 and December 12, protests called to continue promoting the Big Lie and to create a battle-hardened army of proven street fighters who would take up arms in the federal city as the shock troops for Trump's emerging mass insurrectionist army. These events produced many violent injuries and dozens of arrests, as the Proud Boys roamed the boulevards of downtown DC in search of brawls and fistfights with Antifa, BLM, or anyone else who looked at them sideways. Trump's militants marched like Brownshirts down the portion of Pennsylvania Avenue recently renamed Black Lives Matter Plaza by DC mayor Muriel Bowser in response to the paramilitary government riot Trump had released on BLM protesters on June 1, 2020.

Trump issued frequent public calls at events and press conferences, along with incessant social media appeals, for his followers to come to Washington for a "wild" protest on January 6. Tens of thousands came to the District with one objective: to "stop the steal." The day's official festivities for the insurrection "tourists"—as some of my colleagues would come to describe the rioters—started off with Trump's speech right near the White House.

At trial, Madeleine Dean interspersed Trump's tweets of that day with his incendiary incitement to demonstrate how focused he was on building a massive angry crowd: "All of us here today do not want to see our election victory stolen by bold and radical left Democrats, which is what they are doing, and stolen by the fake news media. That is what they have done and what they are doing. We will never give up. We will never concede. It doesn't happen. You don't concede when there is theft involved. Our country has had enough. We will not take it anymore, and that is what this is all about. And to use a favorite term that all of you people really came up with, we will stop the steal . . . We will not let them silence your voices. We're not going to let that happen. . . . Republicans are constantly fighting like a boxer with his hands tied behind his back. It's like a boxer. And we want to be so nice, so respectful of everybody, including bad people. And we're going to have to fight much harder. . . . Because you'll never take back our country with weakness. You have to show strength and be strong . . . We fight, and we fight like hell. And if you don't fight like hell, you're not going to have a country anymore."

Our video, drawn from hundreds of hours of online footage canvassed by our researchers, showed how Trump's various calls to action and invocations to march to the Capitol were received and interpreted by thousands of his followers, who kept drowning him out with responsive screams like "Storm the Capitol" and "Take the fucking Capitol!" Trump's halfhearted and barely noticeable slipping of the word *peacefully* into his remarks was a pathetic attempt to slap a sticker of legitimacy, a laughable alibi, on the speeding train of his militant rhetoric and aggressive incitement of a crowd he had groomed for the fight. At the trial, I likened Trump to a guy robbing a bank who yells "protect private property" on the way out the door.

Zeroing in on the attack itself, Stacy Plaskett and Eric Swalwell slowed down the frenzy of those hours into frames by using a model of the Capitol that Barry's team had developed. They mapped out the movements and organized assaults of the insurrectionists to document where they did damage and how close they came to apprehending and harming Vice President Pence, Speaker Pelosi, law enforcement officers, and members of Congress assembled for the joint session. This heart-stopping presentation immersed the senators in the magnitude of the insurrectionary violence and its almost limitless possibilities for disaster. I remember when Eric showed the now-famous scene of Officer Eugene Goodman running from the mob and heroically turning around Mitt Romney, who was about to walk right into the jaws of danger. Romney turned and ran to safety behind Officer Goodman. In the background, you could see the approaching mob. The senators appeared flabbergasted, and many looked over to Romney sympathetically.

The son and brother of cops, Swalwell did a masterful job preparing the Senate to see the excruciating video of Metropolitan Police Department officer Daniel Hodges screaming in agony as he was trapped in the door of the Capitol. "I'm sorry I have to show you the next video, but in it you will see how blessed we were on that hellish day we had a peacemaker like Officer Hodges protecting our lives."

That scene offered dramatic juxtaposition to the clip Stacey Plaskett showed of the mob angrily hunting for Nancy Pelosi. "They sought out the Speaker on the floor and in her office, publicly declared their intent to harm or kill her, ransacked her office, and terrified her staff," Stacey said. "And they did it because Donald Trump sent them on this mission." She closed with a video of insurrectionists sarcastically hollering out Pelosi's name, which had reminded me, the first time I saw it, of Jack Nicholson, in *The Shining*, and his mad, singsong falsetto: "Where *are* you, Nancy? We're *looking* for you! *Nancy?* Oh, *Nancy?* Where *are* you, Nancy?"

Stacey drove this all home when she reviewed what was being said on right-wing internet sites monitored by the White House in the run-up to the insurrection. One post asserted that "the Capitol is our goal. Everything else is a distraction. Every corrupt member of Congress locked in one room and surrounded by real Americans is an opportunity that will never present itself again."

This petrifying statement struck the Senators hard, as did the pictures of QAnon fanatics rifling through their desks, where they were seated during trial.

Cicilline and Castro made clear that Trump's incitement of the insurrection was accompanied inextricably by his desertion of duty. He abandoned his duties, as commander in chief, to protect the country. He violated his Oath of Office by refusing to defend the Constitution or respond to a violent attack on Congress and the effort to overthrow our election.

This always important section of the argument had turned critical, as some of the Republicans seemed to be buying the idea that Trump's dereliction of duty was wickedly damning and wanted to know more about that element of the offense. Cicilline reviewed Sen. Ben Sasse's statements that senior White House officials with Trump on January 6 described him as "walking around the White House confused about why other people on the team weren't as excited as he was as you had rioters pushing against Capitol Police trying to get into the building." Trump was reportedly "delighted" about the attack and outright rejected urgent pleas to issue a statement calling for an end to the assault on Congress. When he finally did speak, he only made matters worse, further inciting the insurrectionists and inflaming them against Mike Pence. We showed that, even after Trump had learned about all the violence and the threats to Mike Pence's life, he continued to circulate the Big Lie, to lavish praise and love on the insurrectionists, and to castigate Pence—a pattern familiar from his vilification of Governor Gretchen Whitmer, even when her life was being threatened. As I would come to restate later in the trial, Trump's appalling dereliction and desertion of duty were interwoven with the offense of incitement to insurrection—just as they would be for an on-duty police officer who decided to stroll across the street and rob a bank. You cannot be protecting Congress while you are inciting people to overthrow it.

It was during this section that Utah senator Mike Lee interrupted the proceeding by loudly objecting, outside of regular order, to Cicilline's mention of his name. It had come up in connection with a telephone call Trump had mistakenly placed to Lee's cell number on January 6 in search of Alabama senator Tommy Tuberville, to lobby him to keep the objections to the states' electors going as long as possible. Obviously playing to Fox News and other right-wing outlets looking for any diversions from presentation of our case, Lee struck a victim's pose and threw a monkey wrench into the proceedings, demanding that his name be removed from mention. The scuffle briefly distracted the world from the narrative of Trump's crime and gave right-wing TV something to talk about. All I could think of during our momentary break was how we could get Senator Lee to shut up on this completely irrelevant point involving a minor incident that barely involved him anyway. I told Senator Schumer and Senator Dick Durbin, a wise man who was clearly rolling his eyes over the whole thing, that we were happy to withdraw the offending newspaper article, without prejudice to our ability to resubmit it.

When I asked to be recognized, I said that Mr. Cicilline had "correctly and accurately quoted a newspaper account, which the distinguished senator has taken objection to, so we are happy to withdraw it."

"Because it is not true," Senator Lee interjected, although not recognized to speak.

"On the grounds it is not true," I said, "and we are—"

"Castro repeated it, too," Lee interjected again from his seat.

"We are going to withdraw it this evening without any prejudice for the ability to resubmit it, if possible. This is much ado about nothing because it is not critical in our case."

"You are not the one being cited as a witness, sir," Senator Lee shouted out. I ignored him and looked instead at the other senator from Utah, who seemed to evince the slightest bit of a smile.

Meantime, at the end of Wednesday's presentations, Senator Tuberville admitted to the fact that he had told Trump on the phone that Pence had to be evacuated from the Senate chamber—which established that Trump was well aware of the danger Pence was in before he tweeted out yet another slashing attack on him. Ironically, after the self-dramatizing exchange Senator Lee had initiated, he produced for us phone records that were extremely helpful to fleshing out our time line; they helped establish that Trump's continuing pro-insurrection incitements were made even after he had become aware of the bloody violence being executed on his behalf.

Senator Lee's intervention was one of those truly meaningless trial skirmishes staged to distract everyone. It didn't work.

We continued to elaborate our theory of the case. Incitement is not just about what effect the inciter *intended* to produce in his audience; it is also about what effect he *did* produce and what he should reasonably have understood to be the likely consequences of his incitement. From the start, Rep. Diana DeGette wanted to collect whatever information was available about how the protesters heard and understood Trump's drumbeat of exhortation and war cries. She delivered a powerful statement, using the rioters' own words to prove that they were hell-bent on storming the Capitol; that they thought they had been invited there directly by the president of the United States, as they kept pointing out to the baffled Capitol cops; and that the violence was justified to "stop the steal" and protect the election against the mysterious thieving forces who would destroy our democracy.

As I listened to Diana's presentation, I kept thinking about what a Capitol cop from my district had told me about January 7, the gloomy day after, when people kept calling the Capitol Police and the Speaker's Office to ask whether there was a "Lost and Found," because they thought they "might have forgotten" their cell phone or their backpack in the Speaker's office the day before. They were acting as if a violent insurrection to overthrow a presidential election were some kind of rock concert or a Civil War battlefield reenactment. The astonished and quick-witted Capitol officers hurried to get each caller's name, phone number, and address so they could "return their property and deal with any other lingering issues."

But what struck me was that Trump had clearly convinced a lot of people that they had a *right* to be there that day, that they were "guests" in Congress of the president of the United States. Their sense of entitlement at being invited guests was probably reinforced by Trump's casual lie to his rallygoers that he would *be there with them* on the march and by the fact that the police were not shooting rioters (with the sole exception of Ashli Babbitt, as she tried to enter the House of Representatives chamber itself). That remarkable sense of entitlement and impunity, which millions of horrified Americans came to see immediately in racial terms, was doubtless further bolstered by the undeniable fact that thousands of trespassing rioters and hooligans were simply being *let go* by overwhelmed and injured police officers.

Although I loved watching the managers work their magic and felt good about how evenly distributed our arguments were, I did take a turn at delivering a substantive chapter myself. I had originally thought not to take on any of the building-block chapters of the factual proof, so that I could instead focus on the constitutional arguments, our openings and closings, and the Q&A session with the senators and watch the argument unfold in the event that we needed to call any audibles in later phases of trial. But the more deeply we got into the preparation, the more obsessed I became with a chapter relating to prior acts of political violence encouraged and praised by Trump, especially the storming of the Michigan State Capitol in the spring of 2020 accompanied by an extremist criminal conspiracy to kidnap and assassinate Michigan governor Gretchen Whitmer.

This evidence struck me as devastating for numerous reasons. First of all, it showed that Trump knew that his right-wing followers were willing to storm capitol buildings and were capable of doing so. Indeed, he had learned in the process which groups were for-real street-fighting brawlers and not just big talkers. Some of the specific players involved in the Lansing insurrection and familiar to him returned to play key roles in the January insurrection. Trump's network maintained his online political relationships with these elements and brought them together for the "wild" events in DC. As if that weren't enough, Trump had displayed great and public enthusiasm for these darkly absorbing events in Michigan, which helped establish his specific state of mind, his specific MO, and his specific intent in inciting the similar insurrectionary violence that overtook the U.S. Capitol. These January 6 tactics had been road-tested. Nothing that happened on January 6 marked any kind of a *break* from Trump's recent political activities and commitments. On the contrary, the insurrection was an *expression* of his basic mind-set, the *culmination* of numerous episodes of his encouragement to armed and violent forces; and the *fulfillment* of his precise determination to incite mass action to stop the imaginary steal of the 2020 election.

The story began back in March 2020, when Trump attacked Governor Whitmer over her response to the novel coronavirus pandemic. On April 17, 2020, he tweeted, "LIBERATE MICHIGAN," and less than two weeks later, on April 30, his supporters marched on the Michigan State Capitol in Lansing, storming the building. Trump's marching orders on April 17 had been followed by aggressive action on the ground.

The video we played showed how these militant protestors had come to Lansing armed, but more than that, it showed that the Trump-inspired mob in Lansing itself looked stunningly familiar: Confederate battle flags, MAGA hats, weapons, camo army gear—just like the insurrectionists who showed up and invaded the Senate chamber on January 6. Looking at the video from that April siege, one clearly saw that the event at the Michigan statehouse was essentially a state-level dress rehearsal for what was to come in the Capitol on January 6. Further proof of that symmetry could be found in Trump's response to both events. After Lansing, he refused to condemn the attacks or the lawbreakers, choosing instead to pressure Governor Whitmer to listen to the perpetrators' demands, tweeting, "The Governor of Michigan should give a little, and put out the fire. These are very good people, but they are angry. They want their lives back again, safely! See them, talk to them, make a deal."

Perhaps it wasn't surprising that Trump's emboldened supporters stormed the Michigan State Capitol again two weeks later, after Governor Whitmer refused to capitulate to the president's demand to negotiate with them. Energized by his praise, encouragement, and support, the protestors had escalated, and on May 14, Trump's mob again stormed the state capitol.

The political attacks on Governor Whitmer continued. Then, on October 8, the consequences of all these provocations were revealed to the world. Thirteen men were arrested by the FBI for plotting to storm the Michigan State Capitol building, launch a civil war, kidnap Governor Whitmer, transport her to Wisconsin, and then try to execute her. This was a kidnapping-and-assassination conspiracy.

The language the arrested men used was telling. In the charging document, the FBI reported that one of the conspirators said he needed "200 men" to storm the state capitol building and take political hostages, including the governor. The plot was well organized, just like the one that was coming on January 6. The men in Michigan even considered building Molotov cocktails to disarm police vehicles and attempted to construct their own IEDs—precisely something that the terrorists did at the DNC and RNC headquarters on January 6.

Not even these horrifying events were enough to get Trump to stop inciting violence in Michigan. Brazenly, he attacked the governor again the very night this conspiracy became public. He did not condemn the violence. He did not criticize the extremists. He didn't even check on Governor Whitmer's safety. He chose to vilify her, tweeting, "Governor Whitmer . . . has done a terrible job," and then, amazingly, he took credit for foiling the plot against her, demanding her gratitude. Less than a week later, he was back in the state and going after Governor Whitmer once more, driving a crowd in Muskegon to chant, "Lock her up! Lock her up!"

The test run in Michigan showed that many of his followers were prepared to engage in criminal violence on his behalf, with orchestrated attacks and deadly weapons, to neutralize his chosen political enemies. Trump knew precisely what he was doing in inciting the January 6 mob. He had just seen how seamlessly his words and actions inspired actual riots in Michigan. He knew what his true believers were capable of. He knew that, egged on by his tweets, his lies, and his promise of a "wild" time in Washington to perpetuate his grip on power, his most extreme followers would show up, ready to go and "fight like hell" for their hero, just as they had answered his call in Michigan.

Following the Michigan chapter, we proceeded to lay out a series of details about the larger domestic and international implications of Trump's actions, making it clear that the ramifications of January 6 were being felt far beyond our borders, as they undermined our national security, strengthened our authoritarian adversaries, and seeded doubt around the world in the strength of our alliances and durability of American democratic institutions.

At that point, finding ourselves way ahead of the game with regard to time on Thursday, we decided that Neguse, Lieu, and I would address two constitutional arguments we knew the defense would be harping on in the absence of any alibi or defense: free speech and due process. We wanted to put these two things to rest. I notice now in our remarks that, when we got up to speak, a kind of relaxation and self-confidence had set in, the slight easing of formality, which will happen at any trial or any extended event when the participants begin to let their guard down and address the substance of things with more frankness. We connected directly and effectively with the senators on these points and spoke to them less like prosecutors and more like congressional peers explaining the fallacies that might await them from the president's team.

When it was time to close, we could feel it. If they weren't convinced now, they weren't going to be. And if they were convinced now, they had all the arguments they needed to vote and to carry the message to the world. Neguse did his typically admirable and lucid job of synthesizing our principal factual and legal claims and emphasized a passage that stuck in my head, quoting Dawn Bancroft, who said, "We broke into the Capitol . . . We got inside, we did our part. We were looking for Nancy to shoot her in the friggin' brain, but we didn't find her." He had set the stage perfectly for how I wanted to close, on a discussion of the rareness and fragility of democracy and the absolute necessity of our vigilance. I spoke of Tommy's namesake, Tom Paine, and paraphrased his words, taking a few liberties with them for rhetorical effect:

Tom Paine wasn't an American, as you know, but he came over to help us in our great revolutionary struggle against the kings and queens and the tyrants. And in 1776, in The Crisis, *he wrote these beautiful words. It was a very tough time for the country. People didn't know which way things were going to go. Were we going to win, against all hope, because for most of the rest of human history, it had been the kings and the queens and the tyrants and the nobles lording it over the common people? Could political self-government work in America, was the question. And Paine wrote this pamphlet called* The Crisis, *and in it he said these beautiful words. And, with your permission, I'm going to update the language a little bit, pursuant to the suggestion of Speaker Pelosi, so as not to offend modern sensibilities. Okay. But he said:*

"These are the times that try men and women's souls. The summer soldier and the sunshine patriot will shrink at this moment from the service of their cause and their country; but everyone who stands with us now will win the love and the favor and the affection of every man and every woman for all time. Tyranny, like hell, is not easily conquered, but we have this saving consolation: The more difficult the struggle, the more glorious, in the end, will be our victory."

Thursday's trial was over. Our exhausted team met in the war room and took in the TV commentary, which had been excellent over two days of our offering overwhelming factual evidence of Trump's clear culpability in inciting the insurrection. We saw clips of pro-Trump Republicans like South Dakota's John Thune, the number-two GOP senator, saying that we House managers had done an "effective job" and were successfully "connecting the dots" between Trump's statements and actions and the bloody insurrection. We saw James Lankford from Oklahoma saying of our video clips, "It's painful to see. And I still can't believe that there were Americans that smashed their way into the Capitol."

On Friday it would be the Trump lawyers' turn to introduce their case and try to conquer this mountain of evidence of their client's guilt—and after Trump's well-publicized wrath toward Castor and Schoen for their undisciplined stroll down memory lane on Tuesday, we expected them to come out swinging like prizefighters to placate their enraged tyrant client, who accepted no weakness on his team.

We had been talking very quietly among ourselves about the pros and cons of calling various witnesses. We had to decide whether to do it at the close of the four-hour Q&A session, but that key juncture could arrive as early as Friday and would definitely come no later than Saturday. Meantime, I got a phone call from a CNN reporter that began to force the issue. She asked me for comment on a big story that was about to break involving House Republican representative Jaime Herrera Beutler of Washington State, who was prepared to tell the world about an explosive conversation she had had with House GOP leader Kevin McCarthy on the subject of January 6. According to Herrera Beutler, McCarthy told her that he had called Trump in the late afternoon, in the middle of the mob riot in the House, to beg him to call off his followers. Trump responded first by saying that the rioters were Antifa and then, when McCarthy told him that this was false and that "These are your people, Mr. President," Trump replied, "Well, maybe they just care more about a fair election than you do, Kevin." At that point, as I was hearing the story, McCarthy exploded and said to Trump, "Who the fuck do you think you're talking to?"

We had just rested our case, and although I had reserved comment on the revelations, I recognized that the story was pure dynamite. It was not necessarily a game changer that would overturn the minds of the Trump cult's true believers—what would be?—but it confirmed for mainstream America every major claim we had been making about January 6 and exploded all of Trump's evasions and digressions. It demonstrated that Republicans in Congress, like Democrats, were freaking out about the insurrection; that Trump knew exactly what he was doing in deploying violence as an instrument of political leverage to overthrow the election; that Trump, despite multiple appeals, had deliberately refused to do anything to stop the insurrection well into its third or fourth hour; and that McCarthy was angry about Trump's shocking indifference to the dangers surrounding us.

I convened a meeting of the managers to discuss the new context. We had been working back channels to get Republicans, any Republicans, to testify to what they knew about the president's conduct on January 6. This was the evidentiary holy grail of our prosecution team, but the minute it became clear that we were serious about calling them as witnesses, all the Republicans began clamming up. They knew that taking on Trump and the GOP establishment in this circle-the-wagons environment would require extraordinary strength of many kinds and might just be the foolhardiest thing they ever did.

Diana DeGette and I called several House Republicans we knew to ask if they had heard the same story from McCarthy and would be willing to testify to it. Although all of them had heard the story, they had heard it only secondhand, from Herrera Beutler or someone else who had heard it from McCarthy. None of the people who had heard the story retold from Herrera Beutler rather than from McCarthy himself would work as a witness; this would have been double hearsay and a kind of civic taboo. Although the general rules of evidence were merely "rules of reason" in the Senate trial, no one would really accept a witness saying that he or she had heard this story about what Trump had said from *another member* who had heard it from McCarthy. We needed Herrera Beutler herself, because McCarthy had confided in her the whole story and, significantly, because she had made contemporaneous notes of it, which were apparently in her possession. That chain would be quite tight enough.

Getting Herrera Butler to come testify was going to be tricky. She had apparently left her home in Washington State with her family and was seeking privacy, if not sanctuary and anonymity, in California.

The race was on to find her. Diana, who prided herself on having good GOP connections and being our team's "Republican whisperer," insisted that she was "on it" and would make contact with Herrera Beutler. She had left messages for her on her cell phone and was also trying to get in touch through staff members. I said I would try an independent avenue of approach through two GOP friends in the caucus. Barry had already been investigating whether there was anyone within Vice President Pence's circle, including Pence himself, who was ready to testify about his experience fleeing the Capitol and the president's relentless bullying and intimidation, as well as his shocking refusal to aid his own vice president. We would shake the bushes again overnight, as Barry, Aaron, Perry, and the investigators had been shaking them all along, to see whether some as-yet-undetected Republican would choose to become the trans-partisan savior of the republic. We all had high hopes and Hollywood dreams that a righteous GOP whistle-blower would emerge like a redeemer from all the scraping know-nothing sycophancy permeating the Republican Party. But I had my doubts that such a person was out there, and I did not want to build up any expectations on the team, much less in the press, that a star witness was set to emerge. I wanted to make sure we lost no momentum in our original trial strategy, which was working right now, so we would be ready to close the deal with the overwhelming evidence and arguments we had already set before the Senate and the country.

Back home late Thursday night, Sarah had prepared my favorite dinner of pasta with Beyond Sausage and garlic bread, a Tommy favorite, too, which I had loved to make for us together. Tabitha and Ryan were there, along with Sarah's brother Kenneth and Arlene, her mom. So were Noah and Heather and their kids. I wanted to know what they thought about the trial and what the TV pundits were saying. Some commentators were predicting that the Republicans would stand together and do the right thing as a way to rid themselves once and for all of this dangerous abscess on their party, an optimistic prediction I fervently hoped was right. After all, whenever I spoke to Republicans in either chamber, I told them that conviction was an imperative not just for the Constitution and America but for the GOP itself. If Trump were not convicted and disqualified, he would destroy their party. Lincoln would be the first Republican president, and Trump would be the last.

I brought my computer upstairs to work on my closing, planned for Saturday but which could perhaps even come Friday night. As I started working, I got a phone call from former Republican South Carolina congressman Trey Gowdy, whom I had been very friendly with before he left Congress in 2018 and who always cracked me up because the Democrats said he was a doppelgänger for Draco Malfoy in *Harry Potter* and I could never quite erase the image. I thought Trey was leaving Congress to go become a federal appeals court judge, but in truth, it was to make more money as a lawyer to support his growing family. Not just a right-winger infamous for the Benghazi hearings but an old-fashioned gentleman lawyer with a sense of decency, Trey wanted to extend his condolences to me because he said I had been on his mind ever since he read about Tommy. He also said he wanted to compliment me on the job I was doing leading the impeachment effort. He said he was glued to the TV with the rest of America, gripped by the enormity of these events, but that a lot of his Republican friends thought I was doing very well and wanted him to tell me that. I thanked him and said this was high praise indeed, coming from him, given that there were a lot of rumors that Trump had asked him to be the lead lawyer in his defense. Trey didn't exactly let on whether this was true, but I told him that I would be damning him with faint praise if I stated the obvious truth that he would have been doing a better job than Trump's actual team. Trey is a first-class lawyer, and as much as I was hoping for Dershowitz to get the nod from Trump, that's how much I was praying that Trey would have nothing to do with it.

The last time I had seen Trey in Washington, and we were saying good-bye, he told me the problem with me was that I was a liberal. And I told him, "You're damn right I'm a liberal, because the heart of that word is *liberty*, and if we're not fighting for liberty, what are fighting for?" And I said, "I'm a progressive, too, because the heart of that word is *progress*, and if we're not working for progress, what are we even doing in politics?

"But these days," I said, "my favorite thing to call myself is a conservative, because I want to *conserve* the land, the air, the water, the climate system, the Constitution, the Bill of Rights, Social Security, Medicare, the Affordable Care Act, the Fair Labor Standards Act, the National Labor Relations Act, the Clean Air Act, the Clean Water Act, the Civil Rights Act, and the Voting Rights Act. I don't even know why you guys want to call yourselves conservatives anymore, because you want to tear everything down. That's not conservatism. That's nihilism." I think he just chuckled.

I was probably writing for about fifteen minutes when I closed my eyes and fell asleep with my laptop on my chest. Sarah must have thought I was just taking a break or something, because she did not wake me, and then I did not wake up until about four in the morning. It was unpleasant to wake up in the middle of the night to discover that I had not brushed my teeth or washed my face, much less solved the burgeoning crisis over calling witnesses or completed an integrated new draft of my closing statement. I went and brushed my teeth and splashed my face and tried to go back to sleep for a few more hours, but I kept thinking about Tommy imitating Sacha Baron Cohen and insisting, "He's not the legitimate leader, he's not the legitimate leader!"

CHAPTER 13

SPAGHETTI ON THE WALL

Tactics is knowing what to do when there is something to do. Strategy is knowing what to do when there is nothing to do.

—SAVIELLY TARTAKOWER

On Friday morning, I woke up and resolved to live the next several hours like Tommy Raskin, who was always the perfect gentleman in any debate or argument.

I knew that listening to the Trump lawyers' coming polemical arguments would test my ability to keep a straight face and not grimace, recoil, scowl, shake my head, or laugh. It would also test my ability to refrain from madly scribbling down a thousand notes on everything that got under my skin, so I could respond in closing. I knew I would be tempted to try to answer every cheap shot, political hit, character-defaming innuendo, rhetorical evasion, red herring, and logical fallacy that came our way, and I expected dozens of such slams after finding out that Trump was extremely upset not only over the wandering quality of his team's Tuesday opening but by their gentlemanliness and occasional generosity of spirit.

But Barry, who had long ago taught me not to make three or four points where one would do, reminded me frequently to note and respond only to *relevant points of legal or political reasoning*, and even within that category, *only those that might actually move senators in one direction or another*. Barry was a master at focusing on what was relevant and important in the proceedings and not chasing a lot of silly rhetorical rabbits and ideological mirages down the alleyway. Candyce Phoenix, who was my superb staff director on the Oversight Subcommittee on Civil Rights and Civil Liberties, had the same fine lawyer's sensibility and once said to me, "Some arguments just rebut themselves."

Sarah told me in the morning that "You don't need to clean all the spaghetti off the wall. Just let them throw their stuff."

So I would make Tommy Raskin my model. When you go back and look at any of his debates or forensic presentations, you see that he was filled with enormous passion, but that he never lost his cool. Amazingly, he responded only to people's best arguments, not their self-evidently foolish or irrational ones, and he assumed that his audience could filter out the bad arguments on their own. This powerful focus on substance made Tommy's debates an educational experience, as more and more of the nonsense fell by the wayside and the dialogue moved energetically toward truth and clarity. Remembering the precociousness of his wisdom is always heartbreaking to me, because his sparkling goodness operated in a political world of such petty nastiness. I would listen for the next few hours like Tommy Raskin would have, keep him right in my heart, and simulate his optimistic and unemotional affect while in debate.

But I could tell, when Trump's counsel Michael van der Veen uttered his very first sentence, that it would be an act of will to keep my cool in the face of a deluge of Fox News–style right-wing polemics. Van der Veen became famous in the trial for pronouncing his city's name oddly when he threatened to bring any potential witnesses called by the managers to his downtown law firm in *Philly*delphia for a deposition.

But he started by labeling the proceeding a "blatantly unconstitutional act of political vengeance."

Breathe in, breathe out, I told myself.

His opening paragraph was a tissue of lies, distortions, and mischaracterizations. It also had the unmistakable sound of choreographed partisan sound bites, which was inevitable if all the morning's news reports were true: apparently, several Republican senators, including Lindsey Graham and Ted Cruz, had been directly caucusing with the legal team overnight to pepper them with advice.

But there was nothing in the paragraph I'd just heard, Tommy would have told me, that required a response to advance the public understanding. As I listened to van der Veen's angry kitchen sink argument blaming us for essentially everything and distorting our arguments, the wisdom of Tommy's meditative way became plain. Most of the rhetoric began to flow right past me as I searched only for points that actually embarked upon a serious analysis of Trump's crime or that, at least, might lure a reasonable person in with a seductive fallacy. I would move to block only the punches that might land, not all the heckles and jeers brought to stir the blood of far-right cult followers.

For the next several hours I listened, resisted the impulse to fidget, and calmly noted just three points I wanted to respond to in the course of the day:

First, van der Veen argued that on January 6, Trump "devoted nearly his entire speech to an extended discussion of how legislators should vote on the question at hand," the implication being that this somehow proved that Trump, at his rally on the Ellipse, was simply invested in "the proper civic process," as van der Veen called it. Now, this was a serious but utterly deceptive argument, one demanding a response to clarify what had actually happened on January 6.

Trump's team wished to convince us that the president's recruitment and incitement of tens of thousands of political extremists and rabid followers to come to Washington for a "wild" protest to "stop the steal" was actually an effort to make sure that "the democratic process would and should play out according to the letter of the law." On this theory, the rampant criminal violence, personal intimidation, and threats to the constitutional order that materialized were all, then, a radical *deviation* from Donald Trump's plan for the day and, therefore, someone else's fault. The idea of violence resulting from his provocations and diatribes had simply never entered Donald Trump's pretty little head.

The argument defied belief. To begin with, if all the president had been seeking to do was redress his perceived injuries according to the "proper civic process," he would have accepted the rulings of sixty-two federal and state courts across the land, including the U.S. Supreme Court, which upheld the validity of the 2020 presidential election against all his attacks. He would have told his mob that the American judicial system had interrogated and rejected every single one of their claims of massive fraud and corruption and that it was time to concede loss and move on with their lives—instead of telling them to go and "fight like hell or you won't have a country anymore."

Alternatively, if he was furious about the courts *getting the law wrong*, why didn't he unleash the wrath of the crowd against the Supreme Court rather than the Congress, which had never ruled once on the legality of the 2020 election? The answer, of course, is that Trump was not in any way on the side of the "rule of law." He was completely *opposed* to it. He wanted to ignore the towering stack of cases rejecting his fairy-tale claims and run roughshod over the actual constitutional system to prolong his rule. That was his sole objective.

Trump's lawyers were arguing that the entire purpose of Trump's pre-riot speech had been badly misinterpreted by his crowd. When he proclaimed that "you'll never take back your government with weakness," he was simply speaking in the spirit of another great champion of nonviolence, Dr. Martin Luther King Jr., to get his followers to peacefully assemble, like the marchers present to hear Dr. King's "I Have a Dream" speech in 1963. Why, then, had Trump sent his faithful to the Capitol on the exact day and at the exact time we were counting the electors, and with no plan there for speeches, music, or any organized peaceful program? The defense's astounding argument also ignored the fact that Trump had done nothing to *stop* the violence for hours after it began, even after he learned that his own vice president was in serious danger. The whole "I Have a Dream, Part II" framing of Trump's actions and speech was a profound insult to the history and ethos of nonviolent protest in America.

I needed to answer a second point of legal fallacy in the Trump lawyer's diatribe: the claim that a riot could not be both premeditated and incited. As van der Veen argued, "The fact that the attacks were apparently premeditated, as alleged by the House managers, demonstrates the ludicrousness of the incitement allegation against the president. You can't incite what was already going to happen."

But *of course* you can. This argument was plainly false.

For one thing, incitement can be part of the organizing, as it clearly was when the president urged his followers to come to DC and get "wild." For another, people who have been organized to come can be incited once they arrive. Riot organizers can go door-to-door for a month organizing people to come to a "wild" protest at city hall, and then one of them (even the mayor!) can then incite them at the front entrance by passing out Confederate battle flags and telling them to fight like hell or they won't have a city anymore and to never, ever give up the battle. There is no conflict between prosecutors pressing criminal charges for *conspiracy* to riot against certain individuals and *incitement* to riot against others. As beguiling as Trump's argument might be that a riot cannot be both organized and incited, it is plainly false as a statement of both law and of fact.

The Trump lawyers were clearly trying to shift the gaze of the Senate to the conceptually distinct question of who had *organized* the insurrection. Of course, the most likely culprit was Trump and his team, at least from everything we knew, but we simply did not have enough evidence at that point to say for sure that they were the central ringleaders of the conspiracy, and they would of course fight us every step of the way to deny us all the evidence we needed. Undoubtedly, there were dozens, and probably hundreds, of organizers of the violence of January 6, but it would take a much lengthier and more exhaustive investigation to determine exactly the background organizational hierarchy of the attack and the overlapping networks of association and funding that pulled off the siege of the Capitol. But we already knew, from everything Trump had publicly said and done, that he had *incited* the insurrection. That was a matter of public record, and that is what we were starting with.

It would be up to a future independent commission or select committee to investigate the nature of the behind-the-scenes organizing for the official January 6 rally that took place within the core Trump political entourage, among people like Roger Stone, Stephen Bannon, Michael Flynn, and other members of the Trump family; among key fundraisers like Michael Lindell, the MyPillow Guy, Linda McMahon, the former head of the Small Business Administration, and Julie Jenkins Fancelli, heiress to the Publix Super Market chain, who reportedly invested at least $300,000 in organizing the day's official activities; the official "March to Save America" rally organizers and Women for America First. Likewise, any future commission would also be investigating the major extremist groups that descended on the Capitol, like the Proud Boys, the Three Percenters, the Oath Keepers, Aryan Nations, and other private right-wing militia groups. Understanding the various plans for the insurrection was in no way necessary for us to demonstrate Trump's *incitement* of it, and we would never be able to get definitive evidence, without discovery and subpoenas, to show that Trump was also the ringleader *organizing* the insurrection. He may have been that, and a lot of evidence pointed in that direction, but there was still too much we did not know for us to draw up this charge a week after the insurrection took place.

Finally, most of Trump's argument at trial was an effort to negate the meaning of his repeated urgings to his followers on January 6 to go and "fight like hell" to "stop the steal." His argument turned on the idea that countless political leaders (including yours truly) have been caught on tape saying they would "fight like hell" for Social Security or telling other people to "fight like hell" for the Voting Rights Act. If Donald Trump can be impeached, convicted, and disqualified simply for saying "fight like hell," then so can the rest of us.

This superficially alluring argument gave rise to probably the most effective part of the Trump lawyers' presentation, a lengthy propaganda reel, custom-made for reuse on right-wing media, in which dozens of House and Senate Democrats—from Speaker Pelosi to Ayanna Pressley to Elizabeth Warren to me—were captured on video declaring that they were going to "fight like hell" for some cause or bill or community.

The fallacy of this argument, of course, was the idea that Trump had been impeached for uttering a single statement or phrase. In fact, he was impeached for engaging in a lengthy course of conduct and speech constituting incitement to insurrectionary violence and actually leading to it. Nowhere did the managers ever assert that speaking one specific phrase or sentence, or even a dozen sentences, was necessary, much less sufficient, to prove incitement to insurrection. On the contrary, we demonstrated in two days at trial an *entire course* of deliberate conduct clearly aimed at whipping a crowd into a frenzy and then convincing the mob to go fight like hell and take action to "stop the steal," which quickly became a thoroughgoing violent insurrection.

Trump's lawyers may as well have produced a tape showing that Trump had used the word *Capitol* and that a bunch of Democrats had used the word *Capitol*, too. That is not how the law conceives of incitement to insurrection. Trump's repeated use of the phrase "fight like hell" must be understood in context, as part of a larger course of deliberate conduct designed to bring about insurrectionary violence, a course of conduct that succeeded in accomplishing its intended result.

All these points we would quickly come to make in the senators' question-and-answer session, which followed immediately after the break we took when the defense rested. Four hours had been set aside for questions directed at our two teams, but only about half that time was used. I was joined by Castro and Plaskett in fielding the questions. If they were directed at one side, we would have five minutes to answer it, and two and a half minutes if they were directed to both. The questions were not spoken aloud by senators, but were written down on cards and passed forward to be read by the clerk.

Castro did a great job of blowing away the "fight like hell" argument in answer to the first question, which came from Senators Schumer and Feinstein: "Isn't it the case that the violent attack and siege on the Capitol on January 6 would not have happened if not for the conduct of President Trump?"

Representative Castro first answered the question directly by paraphrasing Liz Cheney, saying that "Donald Trump assembled the mob. He assembled the mob, and he lit the flame. Everything that followed was because of his doing, and although he could have immediately and forcibly intervened to stop the violence, he never did. In other words, this violent, bloody insurrection that occurred on January 6 would not have occurred but for President Trump." He went on to emphatically refute the claim still lingering in the air that the prosecution case was based on one speech or one phrase, although that incendiary phrase was definitely part of the process of inciting the mob.

I then got the chance to respond to the "rule of law" argument, the claim that Trump had only been trying, like a good Boy Scout, to see that the "rule of law" was adhered to. Sen. Raphael Warnock, who already seemed to be a powerful force in the Senate in his first few weeks on the job and also just about the kindest person on Capitol Hill, posed a question to our side that was read aloud: "Is it true or false that in the months leading up to January 6, dozens of courts, including state and federal courts in Georgia, rejected President Trump's campaign's efforts to overturn his loss to Joe Biden?"

I rose to recount the story of Trump's devastating losses in sixty-one straight court cases on both the facts and the law: the courts found no fraud, no corruption, and no constitutional violation. I took the time also to note that we had no problem with Trump having pursued his belief that the election was stolen or that there had been fraud, corruption, or unconstitutionality. That was his right. Despite the demonstrably frivolous and vexatious nature of the litigation, the courts were in fact the "proper" arena for hearing Trump's (empty) legal grievances about the election. And all the courts across multiple states and jurisdictions had found no merit to his claims. That was the verdict of the American judicial system. Trump had gotten his "proper civic process" and his due process, and he had lost across the board, fair and square.

Perhaps the most important point I wanted to make was that when Trump crossed over from nonviolent and lawful means in pursuit of his absurd claims to *inciting violence against Congress and the election and Vice President Pence to get him to violate the Constitution*, that was when he entered the realm of insurrection. Meanwhile, the propaganda reels showing that dozens of Democrats had used the word "fight" before only underscored that Trump was impeached not for using a single word but for a long and continuing course of conduct inciting a violent insurrection—that is what the evidence in the trial was about. The propaganda reels showing Democratic politicians speaking on other topics in other contexts would have been excluded in any court in the land and should have been disregarded by the senators as laughably irrelevant to the case.

The questions and answers went on for more than two hours, and I felt that, at every turn, we got the best of them. Senator Warren pointed out that the defense was highlighting the fact that "Democratic members of Congress raised objections to the counting of electoral votes in past joint sessions of Congress" and asked whether "any of those Democratic objections [were] raised after insurrectionists stormed the Capitol in order to prevent the counting of electoral votes and after the president's personal lawyer asked senators to make these objections specifically to delay the certification?"

This trenchant question from Senator Warren gave me the opportunity to point out that there has been a "proud bipartisan tradition" of making technical objections under the "arcane" rules of the Electoral College, but that *none of them before had been accompanied by mob violence designed to shut the whole process down.* I pointed out that Trump's lawyers had had "some fun" at my expense by pointing out that I had objected in 2016 to Florida's electors because some of them had violated the rule against "dual officeholding," but I had never asked a single senator to join me in pointing out this clear technical violation, much less to organize or incite mob violence against Vice President Biden, who was presiding and who properly ruled the objection out of order in a matter of fifteen seconds. None of this, I observed, should be confused with "mobilizing a mob insurrection against the government that got five people killed, one hundred and forty Capitol officers wounded, and threatened the actual peaceful succession of power and transfer of power in America. If you want to talk about reforming the Electoral College, we can talk about reforming the Electoral College. You don't do it by violence."

This answer prompted van der Veen to turn on me: "You need to stop. There was nothing funny here, Mr. Raskin. We aren't having fun here. This is about the most miserable experience I have had down here in Washington, DC. There is nothing fun about it. And in Philadelphia, where I come from, when you get caught doctoring the evidence, your case is over, and that is what happened. They got caught doctoring the evidence, and this case should be over."

Wait a second, we were talking about people getting killed, police officers being smashed in the face, and van der Veen was complaining about participating in a Senate trial as being his "most miserable experience" ever in Washington? I took note of his appalling and self-pitying complaint (which may have inadvertently revealed how he was being mistreated by his own apparently incensed client), and the sharp-eyed Eric Swalwell sent me a note reminding me to deal with this irksome comment in a later answer.

When I rose to respond in two and a half minutes, I took a few seconds to deal with the "worst day in Washington" annoyance.

"Thank you, Mr. President. Counsel said before that this has been 'my worst experience in Washington.' For that, I guess we are sorry, but, man, you should have been here on January sixth." I saw a lot of heads in the room nodding.

When the clerk read Louisiana senator Bill Cassidy's question, I straightened up, because the question indicated that Cassidy had been paying close attention to the time line of Trump's call to Tommy Tuberville and his subsequent tweet about Mike Pence—which suggested that Senator Cassidy's vote might really be in play. His question on the precise sequence and timing of events indicated that he wanted to know how long Trump continued to goad on the insurrectionists after he knew that his vice president was trying to escape from their clutches.

Van der Veen was combative and evasive, hiding behind the rhetorical smokescreen Mike Lee had created around this point while arguing vaguely that there was "a problem with the facts in the question, because I have no idea, and nobody from the House has given us any opportunity to have any idea."

Essentially, his response blamed the House for not doing enough research to determine Trump's movements during the attack, which was rich given that Trump was refusing to testify about his actions and obviously had in his possession and memory all the information the senator wanted to see. That strategic silence had been Trump's decision, not ours, and the Senate could draw a strong negative conclusion about Trump's refusal to testify. This was Michael Anderson's point about the "adverse inference."

I reminded the senators that Trump had been invited to testify to fill in all the gaps of his movements and decisions on January 6, an invitation that his attorneys had quickly and contemptuously rejected. But this was a civil proceeding to protect the people of the United States, not a criminal prosecution that would land anyone in jail. I invoked Justice Scalia, who championed the Supreme Court's position allowing robust "adverse inferences" in a case like this. If an innocent person had an incentive to testify in a civil proceeding to clear up some crucial matter of fact that could lead jurors to draw a negative conclusion about his actions, but that person instead decided to skip out on the whole thing, the jury could hold that refusal to appear against him and interpret every disputed fact against his defense. Scalia likened the situation to parents coming home and asking their teenager, who has fallen asleep in the living room, with his favorite food and drinks spilled out all over the furniture along with a bunch of empty beer cans, what happened, and the teenager's refusing to speak. It would be ridiculous *not* to draw an adverse inference from the silence.

There was an easy way to settle all the questions around what Trump was thinking on that day and what happened in the minutes between when he spoke to Senator Tuberville and when he sent the tweet about Mike Pence. Rather than screaming at us about how "we didn't have time" to get all the facts about what Trump had done, Trump's lawyers should simply "have their client come and testify under oath" about why he was sending out tweets denouncing the vice president while the vice president was being hunted down by a mob that wanted to hang him, presumably on the gallows they had erected outside the Capitol. The facts in this instance lived with one person and one person alone. Speak up, Mr. Trump.

After the question-and-answer portion finally came to a close, we adjourned for the day, and our team gathered back in the war room to lick our wounds, savor our wins, and brainstorm again about the problem of witnesses. Nothing had changed. Essentially, no new doors had been opened. The only member we knew who had had a direct conversation with Kevin McCarthy about his words with

Trump was Jaime Herrera Beutler, and she was completely incommunicado. Diana could not get through to her in California, despite repeated overtures.

I had reached out in the morning to John Katko, the first House Republican to have announced that he was going to vote for Trump's impeachment, and he said he could not testify on that McCarthy issue or on anything comparable to it. He just didn't know anything firsthand along the lines of the Herrera Beutler story. Katko is a brilliant lawyer, one who, I know from personal experience, does not elevate his party loyalty over his integrity and honesty. He later fought effectively to win an agreement for a bipartisan 9/11-style independent commission on the January 6 insurrection, a commission that would have comprised five Republican members and five Democrats, and equal subpoena power—a resolution the GOP had originally sought, but that its leadership then disdained and rejected both in the House and the Senate, where the proposal collapsed. The truth was that the GOP hierarchy simply wanted no independent investigation at all. A real investigation could implicate a huge sector of the GOP power elite and, of course, Donald Trump was trying to put a hex on it.

But Katko did not seem to be in any more of a hurry to testify about Trump than anyone else. Barry reported being similarly stymied by Pence and his staff. No one was forthcoming on the GOP side. The wagons were circling around Trump everywhere except for Herrera Beutler, who was now disturbingly MIA. I reflected on this situation as I made my way downstairs to meet the Capitol Police detail in the pounding, freezing rain. On my way out, I saw, on a TV monitor downstairs in the Capitol, Jake Tapper on CNN saying of van der Veen and the others, "I have never seen a set of lawyers so outmatched as the Trump defense attorneys." But I was starting to feel like all these "gang of lawyers who couldn't shoot straight" stories were providing a useful diversion for Donald Trump from his own blatant culpability and might even be creating sympathy for him on the Republican side. Trump would love nothing more than to say that he was vulnerable in the trial only because his Philly-delphia crew was letting him down. He was expert at swinging back and forth between being a slashing bully and a pitiful, hapless victim of other people's incompetence.

Back home Friday night, I had an amazing conversation with Hannah about the case. She was struck by the evidence admitted about a pro-Trump protester who told his kids before leaving for Washington that he might never see them again. I, too, had been struck by that story, because I thought it confirmed how pervasive the premeditation and planning were for violence at the president's "Save America March" at the Ellipse. But Hannah was wondering how it made the man's *kids* feel, and she was outraged that Trump could lure people into this cauldron of violent bloodshed and then skip away from it all, refusing to testify and disclaiming any respon-sibility. What if that father has to go off to prison and those kids don't get to see him for years? What if he had been killed? Hannah was emphatic that this outcome was wrong. How could Trump be allowed to evade responsibility when all the people he recruited would be prosecuted and facing prison? "He started the whole thing," she said.

Her questions brought me up short. I had been listening to that story myself, but in the mode of a prosecutor eager to make the case for the widespread planning and coordination of violence on that day. Yet my darling Hannah Grace had seen right through the political and legal dimensions of the situation to the humanity and morality of the situation. This was precisely the broader moral context the Senate needed to keep in mind as the day of decision approached. Hannah had given me my closer. What happened on January 6 was unfair to the children of *anyone* present. It was unfair to all the children of America who would inherit the chaos unleashed on that day.

CHAPTER 14

WITNESSES TO INSURRECTION

We shall not cease from exploration, and the end of all our exploring
will be to arrive where we started and know the place for the first time.

—FROM T. S. ELIOT, "LITTLE GIDDING," *FOUR QUARTETS*

On Saturday, the long-simmering debate over witnesses came to a boil. The progressive blogosphere, which Eric Swalwell stayed in close touch with, wanted to know why we had waited so long to make a call about witnesses. It was all the reporters wanted to talk about as they shouted questions at us as we entered the Capitol in the early morning.

But the Senate had adopted a clear rule in its impeachment resolution stating that a motion to call witnesses could come only *after* each side had presented its case and *after* the Q&A period. The timing was set not by us but by the Senate's own rules. I knew progressive America thought we had demolished Trump's position in the first four days of the trial and believed we could put it all away and deliver the whole Senate with one or two breakthrough witnesses, and nobody was more hopeful than I that this was true.

Now was the time to decide.

When I got to the Senate on Saturday morning, I was bombarded by managers, friends, members, and constituents telling me, "You gotta call witnesses. Everybody wants witnesses."

PBS's *Firing Line* had aired on Friday Margaret Hoover's exclusive interview with Republican representative Jaime Herrera Beutler, who offered electrifying revelations about Kevin McCarthy's testy conversation with Trump in the middle of the chaos on January 6.

Herrera Beutler's interview established that, as the Capitol riot was ongoing, the GOP leadership was panicked by the violence; that Donald Trump had already started to lie and equivocate about who was behind the riot, a fact profoundly suggestive of his guilt; that McCarthy had completely rejected Trump's efforts to blame the insurrection on Antifa; that Trump was supportive of the violent actions of the insurrectionists, who were only defending an "honest" election; and that Trump was palpably reluctant to do anything to call off the dogs, even when the violence and destruction were rampant and spreading.

Not only was Jaime Herrera Butler a Republican who had voted overwhelmingly with Trump on the House floor and, thus, obviously had no political axe to grind against him, but we had reason to believe that she had within her physical possession the holy grail for a trial like this: contemporaneous notes she had taken from her conversation with McCarthy.

We had wanted Herrera Beutler as a witness from the start. Although I did not know her at that point (people sometimes forget that there are 435 Members of the House, plus five Delegates, and most people simply don't know everyone else) and although it is possible she could have been flipped during her incommunicado period, I was told by several people, including Diana DeGette and Liz Cheney, that Beutler was trustworthy and would be a lucid and honest witness with a common touch. I was totally sold. Although we had been unable to connect with her over the last week, she could not ignore a subpoena. So our plan was to get a vote for witnesses and then use our new subpoena power to bring her in. If there were any truly undecided senators left, her testimony should push them over to the "guilty" camp.

But this development immediately raised the question of other potential witnesses. I knew that Democrats were gung-ho for the idea of getting witnesses to refute all of Trump's lies point-blank. Americans raised on *Perry Mason* or Hollywood culture carried in their heads a mythic archetype of a breakthrough whistle-blower witness who arrives at the last minute of a trial to dramatically turn the tide and save the day. After several weeks of delving into the field of potential witnesses with Barry, Aaron, Perry, and my minuscule cadre of Republican friends, I had come to the conclusion that no matter how much we shook the bushes and prayed, alas, there really were no witness-stand messiahs out there who had spent the day with Donald Trump and who were now ready to deliver proof of his spectacular criminality and cruelty—just ordinary mortals caught in the normal, depressing vortex of party politics in Trump Land.

To put the point a little more crudely, the potential witnesses in the real world whom people were crying for all offered us as many risks and dangers at trial as potential benefits and breakthroughs. The most commonly demanded potential witnesses were also the most fraught. A lot of people were writing me to urge us to call Pence or McCarthy, for example, on the grounds that either of them could get up and simply put the whole thing away by telling the truth and bringing the entire house of cards crashing down on Trump. I had no doubt that this was true.

But it was equally true that Pence or McCarthy could get up and feign ignorance of any wrongdoing by Trump; explain the whole thing as a big, unfortunate misunderstanding; blame it on a few extremist bad apples and the poor communications inherent in a chaotic day; and then overnight become a hero in the race to be Donald Trump's heir apparent. A witness like McCarthy, who like everyone else would have good reason to believe that Trump was behind the attack and had been agitating all day long for more violence right through the riots, could be converted overnight into the MAGA Man of the Year by keeping the whole GOP ship united and sailing in the same direction, as we managers sank back into our corners deflated, dejected, and demoralized. Who knew what kinds of political deal making were already taking place within the tightly disciplined world of GOP power brokers? As one former Republican senator told me over the phone, "All McCarthy cares about is keeping Trump on his side if the GOP wins the majority back in 2022. He doesn't want Trump to throw him overboard for Jim Jordan."

In other words, what made anyone think that Pence, McCarthy, or any other dark-horse lifelong Republican witness drawn involuntarily from their staff could be remotely trusted with the shattering power of a "star witness" role? If these people were willing to lie about climate change and COVID-19, threats to the safety of the American people and the people of the world, why would they not lie about what had happened on January 6, an event they considered an existential threat to the Republican Party?

It cracked me up a little, as I listened to cell phone messages from the most progressive people I knew (climate activists, union radicals, MSNBC devotees, and others normally skeptical about the ethics of the GOP power structure), to discover that they were sudden converts to the idea that we just "had to" call Pence's chief of staff, or Mrs. Pence, or Donald Trump's personal valet, because they had a "really good feeling" about what one or the other might say on the stand. Talk about magical thinking! If there were closet truth-tellers in GOP Land just waiting for an opportunity to spill the beans on Donald Trump and his fascistic decision making on January 6, why had they not come forward over the last month? Surely, they would have spoken out in the media or at least tried to get in touch with us—I had heard from any number of Capitol officers about their anguish and their anger. In Washington, you are always about 2.5 people away from finding someone's cell phone number. It would not have been tough at all for any secret GOP whistle-blowers to reach us if they felt they had something important they wanted to tell us.

I was most drawn to those who were telling me that we needed witnesses like the Capitol cops and their spouses and children, or renegade MAGA supporters now regretting and repenting their involvement (not that we actually had any of those stepping forward at the time). We did have some amazing police officers I would have loved to put on the stand, but they would only be testifying essentially to their personal experience of the violence we had all just seen for several days. They would strike an extraordinarily resonant chord across America—at least in the America I grew up in—but they would have no more evidence on Donald Trump and his evil deeds that day than what we had already introduced. (I did resolve in my mind to find another opportunity for them to speak, and when we later came to create the Select Committee on January 6, I strongly advocated to our Chair Bennie Thompson that we begin with a hearing of Capitol and Metropolitan Police Department officers, which we did.)

The other major problem I was having was that the "witnesses now" chorus was ignoring the vast and likely prohibitive *price* for calling various mystery witnesses. In theory, we had some kind of slender working majority of senators who would vote to allow us to call witnesses. But it would have been almost impossible to assemble a majority for procuring witnesses for just *our* side without giving the defense team the power to call an equal number of witnesses whom they wanted for *their* side. Some Democrats had already said publicly that they would only support giving *both sides* witnesses, and I knew in my gut that every Republican, including those who had already voted with us on jurisdiction and provided our best hope for winning a conviction, would take the position that they would vote to grant only equal numbers of witnesses to both the House and to Trump. No one I spoke to thought we could get any other kind of arrangement.

So what would that look like? Republicans were already saying that they needed to call Nancy Pelosi, to cross-examine her about this or that intelligence or security failure. Others were saying that if we called McCarthy, their first witness would be Vice President

Kamala Harris, because of this or that issue I could hardly even understand. Bottom line: They were promising a first-class GOP Ringling Bros. and Barnum & Bailey Circus sideshow, complete with the creation of a whole new set of diversionary propaganda talking points that would make Hillary's emails look like small potatoes. It feels quaint even to articulate it, but with the introduction of witnesses, the GOP would be promoting not the quest for truth and justice, but mass entertainment and extreme political polarization.

All these risks and dangers would be undertaken at a time when, according to all the political and legal commentators, we had already blown the Trump team out of the water over the course of the week and when no fair-minded person could any longer raise any doubts about Donald Trump's guilt. Everyone imagined that star witnesses would be a "game changer." But we had pretty much already won the game, and lying witnesses could just as easily change the game *in the other direction*. We had already effectively captured public opinion in a fast-moving and compressed process. All we could do now was squander that momentum and coherence over a period of weeks (even months) while people began whining about how we were distracting everyone from beating COVID-19, rebuilding infrastructure, and doing everything Donald Trump had failed to do for the last four years but that the GOP now blamed on Joe Biden.

I worked through these thoughts and asked Barry privately for his opinion. We converged at exactly the same place: we should go forward to ask for a single witness, Rep. Jaime Herrera Butler, whose deposition could be taken as early as that morning or that afternoon over Zoom. But, of course, we had to *find* her first, I said, looking at poor Diana.

We brought Diana, who had been stalwart in trying to chase down Herrera Beutler, into our huddle along with all the managers and the staff, so I could tell them where I was at that point. They all agreed with our analysis. Eric told me that the progressive blogosphere was still blowing up with demands that we call witnesses and issue a statement explaining where we were. I told him we needed to hang tight on that, and they would find out soon enough. We had to play our cards very close to the vest. But I needed to call someone before making my final decision: Speaker Nancy Pelosi.

The Speaker and I had spoken every two or three days in the run-up to the trial, but mostly because she was checking in on how Sarah and I and the girls were doing. She was remarkably hands-off in terms of management of the case—I understood it had been the same with Adam Schiff and Impeachment 1.0—and I consulted her only at a few critical junctures, where I felt her surpassing political judgment was needed. She had shown great trust and confidence in me at a time of extraordinary personal and political tumult, which is something I grew to appreciate immensely. Sometimes when people asked me how I had the strength to say yes to Nancy Pelosi after what my family had just experienced, I said that it was my family and Nancy Pelosi who gave me the strength and power to do it.

Pelosi grasped the dynamics of the situation the way she always did, at the speed of democracy. When I explained where I was on witnesses—deeply skeptical of opening things up generally, despite the clamor from our friends, but determined to get Herrera Beutler's testimony in as quickly as possible, once we knew what she had to say—Pelosi agreed immediately and asked why we would even think about reaching for more witnesses.

I told her that there was apparently a sense out there that Mc-Carthy might be a good witness for us.

"McCarthy?!" she said. "I wouldn't trust him any more than I would trust the guy with ram's horns," she said, referring to the infamous "QAnon Shaman" who had taken part in the Capitol riot. She always made me laugh. "McCarthy," she said again. "Why Mc-Carthy?"

"Madam Speaker, you're reconfirming my instincts. These guys will be back at Mar-a-Lago asking for money and endorsements in a matter of days if they're not already there," I said. "I don't know why McCarthy. They think he might tell the truth because he said to Trump, 'Who the eff do you think you're talking to?' They were impressed by his profanity."

"They don't understand politics *or* profanity, then," she answered, speaking a mile a minute now. "When McCarthy was saying, 'Who the eff do you think you're talking to?' he wasn't acting like a tough guy, like a mobster. He was basically saying, 'Mr. President, no one has been carrying your water in this election business more than me, who the eff do you think you're talking to? No one has shown their devotion to your election more than me, and now all of our lives are in danger because of this mob you sent over, which is out of control, and how dare you say I haven't been loyal enough? Who the eff do you think you're talking to? I run your whole operation here.'"

"I got you," I said, everything having been clarified by the Speaker in her inimitably insightful way. "He was saying, 'How dare you give someone else the award for Sycophant of the Year when that's my job, Mr. President'?"

"Yeah, but these other people were willing to *kill* for him," she said.

"It's a race to the bottom. McCarthy is not our guy," I said. "After all, it may take a racist fanatic to kill for Trump but any old Trump sycophant will *lie* for him."

"Right. He's not your witness," the Speaker said. "Stick with Herrera Beutler."

We went to the Senate floor with that strategy. We would call one witness, Congresswoman Jaime Herrera Beutler of Washington State, and I introduced my motion to subpoena witnesses with Sen. Patrick Leahy at the desk.

In my speech, I referenced the statement Herrera Beutler had made on television the night before, which confirmed that, right in the middle of the insurrection, when House Minority Leader Kevin McCarthy called the president begging for his help, President Trump had responded, "Well, Kevin, I guess these people are more upset about the election than you are." As I explained, we felt that this was a critical piece of new corroborating evidence that confirmed the House Article of Impeachment in many particulars, showing Trump's horrifying dereliction of duty, guiltier-than-sin state of mind, and continuing incitement to insurrection even after it was clear the hordes had overrun the Capitol.

"For that reason," I said, "and because this is the proper time to do so under the resolution that the Senate adopted to set the rules for the trial, we would like the opportunity to subpoena Congresswoman Herrera Beutler regarding her communications with House Minority Leader Kevin McCarthy and to subpoena her contemporaneous notes that she made regarding what President Trump told Kevin McCarthy in the middle of the insurrection. We would be prepared to proceed by Zoom deposition of an hour or less, just as soon as Congresswoman Herrera Beutler is available."

I went on to underscore something that Congressman Herrera Beutler had said: that she hoped her willingness to speak out would lead other material witnesses to come forward. If that happened, I made it clear that we would also use the opportunity to take additional depositions.

Needless to say, Trump's lawyers were displeased and struck a tone of outrage, despite the simplicity of our motion. Van der Veen objected fiercely, saying, "If they want to have witnesses, I'm going to need at least over one hundred depositions. Not just one." He also rejected the idea that his own parade of "at least over one hundred" witnesses would be allowed to testify via Zoom. "These depositions should be done in person in my office in *Philly*delphia," he said, reviving his special pronunciation of the city's name. "That's where they should be done."

The odd ferocity and sweet provincialism of his response provoked huge laughter in the chamber, even among some Republicans, which then caused van der Veen to upbraid the senators in an angry tone: "I don't know how many civil lawyers are here, but that's the way it works, folks. When you want somebody's deposition, you send a notice of deposition, and they appear at the place where the notice says. That's civil process."

With more laughter in the air, he said, "I don't know why you're laughing."

At that point, things in the Senate chamber could really have headed south, but the presiding judge, Senator Leahy, who had by now endeared himself to all of us with his awesome decency and old-fashioned New England good manners, intervened both to rescue Senate decorum and bail out the floundering defense counsel: "I would remind everybody we would have order in the chamber during these proceedings," Leahy said.

When van der Veen resumed his place at the lectern, I was hoping he would just take a moment to breathe and then return to his argument. But he kept the beef going: "I haven't laughed at any of you, and there's nothing laughable here."

AWKward, as my kids would have said.

In any event, van der Veen insinuated that our request to call a single witness for less than one hour reflected poorly on the quality of the House managers' investigation, and he remarked that we "need to live with the case [we] brought."

I answered that this moment had already been set as the right time and place for calling witnesses, and that Rep. Herrera Beutler's statement had just come to our attention and was urgently relevant to bolstering our claim that, after the breach and invasion took place, President Trump did not work to defend the Capitol in any way but, rather, pursued his political campaign to overthrow Biden's Electoral College majority (and then force the contest into a contingent election in the House). This was a conversation van der Veen obviously did not want to have.

The major substantive point van der Veen made that I felt compelled to respond to was that any testimony about what Trump did *after* the violence came to the Capitol was irrelevant to determining whether he had incited the violence in the first place. This was a clever but deeply flawed argument, one I needed to debunk for all the Senate.

Here is what van der Veen said: "It doesn't matter what happened after the insurgence into the Capitol Building, because that doesn't have to do with incitement. Incitement is—it is a point in time, folks. It is a point in time when the words are spoken, and the words say, implicitly say, explicitly say, 'Commit acts of violence or lawlessness.' And we don't have that here."

But he had missed two key points of relevance that made his argument nonsense.

In the first place, the statements and tweets Trump made after the storming of the Capitol Building would clearly be relevant if they constituted a continuation of the criminal conduct itself. Trump was charged with incitement to "insurrection," and nowhere was it written that the insurrection was over and complete once the physical breach and trespass of the Capitol had taken place. Trump continued to promote and facilitate the insurrectionists' violence and lawlessness long after the building had been stormed, and the insurrectionists were armed with cell phones which enabled them to read the president's encouraging tweets while they were rioting. The insurrection lasted for several hours before it was put down, not several minutes.

Second, evidence of the president's comments about insurrectionary violence after the beginning of the siege were clearly relevant to exposing the purpose of his having invited a "wild" protest in the first place and the intent of his statements urging his followers to "fight like hell" to "stop the steal" or "you won't have a country anymore." If the president had reacted with horror and shock to the violence overtaking the Capitol, that reaction would support the claim that he had had no intention to incite violence and that his nonviolent purposes had been misunderstood and betrayed by his followers. However, if he reacted with delight and redoubled support for his followers even after they had engaged in widespread violence, that evidence would clearly help establish his specific intent to incite insurrectionary violence. I would return to this theme in my closing.

The motion to grant us the Zoom witness testimony of Representative Herrera Beutler passed 55–45, with Lindsey Graham, one of Trump's most ardent cable news defenders, voting with the majority—which caused quite a stir of mystery on the Democratic side and consternation among the GOP. Was Lindsey Graham, of all people, now in motion? Were the partisan lines becoming unstuck?

While speculation ran wild, the practical meaning of our successful motion, like everything else in the Senate, now had to be worked out in negotiation between the two sides. Barry Berke and his team went to work. The defense team was prepared to push a witness motion of its own—not for "at least one hundred witnesses" but for a single witness of its own against Rep. Herrera Beutler. This was reasonable, and fine with me, but only if we could cross-examine their witness in deposition (and they could likewise cross-examine Representative Herrera Beutler).

There was about an hour of kibbitzing and caucusing to reach a compromise, and we received a surprise visit to the war room from Delaware senator Chris Coons, whom I had met long ago, before either of us had been elected to Congress, through the Rodel Fellows program. (He was a county executive in Delaware then, and I was a state senator, and I took an immediate liking to him.) Huddled together with the managers, Coons said he spoke "only for myself" and not for President Biden—a remark that kicked off days of discussion among the managers about whether that actually meant he spoke just for himself but wanted us to *think* he was speaking for President Biden, or that he was *really* speaking for President Biden but wanted us to *think* he was speaking just for himself. In any case, he wanted to remind us to be reasonable in our negotiations with the defense team, because the next day was Valentine's Day, and all the senators had immediate travel plans. I told him we would definitely keep that in mind and that we had no interest in unnecessarily prolonging things.

We soon settled on an excellent, rather thrilling compromise: a stipulated admission to the record of Herrera Beutler's sworn and uncontradicted statement about Kevin McCarthy's account of his interactions with President Trump on January 6.

We managers considered the admission of Herrera Beutler's statement into the record a huge win for our side. We had focused the attention of the nation on the fact that, in the middle of the insurrection, House Minority Leader McCarthy had implored the president to call off the wolves, but that Trump deliberately kept the heat turned up high to advance his political objectives. In doing this, he had completely abandoned his duties and abdicated his role as commander in chief. The whole world could see the plain truth of what had happened: that Donald Trump had a plan to stay in the White House and, as Joaquin Castro put it so perfectly at one point, he "left us for dead."

Any other calls for witnesses I considered unnecessary and risky. We simply were not going to be able to end on a higher note than this, or with a vaster distance between the weight of the case made by the House and the weight of the case offered by the defense than we would at this very moment.

It was time to close, and I started by underscoring the importance of the newly admitted and uncontradicted Representative Herrera Beutler statement, which I used to refute van der Veen's breathtaking claim that a defendant's conduct *after* the alleged commission of a crime was irrelevant and inadmissible.

I wanted them all to see, and to figure out for themselves, why this made no sense:

"Say you light a fire, and you are charged with arson," I proposed. "And the defense counsel says, 'Everything I did after the fire started is irrelevant.' The court would reject that immediately and say, 'That is not true at all. It is extremely relevant to whether or not you committed the crime.' If you run over and try to put out the flames, if you get lots of water and say, 'Help, help, there is a fire,' and you call for help, a court will infer that—or it could infer that—

you didn't intend for the fire to be lit in the first place. They would accept your defense, perhaps, that it was all an accident. Accidents happen with fire.

"But if, on the other hand, when the fire erupts, you go and you pour more fuel on it, you stand by and you watch it, gleefully, any reasonable person will infer that you not only intended the fire to start but that once it got started and began to spread, you intended to continue to keep the fire going. And that is exactly where we are, my friends.

"Of course, your conduct, while a crime is ongoing, is relevant to your culpability, both to the continuation of the offense but also directly relevant, directly illuminating, to what your purpose was originally: What was your intent? And any court in the land would laugh at anyone—would laugh out of court any criminal defendant who said, 'What I did after I allegedly killed that person is irrelevant to whether or not I intended to kill them.' I mean, come on. Donald Trump's refusal not only to send help but also his decision to continue to further incite the insurgents against his own vice president—*his own vice president*—provides further decisive evidence of both his *intent to start* this violent insurrection and his *continued incitement once the attack had begun* to override the Capitol."

I wanted to then reinforce in their minds that Trump's lack of official response to the violence told us everything we needed to know about whether he had incited the violent insurrection purposefully or had just accidentally said some things that were badly misinterpreted by his followers. I read the Senate the words of Republican representative Anthony Gonzalez of Ohio, a former pro football player, who said on January 6, "We are imploring the president to help, to stand up, to help defend the U.S. Capitol and the United States Congress, which was under attack. We are begging, essentially, and he was nowhere to be found."

Trump was indeed "nowhere to be found," a ringing and authentic statement on his abandonment of his official duties. Trump had traded the position of commander in chief for the position of inciter in chief, which meant that his dereliction or desertion of duty was intertwined with his advocacy of violent insurrection. This was an important point to make, because the press was suggesting that some GOP senators, especially former military, felt we had proven Trump's dereliction of office yet had not charged it. Thus, I forcefully argued that "dereliction of duty" and "desertion of duty" were "central to his incitement of insurrection and inextricable from it." It was like a police officer who walks into a bank and robs it. He would be guilty of bank robbery, and his dereliction and desertion of duty would be essentially folded into the offense itself. You can't defend a bank against robbery if you're the one robbing the bank.

After an interlude in which Cicilline vividly refreshed the factual proof for a final round, I returned to the lectern. I was moved to turn away from Trump for a moment—I always felt that both speaking of him (having to use his name) and listening to people talk about him were overwhelmingly draining exercises, best accomplished intermittently and in tiny doses.

My precious daughter Hannah Grace was in my heart and much on my mind. I could not stop thinking about what she had said the night before, or of her anger over Trump's having lured in thousands of kids' parents to participate in that bloody nightmare and then waltzing away scot-free as they faced months or years in prison for their crimes.

And I said: "Mr. President, senators, my daughter Hannah said something to me last night that stopped me cold and brought me up short. The kids have been very moved by all the victims of the violence, the officers and their families, but Hannah told me last night she felt really sorry for the kids of the man who said goodbye to his children before he left home to come and join Trump's actions. Their father had told them that their dad might not be coming home again, and they might never see him again. In other words, he was expecting violence, and he might die, as insurrectionists did.

"And that shook me. Hannah said, 'How can the president put children and people's families in that situation and then just run away from the whole thing?' That shook me, and I was filled with self-reproach because, when I first saw the line about 'your father going to Washington and you might not see him again,' I just thought about it, well, like a prosecutor, like a manager. I thought, 'What damning evidence that people were expecting lethal violence at a protest called by the president of the United States in saying their final good-byes to their kids.' But Hannah—my dear Hannah—thought of it like a human being. She thought of it—if you will forgive me—like a patriot, someone who just lost her brother and doesn't want to see any other kids in America go through that kind of agony and grief."

I continued: "Senators, when I say all three of my kids are better than me, you know that I am not engaged in idle flattery. Maybe some of you feel the same way about your kids. They are literally better people. They have got a lot of their mom inside of them. They are better than me. And Hannah saw through the *legality* of the situation. She saw through the *politics* of the situation, all the way to the *humanity* of the situation and the *morality* of the situation. That was one of the most patriotic things I ever heard anybody say. The children of the insurrectionists, even the violent and dangerous ones—*they are our children, too*. They are Americans, and we must take care of them and their future. We must recognize and exorcise these crimes against our nation, and then we must take care of our people and our children—their hearts and their minds. As Tommy Raskin used to say: 'It's hard to be human.'

"Many of the Capitol and Metropolitan Police officers and Guardsmen and -women who were beaten up by the mob also have kids. You remember Officer Fanone, who had a heart attack after being tased and roughed up for hours by the mob, and then begging for his life, telling the insurrectionists that he had four daughters, and that just about broke my heart all over again. We talked about this for a long time last night. My kids felt terrible that other kids' fathers and mothers were pulled into this nightmare by a president of the United States."

As I wound down my remarks, I returned to the question I'd borrowed from Officer Dunn and Fannie Lou Hamer to pose in my opening: Is this America? Does January 6 provide the image of what we want to become as a nation? Will the quest for a more perfect union be replaced by authoritarian demagogues unleashing mass political violence to overwhelm the rule of law and democracy?

To me, that's what this trial was all about. Will we allow political power to be determined by violence and lawlessness rather than the democratic process?

As I told the senators, this trial was not about Donald Trump, for the whole world knew exactly who he was and what he stood for. The trial was about who *we* are.

I had a long, intricate closing that I am pretty proud of, so I hope you will go watch it, if you have not already done so, but there is one thing that I wanted to say that I did not, out of a superabundance of caution over inflaming partisan feeling. But I have spent many restless nights wondering whether I should have said it and whether it might have made any difference if I had.

I wanted to grab the question of partisanship and impeachment by the horns and tell the senators that we needed to have some honest conversation about it. Most politicians run around denouncing partisanship, but then lead their legislative and political lives within totally partisan blinders and strictures. It would be better if we denounced partisanship a lot less and broke from it a lot more.

But I think we need to give two cheers for partisanship because, when you think about it, partisanship is just a reflection of the First Amendment: we have the right to think, speak, associate, and assemble freely. Our votes are equal, and we can form parties with other citizens to advance our own political views and philosophies. There is, of course, an easy way to get rid of "partisanship," and that is to get rid of all parties except for one, and that of course is the situation in an authoritarian or totalitarian state, a one-party dictatorship. When you think of it that way, the constant political conflict and competition between parties may be a small price to pay for actual political freedom. It may indeed be the sign of civic and constitutional health.

Moreover, once they are formed, political parties help define policy agendas and issues, educate and organize the public, clarify conflict, mobilize and turn out voters and send them signals about what different candidates stand for. When elections are over, parties help translate platforms and programs into legislative priorities and agendas.

So far, so good. But once we are elected, we must remember that we have a responsibility to represent *everyone* in our districts and our states, the whole people, not just the members of our parties. The word *party* comes from the French word *partie*, which means "a part," and our party is just a part of the whole, not the whole itself.

Although so much of congressional politics is partisan polemic and invective, we all know pretty well how to act in nonpartisan ways, and my proof for this is the dimension of our work we call "constituent service" in our district offices. If you call my excellent office in Rockville, Maryland, and ask for help on Social Security, Medicare, Medicaid, passport or visa assistance, VA benefits, unemployment, or any other of a number of services, we will never ask you whether you are a Democrat, a Republican, a Libertarian, a Green, an Independent, or something else. It is totally irrelevant to us. We ask only if you live in the Eighth District, and then we go to work for you to navigate the twists and turns of federal bureaucracy and get you exactly what you need and deserve. (My team is amazing.)

We obviously have serious and honest differences in legislative agendas among the parties, but all of us can strive to bring that same "constituent service" philosophy to our legislative duties—for, as legislators, we also are acting for the whole. This is tough when it comes down to bedrock value differences: some people just think we owe all kids an equal fair shot at success in life and therefore an excellent and equal education, while others just feel they have a right to keep all the money they have made and not to have to share it with the government to help the children of strangers.

But when it comes to impeachment and government accountability, it is urgently necessary for us not to think in partisan terms but, rather, to act as *constitutional patriots* and place our oaths and commitment to constitutional processes and values way above the narrow and short-term interests of our political parties and way, way above the ambitions of specific power-seeking individuals. That is why Liz Cheney and Adam Kinzinger and all the other eight House Republicans who broke ranks and voted for impeachment will always be heroes in my book, first-class constitutional patriots who set aside party discipline and electoral calculation and focused on the absolute necessity of presidents defending political democracy and protecting the rule of law rather than undermining and trampling them. I would have quoted Thomas Jefferson, who said, "If I could only go to Heaven with a political party, I would prefer not to go," and in his first Inaugural Address, "We are all republicans. We are all federalists."

Reading it now, I guess it seems kind of obvious and not really like much of a breakthrough argument. But I did want to say something at the end that would be uplifting and just a little bit tart about rank partisanship or, even worse, our new cult of political personality worship.

This I said, after reminding them that, whatever else they came to do in Congress, whether on farming or space exploration or higher education, nothing would define their legacy more than this vote: "None of us can escape the demands of history and destiny right now. Our reputations and our legacy will be inextricably intertwined with what we do here and with how you exercise your oath to do impartial justice—impartial justice. I know and I trust you will do impartial justice, driven by your meticulous attention to the overwhelming facts of the case and your love for our Constitution, which I know dwells in your hearts.

"'The times have found us,' said Tom Paine, the namesake of my son. 'The times have found us.'

"Is this America? What kind of America will we be?

"It's now literally in your hands."

I gave the remainder of my time, around twenty-seven minutes, back to the Senate. But following the defense's closing, there was one final intervention required, so I got back up briefly. Trump's lawyers had argued that there was no well-developed body of law defining incitement of insurrection that the Senate could use as precedent for trying and convicting a president in Trump's situation. Therefore, they argued, impeaching a president for committing incitement to violent insurrection against the union was unfair, a brand-new surprise offense offending due process that they called the new "Raskin Doctrine."

Well, it was true that there was no well-developed body of precedent defining constitutional incitement to violent insurrection, because it had never before occurred to *any* president of the United States of America—from George Washington to John Adams, to Thomas Jefferson, to James Madison, to James Monroe, to Abraham Lincoln, to Jimmy Carter, to Ronald Reagan, to George W. Bush, to Barack Obama—to incite a violent insurrection against his own government. So it was true that Trump had blown the doors of history off the hinges with his incitement to insurrectionary political violence and that the Senate would have to stand up for the very first time to define his unprecedented and original constitutional crime as a high crime and misdemeanor. This should not be difficult. After all, if this was not a high crime, what on earth would be?

So I jumped to accept and embrace their new so-called "Raskin Doctrine" as a badge of honor. I said I would wear it with pride. "If that is the Raskin Doctrine," I said, "that a president of the United States cannot incite violent insurrection against the union and the Congress—then I embrace it, and I will take it as an honor. Most law professors never even get a doctrine named after them, so I will accept that."

The moment I said this—totally off the cuff, mind you—I realized that, although I succeeded in making the point that it is perfectly obvious that a president may not organize and counsel violent insurrection, I had also just unwittingly defined myself at this climactic moment not just for the U.S. Senate, but for the entire world, as a *law professor* rather than as a *member of Congress*. What kind of giant faux-pas was this? I suppose I was feeling more like a law professor that week than I had in years because I had the chance to talk uninterrupted about the Constitution with a captive audience for huge stretches of time. In the House, of course, we became masters of the famous "One Minutes" on the floor and our luxurious "five minute" blocks for questioning and peroration in committee hearings. This was a radically different experience, prosecuting an impeachment trial. No one really seemed to notice my faux-pas but I was there representing the House of Representatives and the people we represent, not an academic idea. The dangers of rhetorical improvisation.

When I finished speaking, I experienced a surge of emotion within and around me. It was dumbfounding to think it was over. I suddenly felt crestfallen, as if Tommy, whom I had felt with me as a physical and spiritual presence, would leave me now. But he did not. My chest and my heart were full. I saw Sen. Alex Padilla smiling at me—he had passed me a note on Thursday saying, "You lead with *your heart*," a note that I still keep in my wallet. I even exchanged a kind of meaningful look with Mitch McConnell—was it too much for me to keep hoping he would vote to convict? For a passing moment, I felt it was not. Barry passed me a note saying, "That was great."

I saw a sense of vindication and even pride on the Democratic side of the aisle, as if to say: We may have won or we may have lost, but we have definitely redeemed faith in democracy by telling the truth unvarnished. Sen. Sheldon Whitehouse of Rhode Island, seemingly spellbound by the proceedings, smiled and nodded at me. I saw Mitt Romney, his eyes glistening, looking at us. Some people seemed moved in their seats—Senator McConnell, whom I still hadn't met, continued to have that ineffable about-to-cry look; as did the staffer with the kind face sitting next to him, whom I still would love to meet one day. I saw in Nevada Sen. Catherine Cortez-Masto a kind of visionary and dreamy look, as if she could imagine a day when all of this would be just a dismal memory and democracy would be riding high again. Certain people caught my eye as we waited for the vote to begin: Senator Gillibrand; the two new senators from Georgia, Raphael Warnock and Jon Ossoff, who had a glow of destiny around them; Elizabeth Warren; and even Bernie Sanders looked connected. I thought about Tommy and Hannah and Tabitha. I thought about Sarah. I thought about my parents, neither of them with us for so long now.

I was flooded with feeling. I began to think of all the more comprehensive things I had not said because I had been improvising at the end and not following any of my various scripts and drafts still floating around—about the practice of nonviolence and democracy; the demagogic disrespect for the primacy of Congress; and all of Michael's finer points about taking adverse inferences seriously when a witness skips town. Tommy would have gotten all of it in. A poet and a philosopher, he would have been more inspired and more systematic, too. His own toughest critic, he also probably would have been castigating himself for anything he had left out, just as I had already begun to do to myself. Well, it would all become part of tomorrow's work and I would have plenty of time to think of it during sleepless nights ahead.

Like son, like father. Like father, like son.

We stayed seated, and the Senate readied to vote. I was still thinking that a cascade of shock and horror over what Trump had done could wipe out his entire operation. This was the GOP's big chance to get rid of a constitutional criminal in their midst, one who was seriously alleged to be a violator of women, a con man, a compulsive liar, a corrupter of elections, an inciter and cheerleader for violence, and an enemy of the constitutional order and the common good. In my dream—in my reality now, perhaps—the vote would come up 100–0. America would stride boldly into a future of strong democracy, pluralism and equality, justice and freedom.

Patrick Leahy, with his great and sonorous senator's voice, asked, "Senators, how say you, is Donald Trump guilty or not guilty?"

This was it.

The voting began. It ran in alphabetical order for the 100 Senators. We were hearing the pundits estimate, on the ride down, that it would be 53–47, while I still believed it might end up 100–0.

It started off with Tammy Baldwin from Wisconsin, who voted to convict the president. The very next senator, Barrasso of Wyoming, voted "Not guilty." It would not be 100–0. It see-sawed back and forth. Then some completely unexpected, early-alphabet Republican senators voted to convict, sending me on a mental joyride again—this really might happen: Richard Burr of North Carolina pronounced him "Guilty," and David Cassidy of Louisiana, who had seemed more engaged during the proceedings than almost any other Republican on the floor, stunned the Senate by saying loudly and clearly, "Guilty!" Alas, it was no mass movement, but we built a sizable majority by the end. When all was said and done, the vote to convict was 57–43. Seven Republicans—representing New England, the Mid-Atlantic, the South, the Midwest, and the West—had joined every Democrat in the chamber to vote to convict. Burr, Cassidy, Collins, Murkowski, Romney, Sasse, and Toomey all came our way. That made it 57–43.

I glanced over at van der Veen and Castor. They were celebrating, seemingly exultant in "victory," perhaps more ecstatic than anyone who just barely wins a fixed game ought to be. As I sat in my chair in the Senate and imagined talking points about what we had achieved for a press conference our team had planned—*the most bipartisan presidential impeachment vote in history, seven Republicans from every part of the country joined every Democrat, robust bipartisan majorities to impeach and convict in both houses, Trump convicted in the court of public opinion and the eyes of history*—I was shaken now by a sharp sense of having let everyone down. I worried that I should have pressed for the nonpartisan alphabetical seating arrangement I had always favored, or for a closed ballot. I should have made it far clearer to the Senate that they could draw adverse inferences from Trump's absence. I should have used all our argument time rather than returning any minutes.

We won 14 votes more than Donald Trump, but he had "beaten the constitutional spread" and was naturally already out and about claiming vindication. The incorrigible corruptor and twister of all our values had eluded the complete grasp of the law once more. Impeached, discredited, disgraced, a national pariah, he would remain unbowed and unrepentant, a lying, sinister, and controlling force over one of the nation's two major political parties.

Joe Neguse asked if I was all right. I told him I was. I told him, "Great job, Joe. You were amazing for America."

We headed back to Room 219. The C-SPAN cameras caught me and Joe walking out while Michael van der Veen—I kid you not—was picking up U.S. Senate coasters left out on our desks and putting them in his pocket. My niece Mariah sent me a meme of that special moment, with the label "Stop the Steal!" I texted her, "You can take the lawyer out of *Philly*delphia but not *Philly*delphia out of the lawyer," teasing her, as that is where her mom Mina's family is from. My Philly friends were already sending me petitions and letters distancing themselves from Trump's legal team. A reporter called me a bit later to ask my reaction to van der Veen's posttrial statement that they were "going to Disney World!" To which I replied: "They were always at Disney World."

Back in the war room, we saw Senator McConnell on TV making his speech from the floor, stating that there was "no question" that Trump was "practically and morally responsible for provoking the events of the day," and he kept invoking the impeachment managers and what we had proven about Trump's complicity in the insurrection. Yet, in voting not guilty, McConnell had hung his hat on the discredited jurisdictional argument that a former official could not be tried in the Senate, the argument we had defeated and dismissed on Tuesday, the first day of trial, which seemed like a month ago.

Everyone was gathering around watching.

"He didn't have the votes to convict, and he didn't have the votes to survive if he voted to convict in the minority," I said. "So he's saying Trump's factually guilty, but he lets him off on this phony technicality."

"Profile in courage," Swalwell said.

"Profile in politics," I said.

I was eager to get back to Takoma Park, to Sarah and the girls, to our family. On the way home, we stopped at Caruso Florist near DuPont Circle. The freezing air was bracing, but it felt good to the touch. I found a beautiful bouquet of white roses for Sarah for Valentine's Day. The masked guys at the store wanted to give it to me for free. I told them I couldn't accept their generous gift but that I appreciated it; I wanted to buy flowers for my wife, my love. They said they had watched every minute of the trial and that we had made them proud as Marylanders for standing up for democracy. When they gave me a bunch of red roses to go with my bouquet, I shared them with my security detail, so they wouldn't show up at home empty-handed. You know we've all got to stick together on Valentine's Day.

Back in Takoma Park, we drove up to a remarkable homecoming. Sarah and Kathleen had organized a whole bunch of neighbors to greet us as we arrived. There was a freezing rain coming down, but dozens of masked people had assembled. Alas, between the masks and the misty cold weather and my sleepy eyes, I couldn't make most people out clearly. But I did pick out my friend Tom Cove, Kathleen's husband and a great father whom I admire. I recognized our close friends and neighbors, city councilman Peter Kovar and Paula Kowalczuk, who came to join me at our house on the dreadful day of December 31 when we lost our beloved Tommy, and now here they were again by my side at the end of our journey. I saw our friends Tebabu and Sara Assefa.

People brought chocolate chip cookies and carried homemade signs: "Thank You, Jamie," "Tom Paine Says Thank You, Jamie," "We Love the Managers." "Tyranny, Like Hell, Is Not Easily Conquered." "Real Patriot." Amazingly, the signs were also sprouting up in people's front yards and would greet Sarah and me on our dog walks for many, many months to come. The feelings of solidarity, gratitude, and tenderness we experienced from those signs were indescribable.

As I got out of the car, people clapped, and Drew and the team gave me the go-ahead to address everyone. I went up on our porch, the place where fifteen years ago I had announced my intention to run for the state senate, where my son, Tommy, had kindly introduced me to the world. It felt like I had been gone a long, long time, though I had never really left home.

Before I spoke, I thought about the last line in *Catcher in the Rye*, when Holden Caulfield says, "Don't ever tell anybody anything. If you do, you start missing everybody."

I thought about my mom, Barbara Raskin, the writer, whose worship of language dwells in my heart and whose sentence rhythms still ring in my ears. I wondered whether this partial 57–43 victory we had won could be counted also as a minor victory for the English language, which had been so trashed and degraded along with everything else in these dismal years. I thought about how much my mom loved Tommy.

I looked out at all these wonderful people and thanked everyone for coming and told them I now expected "a standing ovation every time I get home from my office."

They laughed merrily.

"It is hard to know how to thank one's family, one's friends, or one's community for something like this," I said. "I just want you to know that this place, our town, was in my heart the entire time. Thank you, Sarah and Hannah and Tabitha. Thank you to my staff, to Julie and Kathleen and everyone else. Thank you, Takoma Park. Thank you, Maryland. Thank you, America. I love you all."

It was a cold and tearful moment, but I could feel the warmth of this community, the love of my family, and the memory of my precious son all gathering me up now.

People watched me for a moment, but I had nothing left to say.

I blew them all kisses.

And then Sarah and I went back inside to start looking for our future.

EPILOGUE

The summer soldier and the sunshine patriot will, in this crisis, shrink
from the service of their country; but he that stands by it now, deserves
the love and thanks of man and woman.

—THOMAS PAINE, *THE CRISIS*, DECEMBER 23, 1776

Family and friends assembled by the carful on a cold Saturday, April 3, 2021, for a beautiful and wrenching three-hour memorial service for Tommy at Robert F. Kennedy Memorial Stadium in Washington. While Sarah and I had resigned ourselves to a Zoom service in the stifling COVID age, Hannah and Tabitha conceived of an old-fashioned drive-in at the RFK parking lot, and Ryan organized something spectacular. Hundreds of mourners tuned into 98.7 FM on their car radios for the audio and honked at the large screen to applaud.

The program was enthralling. Guided by weekly family brainstorming sessions, Ryan had brought together a constellation of clips and images of Tommy in action: in his poetry and prose, in generations of family photos, and in a beautiful mélange of in-person eulogies and live music, mostly Tommy's favorite Bob Dylan songs, sung by Dar Williams and Lucy Kaplansky, who traveled long distances to join us. Tabitha welcomed everyone, Hannah spoke about how she and Tabitha and Tommy's cousins and friends were working to keep his ideals alive through his memorial fund, and I closed the ceremony.

I recalled a passage Tommy loved, from Ludwig Wittgenstein, who said that the truth of empirical propositions is whether they correspond to reality, the truth of analytical propositions is whether they follow the rules of logic, but the truth of ethical propositions is determined by the courage with which you act to make them real in the world. On this standard, there have never been truer ethical claims than the ones made by Tommy Raskin, because he was all courage and engagement with his moral convictions.

Tommy's friends invoked his qualities as a careful listener, a creative consoler, a leader unafraid to rock the boat, a visionary of urgent moral change, the joy of the party. Sarah told a hilarious story about driving between Maryland and Boston on I-95 with Tommy when he was a sophomore in college. As they passed through Baltimore, Tommy quietly read an article by Professor Walter Block and took exception to several derisive things Block had written about animal rights. Reaching out to him in that moment via email, Tommy challenged Block to a debate on the subject—only to be told that the professor would not debate him unless Tommy first published his critique in a peer-reviewed journal. After asking Sarah what a peer-reviewed journal was, Tommy spent the ride through New Jersey writing his own article, which he then—still in the car, mind you—submitted to numerous online peer-reviewed journals. It was promptly accepted for publication by one of them, and Tommy then scheduled an online debate with Professor Block—all before arriving in Boston.

Most eulogists were friends from each phase of Tommy's life—childhood, high school, college, law school—but we also heard from Michael Anderson, he of "adverse inference" fame and my best buddy from college and law school ("Uncle Michael"); he and his wife (and our law school roommate) Donene Williams, were close to

Tommy, who lived at their house most of his first year of law school. Michael explained why he thought Sarah and I and everyone else at the memorial kept referring to Tommy as "magical." Invoking Mikhail Bakunin, who said that God would send angels to the earth not in the form of priests or police officers but as fortune-tellers and "anarchist magicians," Michael observed that Tommy had breathed in all the pain of the world and breathed it out magically as love.

It is fall 2021, usually a time of boundless high hopes and romantic dreams in our education-obsessed family. But this is a melancholy back-to-school season. Tommy would have been starting his third and final year at Harvard Law School, hanging out with his wonderful law school friends, probably teaching college students, mapping out his next steps in life.

Every day, our dear boy is still gone.

Still gone.

In the semi-light of his bedroom, I touch his comforter, his pillow, a notebook.

No longer drowning in agony and grief all the time as at the start of this trauma, we still miss him terribly throughout the day. The circumstances of his departure are far less important to us now than just the unyielding fact of not having him here. We want him back. We miss him. I would give ten years of my life to have one more minute with him.

I remember Tommy doing his homework one day when we lived in France during my sabbatical year. He lifted his head to look skyward and then pointed out that while we said "I miss you" in English, in French they said "Tu me manques," the seeming reverse to express the same feeling.

"'I miss you' and 'you miss me' mean the same thing!" he said.

We have blessings to count on these days, too. Hank was deeply affected like the rest of us by the loss of Tommy and then by the Trump mob's "wild" attack on American democracy to which he had a front-row seat—a seat under a desk, that is. A creative online entrepreneur with daring ideas, Hank has joined forces with Hannah to help Democrats in western Nevada fight to defend democracy in America, something that, before January 6, he had never thought he would be doing. But as he has expressed to me, none of the things that make life beautiful and wonderful in our country will continue if we don't have a democratic infrastructure to back it up. He and Hannah are definitely in the fight.

Hannah graduated with an MBA from the Haas School of Business. She has continued her work with start-ups at the Silicon Valley Bank but has been spending a lot of her free time chairing the family-and-friends board of the Tommy Raskin Memorial Fund for People and Animals, which we launched in January of this year, and doing a sensational job at that. The fund has now collected $1.2 million from more than eight thousand donors around the world. Composed of a dozen of Tommy's loving cousins and siblings and four rotating friends (Jason Altabet, Sam Dembling, Nurah Jaradat, and Isabel Thompson to begin), the board has made significant grants to Mercy for Animals, Oxfam, Amnesty International, and a few groups Tommy didn't know about, to respond to the humanitarian crises that have arisen since he left us—such as the debacle of the U.S. exit from Afghanistan, the earthquake in Haiti, both in August, and the proliferating calamities of climate change. Hannah's beautiful steadiness and leadership are a solace to me.

In addition to the fund, more than a dozen new projects have been launched in our community and around the world by people moved by Tommy's life and spirit, including:

- The Thomas Bloom Raskin Act, passed by the Maryland General Assembly and signed into law by the governor of the State of Maryland, has established a telephone hotline for people in crisis and a program by which counselors will call them back on a regular basis to check on their emotional and mental health.

- The Humane Society of the United States has created the Thomas Bloom Raskin Farm Animal Policy internship.

- Arena Stage, a not-for-profit regional theater in Southwest Washington, DC, has created an annual Tommy Raskin arts camp scholarship for a child in need.

- The Marshall-Brennan Constitutional Literacy Project has created a Tommy Bloom Raskin Young Activist Award.

- The Berg Society in Madrid has established the Tommy Raskin Scholarship on Human Rights and Sustainability.

- The Women's Democratic Club of Montgomery County, Maryland, has created a Tommy Raskin Speaker's Series.

- Tom and Mary Snitch have purchased five new sewing machines for Damien Center, a village in Luangwa, Zambia, to establish the Tommy Raskin Sewing Center, where village women can produce bags, tablecloths, school uniforms, and COVID masks.

- Mercy for Animals has created a Tommy Raskin Award for Animal Welfare.

There are many other such projects in development. Hannah wants Tommy's birthday, January 30, to be a global day of service in Tommy's spirit, and we receive messages frequently about new honors, gifts, and remembrances made in Tommy's name. We rejoice in the fact that, in a world of noise and chaos, so many people are still taking the time to appreciate and honor Tommy's life.

Tabitha has moved home from Philadelphia to be closer to us and the dogs (or maybe to the dogs and us), and we are ecstatic to have her back. Although she had to leave Teach for America and the Murrell Dobbins Career and Technical Education High School, where she had been teaching for a year, she has found a great new middle school math teaching job in Washington, DC, and is in her element. We see her often. She and Ryan found a cool house in a great neighborhood in the disenfranchised capital city; next Fourth of July, they will enjoy a breathtaking rooftop view of fireworks on the National Mall. And in case you're wondering—no, Tabitha still hasn't been back to the Capitol since January 6. I feel that's my challenge: to change American politics enough to win her back. But Ryan, who always thought he would use his awesome organizational and planning skills as a financial adviser, has thrown himself, for the time being, into politics as an organizer and fund-raiser. A new generation arises.

Sarah has been healing well; seeing her many friends; tending to her garden and houseplants as her father, Herbert Bloom, would have done; taking care of me and everyone around her—as I write this, she has just gone off to the post office to send my sister Erika something for her birthday. Sarah has resisted invitations to go back into government and instead has accepted an offer to become a law professor at Duke, where she spends the middle of the week teaching about corporations, climate policy, and financial regulation, leaving me alone with my politics, my writing, my thoughts, my past. I love her very much and am proud of the way she continues to make our lives beautiful while making our economy more just.

What about me? Well, I can't tell you much you don't already know by now. After throwing myself into the trial in the Senate, I threw myself into this book—something unlike anything I have ever written before. I have always felt that my academic and political careers drew on the side of my brain shaped by my father. But writing this book has provided an astonishing reconnection not just to Tommy but also to my mother, Barbara Raskin, whom we lost in 1999, more than two decades ago. As a kid, if ever I awoke in the middle of the night, I would hear her click-clacking away on her typewriter, hitting that chiming Return bar to start a brand-new line, that fresh, emphatic sound of creativity on a blank page. Even now, in my mind, I can hear the clatter of that typewriter as if it were yesterday. But now *I'm* the midnight writer—another scribbler searching for the truth that lies not just outside of us but inside of us. Actually, my mother's rightful heir in prose is my big sister, Erika, the one who got teargassed when the alt-right came to Charlottesville in 2017. Erika is a seriously funny and engaging writer, and I am her rhapsodic fan. But now I, too, can taste a little bit of that manic writer's spirit when night falls.

The boring truth is that my day-to-day life is not just different in every way now but also very much the same—the glorious grind of politics, the halting pursuit of justice, the zealous defense of democracy. Still, everything I do today is consciously infused with my love for Tommy, informed by his beautiful, elusive values. My friends are still my friends, though I treasure every minute with them so much more. I still talk to Larry Tribe about constitutional ideas; I still get great life advice from our friend Kate Bennis, a communication coach; I check in with my mom's loving and brilliantly zany best friend, Helen Hopps, a hero to little kids in our family; I usually start my day with a call from Noah and Heather, in which he spouts his sublime hilarities and she tells me all her elaborate political theories. Democracy Summer is now a surging nationwide program allied with the Democratic Congressional Campaign Committee, and I am working to recruit no fewer than a thousand fellows to build strong democracy in America in the summer of 2022. Democracy badly needs the idealism, energy, and passion of young Americans, and I am out to recruit them.

And so, I work. I work with my colleagues and friends. And because our work is to make democracy work and to save humanity from climate chaos, *the times have found us*, as my old hero Tom Paine once said—a refrain I stole from him and that Nancy Pelosi stole from me and that the Democratic Caucus stole from her. All over the world, the bullies and strongmen, the despots and tyrants, the kleptocrats and dictators, are on the march, and liberal democracy is under siege. We glimpsed our potential dystopian and authoritarian future on January 6, 2020, a bloody day of sedition, insurrection, and attempted coup that constitutes an action-packed advertisement for what government in America could easily become in this century.

The first and most elemental fight is for the truth. Naturally, Donald Trump and his enablers acted quickly to whitewash the violence of January 6, claiming that the protesters at the Capitol were a gentle, "loving crowd" who represented "zero threat" to members of Congress. Contradicting what we all saw with our own eyes, they invite us to believe that these "patriots" warmly greeted the Capitol Police officers on duty that day with "hugs and kisses" as they entered the building. And now it is an article of faith in the right-wing media that government lawyers are "persecuting" hundreds of "innocent" rioters. Trump's forces are also working to propel Ashli Babbitt, the insurrectionist who was fatally shot as she tried to break into the House chamber after storming the Capitol, into the ranks of American martyrs.

As a society, we have not fully reckoned with the explosion of antidemocratic violence that shook the Capitol that day, nor with Trump's dangerous plan to stage a political coup in the Electoral College. At the close of the Senate trial, a commanding bipartisan majority in Congress found that Trump had incited insurrection, but he managed to escape conviction by two thirds of the Senate. The nation is, therefore, in a fundamentally uncertain and perilous place: the twice-impeached Trump should be a pariah and outlaw in mainstream American political culture, but instead he is the undisputed master of one of our two major political parties and continues to dominate many of the levers of political power in the country. Will the popular majority be able to govern effectively in the face of political manipulation and violence by Trump's mob or will the minority Trump party be able to rule through antidemocratic mechanisms like voter suppression and gerrymandering backed up by political violence?

This uncertainty mirrors the essential underlying conflict in America about whether we're capable of putting the pathologies of political white supremacy behind us once and for all. More than 150 years after the end of the Civil War, will we finally reject and transcend racism in this century? Or has it become such a necessary tool for maintaining right-wing political power and social control that we are bound to see it wielded in perpetuity through dangerous new forms of mass propaganda and self-deception? The dominant pro-Trump elements in the GOP traffic in every form of irrationality and disinformation. On the events of January 6, they circulate conspiracy theories and lies about what happened at the Capitol that day, from hoisting the go-to canard that "Antifa did it," to likening the whole episode to a "normal tourist visit," to claiming that the insurrectionists charged with assaulting federal officers and interfering with Congress are "political prisoners" being held for their beliefs as opposed to their actions.

The violence of January 6 and the unsettling constitutional implications of Trump's attempted inside coup obviously demanded the formation of an independent public commission outside Congress to investigate the events of that day. Coming out of the Senate trial, all the managers strongly agreed that a commission was necessary. During the trial, we had been focused on just one man and one crime, but America needs to understand the whole architecture of the January 6 attack: who organized it and paid for it, who coordinated the inside political strategy with the violent elements running amok outside, and what we need to do to stop the next assault, or at least prepare ourselves for it.

House negotiators—Rep. Bennie Thompson, Democrat of Mississippi, and John Katko, Republican of New York—converged on an obvious solution: an independent 9/11-style commission made up equally of five outside Republican and five outside Democratic appointees (no members of Congress allowed) with equal subpoena power and charged with determining the events and causes of the January 6 riot. But then Minority Leader Kevin McCarthy turned the Republican conference against the very deal the GOP had asked for. On May 18, the House passed the bill by a vote of 252–175, with only 35 Republicans (double the number of Republicans who voted to impeach Trump in the House but still a pathetically small group) joining every Democrat to vote yes.

Yet, in the Senate, Mitch McConnell successfully blocked the search for the truth. One day after the House vote, Trump called on McConnell to reject it. McConnell relented and proceeded to use the filibuster to kill the legislation, calling it "slanted and un-balanced" and lobbying Republicans to vote no as a "personal favor" to him. A group of police officers and their families asked to meet with GOP Senate leaders. Gladys Sicknick, the bereaved mother of Officer Brian Sicknick, who was involved in the bloody fighting on January 6 and died the next day, begged the Republicans to pass the bill for the bipartisan investigation. Mrs. Sicknick wore her son's ashes in a pendant around her neck, but her eloquent appeals fell on deaf ears. McConnell's seething posttrial diatribe against Trump was ancient history now.

Only six Republican senators joined the Democrats in voting to move forward, creating a 54–35 margin, not only six votes short of 60, the magic number in the Senate, but three votes fewer than we won in the impeachment trial. Once again on top as undisputed master of the GOP, Trump continued to vaporize remnants of op-position in his party.

Speaker Pelosi announced that the House would proceed on its own to create a thirteen-member select committee to investigate and would not allow the GOP to play any more games. There would be eight Democratic appointees and five Republican appointees, the regular partisan ratio, and the Speaker would have final approval over all appointments, the precise structure that was in place for the GOP's House Benghazi Committee. The vast majority of Re-publicans voted to oppose the select committee, with only Reps. Liz

Cheney and Adam Kinzinger having the courage to break ranks and vote yes.

Pelosi went on to name the estimable Rep. Bennie Thompson chair of the committee and appointed not only several Democrats who were veterans of the two Trump impeachment trials (including Zoe Lofgren, Adam Schiff, and me), but also the former chair of the House Republican Conference, Liz Cheney. I thought this was an inspired cross-partisan appointment and endorsed it strongly. In our struggle on Capitol Hill to stop Trump's authoritarianism and incitement of right-wing violence, no one has been more emphatic than Liz in pointing out that American democracy itself is on the line.

When McCarthy named his picks, Speaker Pelosi accepted three of them but refused to approve Jim Banks of Indiana, who would have been their ranking member, and Jim Jordan of Ohio. This, too, was the palpably right decision. Banks had been trashing the idea of a select committee, saying that we should instead look at the violence that broke out around the Black Lives Matter movement in the summer of 2020—giving us a preview of the diversionary diatribes he would bring to the committee. (Much of the violence he referred to came from right-wing sources, like Kyle Rittenhouse, a seventeen-year-old vigilante from Antioch, Illinois, who traveled to Kenosha, Wisconsin, with an assault rifle and allegedly shot and killed two people participating in a Black Lives Matter protest.) Jordan, of course, was a political intimate of President Trump's and was present in at least one strategic session with the White House to plan the January 6 offensive. He had even spoken on the phone with Trump on January 6, as the insurrection was unfolding, and was thus a potential witness. An unrepentant defender of the effort to nullify Biden's electoral vote majority and a nimble fellow traveler of Trump's Big Lie, Jordan in my view would only have used his seat on the committee to promote the daily spin cycle of Donald Trump's propaganda machine.

Given that the controlling right-wing elements of the GOP

(including Jordan and Banks) had already sabotaged our plan to conduct a truly bipartisan outside investigation, Pelosi was under no moral duty to allow them to ruin this select committee, too, our final chance to get at the truth. In retaliation, McCarthy pulled all his Republicans off the select committee, and Pelosi proceeded to add in their place the other Republican who had voted in favor of creating it, Illinois representative Adam Kinzinger, a veteran and another rock-solid constitutional patriot.

The early work of the U.S. House Select Committee to Investigate the January 6th Attack on the United States Capitol has been promising. It is the first committee I have served on in Congress where all the members and staff are laser-focused on our common project, to determine the truth and to protect democracy, rather than on partisan crossfire and showboat polemics. For this reason, we have been effective from the start under Chairman Thompson as we have worked to deliver a comprehensive report to America both about what happened on January 6, 2021, and about how we can protect future elections and democratic institutions against fascist sabotage.

Our first hearing featured four police officers from the front lines of the battle on January 6. I had urged Chairman Thompson to call this hearing right away—I had a strong sense from the Senate trial that America needed to hear from the people on the ground that day who had guarded democracy with their valor, toughness, and very lives. The cops who testified would remind us that the actions of the rioters on January 6 were no "normal tourist visit" and that we had come close to losing not just the Capitol but the processes of democratic self-government—not to mention the lives of who knew how many more people. Speaking with eloquence and quiet indignation, all four uniformed officers acted with as much strength in testifying on July 27 as they had in fending off the attack on January 6.

Sgt. Aquilino Gonell and Officers Harry Dunn, Michael Fanone, and Daniel Hodges described "medieval-style" combat, complete with eyes gouged, fingers broken, steel pipes and baseball bats brought down on heads, bear mace and toxic liquids sprayed directly in faces, and hours of sadistic abuse by a fanatical mob. An army veteran, Gonell said that what he went through on January 6 was more terrifying than anything he had experienced as a soldier deployed to Iraq; he was convinced he was going to die at the hands of fellow Americans. Fanone, who was Tased and punched, repeatedly begged for his life by invoking his four daughters; he was beaten unconscious and suffered numerous severe head injuries before being rescued by fellow officers. The mob attacked Hodges without mercy, crushing him in a doorway in a scene of torture and agony that rocketed around the world.

My constituent Officer Dunn made the mistake of trying to reason with the rioters. When they yelled out that no one had voted for Joe Biden, he told them that, actually, *he* had voted for Joe Biden. This prompted the mob to hurl a torrent of racial epithets at him, stirring Officer Dunn to shout in frustration later in the day, after hours of abuse, "Is *this* America?"

All four officers bristled over the craven ingratitude and selfishness of so many Republican members of Congress. Most Republicans in the House refused even to lift a finger to protect democracy by voting to establish an independent commission on the January 6 attack. But when they were in danger themselves on that day, they freely availed themselves of all the police security we had, taking for granted the blood sacrifice of the officers, treating them like personal servants and bodyguards rather than fellow citizens defending our common political democracy.

"I feel like I went to hell and back to protect them and the people in this room," Fanone said. "But too many are now telling me that hell doesn't exist, or that hell wasn't actually that bad."

Our victory over authoritarianism is not assured—far from it. While radical optimism, which has returned to me body and soul, is one thing, self-delusion is quite another. If democracy is to survive and flourish, we must keep our eyes open and our ears to the ground.

At Trump's second impeachment trial, we had a chance to tell (and *show*) the truth of Trump's conduct and to mobilize the majority of Americans behind constitutional democracy—two goals we succeeded in accomplishing. But we also had the chance to convict and repudiate Donald Trump for his high crimes against constitutional democracy and then to disqualify him from ever holding office in the United States again. By doing so, we could effectively have purged Trumpism from the American body politic. This we failed to do.

Claiming vindication and exacting revenge, Donald Trump consolidated his power and increased his stranglehold over the Republican Party, demonizing and ostracizing critics like Rep. Liz Cheney and Sen. Mitt Romney, controlling the flow of endorsements and political money to Republican candidates, and attempting to oust and replace Georgia secretary of state Brad Raffensperger and any other GOP elected official, at any level, who rejected the Big Lie and defied Trump's orders in 2020.

We can now say that the Democratic Party, whatever its faults, *is* the party of democracy and that the Republican Party is the party of Trump, authoritarianism, corruption, and insurrection. If they can't control the government, they will do whatever they can to ruin the prospects for the rest of us to make any social progress at all. The Grand Old Party has become the Rule-or-Ruin Party.

Democrats are by far the majority in this country, and we are growing in strength all the time. Hillary Clinton beat Trump by more than 3 million votes in the popular tally in 2016, and Joe Biden beat Trump by more than 7 *million* votes in 2020. Younger Americans are registering as Democrats in hugely disproportionate numbers, and they aren't buying anything the GOP has to sell. In the 2020 election, voters under age thirty chose Biden over Trump by 25 points (61 percent to 36 percent), and if the Electoral College were controlled only by Americans age thirty and under, Trump would have won fewer than five states in 2020. All the fastest-growing demographic segments of the American population—young people,

Latinos, immigrants, unmarried women—trend blue. The future is indeed ours.

However, in the present, we face a serious problem: GOP power brokers are making sure that our political institutions work at every turn *against* majority rule. Controlling many more state legislatures than Democrats do, Republicans gerrymander Democrats into oblivion in congressional and state legislative elections. They then block Democratic efforts at redistricting reform in the Senate with the use of the filibuster, another essential antidemocratic instrument. To further obstruct voting and thwart the will of popular majorities in the administrative machinery of government, their gerrymandered legislatures pass laws expressly designed to suppress voter participation, making it harder (or, in the case of baseless voter roll purges, impossible) for some groups to access the ballot. And then they cement their hold on the whole system by packing the courts with right-wing judges to enforce all the exclusion. They are fortified in all this self-entrenchment by Supreme Court decisions like *Shelby County v. Holder* (2013) and *Brnovich v. Democratic National Committee* (2021), which have dismantled the Voting Rights Act and weakened the Constitution, which had both been for a brief shining moment effective guardians of the voting rights of the people. In short, the leaders of the GOP—not just Trump, mind you—are using every trick in the book to stifle majority rule and to erase popular democracy in the same way they have been working to erase science and history.

The majority of Americans are caught in a vicious circle of antidemocracy.

In 2024, we will likely face another onslaught against the will of the people in the Electoral College. This obsolete system is bad enough, with its reduction of the entire general election to six or seven swing states before the contest even begins and with its strange power to propel to victory the *loser* of the popular vote (most recently in 2000 and 2016). But now that Trump has let the coup-and-insurrection genie out of the bottle, we have entered an even more dangerous new period in the haunted history of this flawed institution. In the twentieth century, the general election was essentially among the people in the states, and the key to victory for a candidate was winning the popular vote in enough states to clinch enough electors to reach 270. The Electoral College winner was usually known on Election Night (or soon thereafter), when the states' votes were added up.

But now, with Trump's dangerous approach to elections, all the formerly invisible administrative stages of the Electoral College process that used to be considered routine and pro forma—governors preparing "certificates of ascertainment" for the congressional archivist, electors meeting in the state capitols to record their votes, Congress in joint session counting and accepting certified electors from the states—have been transformed into new arenas for no-holds-barred partisan struggle and conflict. They have all become levers to be manipulated, pressure points in the body politic ripe for unprecedented partisan probing. Trump has ushered in a system whereby, at every stage and in every nook and cranny of the Electoral College, the campaign *just keeps going*. As far as our democratic norms are concerned, nothing can be taken for granted anymore.

What Trump's actions revealed is that our presidential election system is riddled with potential vulnerabilities and booby traps. In 2020, he trampled over all of them, and they held—just barely, by the slimmest of margins. Why? The system survived because, in different parts of the country, a handful of public servants—Republicans in most cases—nobly upheld their oaths of office and refused to destroy the integrity of the election. We can't count on that safety valve next time. Since his second Senate trial, Trump has been engineering the purge of apostates, trying to replace serious public servants in positions of influence in the electoral system with opportunists and sycophants. In 2024, Trump loyalists will be running the electoral machinery in a lot of places, ready and willing to follow their leader.

Even worse, Trump's devotion to the Big Lie and his continuing polemics against imaginary voter fraud in 2020 have brought into public disrepute the central facet of the electoral process, the part that gives it democratic legitimacy: the act of citizens voting in their states. Obsessed with his loss and evidently convinced that challenging old election results will be his springboard to political resurrection, Trump has deployed his loyalists and operatives to try to discredit swing state results from the 2020 election, the same results he challenged after the election but that were upheld in all sixty-one courts that examined them. In Arizona, Trump partisans in the state senate in 2021 commissioned an expensive hatchet job on the 2020 election in Maricopa County by a private pro-Trump outfit called Cyber Ninjas. To my mind this was a sloppy investigation designed to confirm Trump's delusions but still it could not deliver on that objective. This "fraudit," as Arizonans against the process call it, was denounced by every serious election official in the state, ranging from Maricopa County recorder Stephen Richer, a Republican, to Secretary of State Katie Hobbs, a Democrat. And yet, Trump convinced Republicans in the Pennsylvania State Senate to follow suit and pursue a similar "forensic investigation" of the 2020 election results in *that* state, prompting the Philadelphia city deputy commissioner, Nick Custodio, to state that "the repetition of baseless claims by elected officials poses a real challenge to our democratic processes."

By expanding the scope of the Big Lie to cover almost everything in our politics and by building up an infrastructure of cult followers as key personnel in the electoral and legislative machinery of the states, Trump appears to be preparing to institutionalize political minority rule in America—specifically, that minority of Americans which continues to refuse to accept the end of white supremacy and that is readily mobilized by political lies and racial grievances. What will this mean in 2024? I see numerous scary scenarios involving Trump as the GOP nominee.

Imagine that gerrymandering and voter suppression produce GOP majorities in the House and Senate in 2022 but that we have a 2024 general election in which Biden wins by more than 7 million votes and in which the popular vote in the states breaks down exactly as it did in 2020. Now assume (not hard to do) that Trump again cries fraud and asserts that Biden electors should be "returned" to their state legislatures and the contest thrown into a contingent House election. What would new Republican majorities in the House and Senate do? My fear is that GOP leaders—having been totally whipped by Trump and the violent insurrectionists, and having effectively purged themselves of dissenters like Liz Cheney, John Katko, Richard Burr, and Bill Cassidy—would snap to attention and grant Trump's wishes, rejecting enough Biden electors to activate a contingent election and get Trump elected "immediately" in the House. This works even without a violent insurrection.

A variation: Imagine the GOP succeeds in purging its ranks of uncooperative state election officials (like Georgia's secretary of state, Brad Raffensperger). When Biden wins an honest majority of the popular vote in Georgia in this scenario, imagine Trump cries fraud, and the election returns are nullified and reversed at the state level by a newly pliant Georgia secretary of state (probably my House colleague Jody Hice, the Trump loyalist who has been recruited to run) and swiftly replaced by a slate of Trump electors named by the GOP-run legislature. Imagine this course of events taking place in several states, and then imagine the House Democrats, in the joint session of Congress, trying to challenge these state-certified electoral slates because the popular vote has been canceled out in violation of the Equal Protection Clause (one-person, one-vote) and the Republican Guarantee Clause, which guarantees the people "a Republican Form of Government" (*not* a Republican *Party* form of government). Would concurrent GOP majorities still accept electors substituted by GOP legislatures in place of the electors chosen by the people? Almost undoubtedly, they would. Would concurrent Democratic majorities try to reject those substitute electors and reinstate the electors chosen by popular vote? Almost certainly. If we permanently break from the tradition of respect for the will of popular majorities, there will be no end to conflict over where political legitimacy truly lies. The permutations and possibilities of a full-blown presidential election breakdown are infinite in the haunted house of the Electoral College.

The shameful GOP whitewash of the insurrectionary violence we all witnessed on January 6 introduces another combustible element. Will Trump's newly empowered election mob accept anything other than total victory for their leader? Certainly not. That means that street violence and Electoral College violence are both quite certain to reappear, and who knows how far the insurrectionists will be willing to go next time. It seems obvious to me that Donald Trump used all the right-wing extremist groups for his own purposes on January 6, 2021, but no one should overlook the fact that they used him, too. Trump's multimillion dollar investment in mobilizing and valorizing a movement against the constitutional order unified diverse white supremacist groups to work together toward common goals. They went from being a band of 500 isolated extremists at the "United the Right" Rally in Charlottesville in 2017 to the front lines of a mass right-wing street movement of 40–50,000 people in 2021 that interrupted the presidential election process and came close to overthrowing the American constitutional order itself. Long after Trump has cashed in all his ill-gotten chips, we will be living with the violent street movement he breathed life into.

And will one major party's tacit embrace of strategic political violence teach "bloody instructions," as Macbeth would say, to the other side and the movements surrounding it? Can electoral violence be contained? We can only hope that the American majority will follow the nonviolent democratic path of John Lewis, Dr. King, and Bob Moses. But if we do not also defeat the forces of domestic violent extremism, no one knows where this spiraling political fanaticism may lead us as a country.

Just as protecting the voting rights of the people against gerrymandering and against the new wave of voter suppression is imperative, the Electoral College system needs to be abolished and replaced with a National Popular Vote or, at the very least, be seriously reformed in the Electoral Count Act—and quickly. American people who value democracy need to rise up beyond the parties to solidify democratic political arrangements in order to prevent another electoral debacle or a complete constitutional collapse in the form of an Electoral College coup, a bloody insurrection, or some kind of social media–inflamed civil war. Do we have, deep within our national politics and deep within our souls, the political vision and social cohesion necessary to make the democratic structural changes we need? I wonder. If we don't, then we had better find them soon. The alternative of proceeding along the present path of chaos is dangerous to us all.

In an authoritarian society, a dictatorship, or a failed state, the mental and emotional health of the population is, at best, irrelevant to the government, and it may in fact be dangerous to the government. The ruling class is no more interested in uplifting the mental and emotional health of the population than it is interested in promoting mass education, public literacy, or the physical health of the public. Healthy people might have the strength to challenge injustice and corruption. Better that everyone be submerged in fear and cruelty, lying, illness and plague, public health disinformation, cynicism, and self-loathing.

But, in a democracy, the mental health of the population is essential and vital to the success of the society. Democracies thrive on participation and involvement of everyone; we all benefit when we are all at our physical and mental best. Strong democracies thus not only promote mental health to empower people to pursue their own individual happiness but also promote healthy social dialogue, compromise, and decision making. A mentally healthy society will take care of its people, generate patterns of healthy interaction in social and political life, and produce healthy leaders; a mentally unhealthy society will ignore the needs of its population, create dynamics of infuriating corruption, dysfunction, and despair in social and political life, and produce unstable, fanatical, and narcissistic leaders.

I have come to believe that good memory is a building block of mental health and balance—not just for individuals but also for families, societies, and political movements. What do we remember and how do we remember it? What do we choose to forget or to lie about?

Memory requires active engagement with the complexities of the past. It is not an unthinking or passive process, like breathing or (for most people) sleeping. I have found that good memory, like good history, requires disciplined and focused attention, an honest effort to overcome one's perceptual and cognitive biases, and sustained effort. At a time when we rely on the internet to remember facts and dates for us, there is a serious danger that we will begin to forget essential parts of the human story and that ideologues and propagandists online will supply the missing "information" in a way that destroys the Enlightenment project of truth and justice.

We live in a time when political actors are willing to make war on scientific facts relating to climate change or COVID-19, and on historical memory, denying, obscuring, or lying about specific historical events—like the Holocaust, slavery, the 2020 presidential election or the January 6 insurrection. The perpetrators of these Big Lies know that human memory is a fragile instrument and that what we "remember" of an event can easily be influenced by what others tell us to believe about it or by the constant repetition of lies about it.

The lies people tell about the past obviously have profound importance for the politics of the present, and for that reason, historical lies are often deliberate and motivated. In the immediate aftermath of the Civil War, unrepentant champions of the Confederacy and apologists for slavery promoted what came to be called the Lost Cause myth, which glorified Confederate generals and soldiers; romanticized life in the antebellum South for both slave owners and the people they enslaved; demonized Abraham Lincoln, Frederick Douglass, and

Union forces; and falsely claimed that the Civil War had nothing to do with slavery but was a simple constitutional disagreement between the states and the federal government over the proper allocation of powers. The Lost Cause myth became ideological cover for the country's abandonment of Reconstruction and its embrace of Jim Crow apartheid, with all its ruinous consequences for the progress of millions of African Americans and the nation itself.

Donald Trump's outrageous lies about the 2020 election mirror the basic form of the Lost Cause myth. According to Trump, he won a landslide victory, but it was stolen from him and from his noble, long-suffering supporters after he valiantly led the country to record prosperity and well-being—a claim that conveniently omits the tens of millions of people who lost their jobs and the hundreds of thousands who lost their lives under his administration's lethally reckless mismanagement of the COVID-19 pandemic.

Moreover, Trump's political purposes and propaganda connect directly to the substance of the Lost Cause myth. He purports to speak for an aggrieved, downtrodden (and obviously white) majority of the country whose votes were canceled out by suspiciously large numbers of votes coming out of big cities like Baltimore, Detroit, Philadelphia, and Chicago and whose candidates (who had been "winning") were denied their rightful victory by radical Democrats, RINOs, and shadowy international forces. For hours on January 6, while his supporters rampaged, Trump continued to send out messages flattering the wounded pride and righteousness of indignant pro-Trump rioters—such as his remarkable tweet declaring, "These are the things and events that happen when a sacred election victory is so unceremoniously & viciously stripped away from great patriots who have been badly & unfairly treated for so long."

It would be hard to surpass the Lost Cause sentiment embodied in those words, but the following tweet (sent at the end of the day, hours after the violence began and when military reinforcements were arriving) also drowns its intended readers in the romantic language of Trump's gargantuan election lie, a lie saturated in racial grievance: "There's never been a time like this where such a thing happened where they could take it away from all of us, from me, from you, from our country. This was a fraudulent election. But we can't play into the hands of these people. We have to have peace. So go home. We love you. You're very special. You've seen what happens. You see the way others are treated that are so bad and so evil. I know how you feel."

It is no coincidence that a Confederate battle flag was waved inside the Capitol on January 6, nor that lots of Trump's marauding partisans carried a Confederate battle flag throughout the day. The "stolen" election is the new-and-improved Lost Cause myth.

Our work in Congress to determine the facts of January 6 is a struggle to protect social memory and historical facts in the face of ideological derangement and disfigurement. January 6, 2020, was—and forever will be—the day in history when Donald Trump and the racist forces he helped draw to Washington tried to *steal* an election, nullify the rule of law, and cancel out the voting rights of the American people. The lie that Republicans promote that Trump and his mob were themselves working to stop an election from being stolen is merely propaganda designed to justify the new wave of vote-suppressing legislation the GOP is trying to pass all over the country. That's why the work of the Select Committee to Investigate the January 6th Attack, and how we as a country interpret and act upon the committee's findings, becomes such a critical test for us. If we cannot get the past right, we will get the future all wrong.

Most of us navigate around the unthinkable, whether it's fascist violence in our society or the escalating dangers resulting from climate change. We blink it out. We proceed *unthinkingly*, which may mean also *unfeelingly*. We may even be *thoughtless* about the horrors that came before our time on earth. Not Tommy Raskin.

When it came to facing the unthinkable, Tommy was unflinching. At a young age, he confronted the shocking reality of genocide against the Native Americans; the enslavement of African Americans; the Nazi Holocaust against the Jews, Gypsies, gay people, and other minorities; the bombing of Hiroshima and Nagasaki; and the brutal meaning of war for its victims and their families. He made himself think about these things in a way few people ever do. As a boy, he would often excuse himself to go to his room because he "wanted to think," and then he would go think about things the way other kids might have played Monopoly or watched a movie. Then he would come back with an essay he had written, or a new argument, or questions for discussion.

Cicero said that "to be ignorant of what occurred before you were born is to remain always a child." My son, Thomas Bloom Raskin, did not die as a child. He died as a man, as full in his knowledge of history and his appreciation for the human condition as any man who ever lived. He dared to think, and he dared to feel.

But Tommy Raskin dared to think the unthinkable also when it came to *transforming* the human condition. He thought about ways to resolve intractable international problems, about policies that could abolish war and hunger and empower the global poor. He thought all the time about how social institutions could be structured around not hierarchy and inequality but the empowerment of people and liberation of their creative energies. He dared to think about not only what was unthinkably dreadful in the human experience but also what might be unthinkably beautiful in our potential future, a future he hoped for and worked to bring about so passionately.

By daring to think about unthinkable evil, by daring to imagine an unimaginable future, Tommy became a visionary saturated in feelings most of us never choose to experience. While he was with us, he shared his vision of the world with everyone around him, permanently altering the ways in which *we* see it, making the evil in the world more vivid and the good all the more precious and tantalizingly within reach.

The weekend before we lost Tommy, I misplaced my glasses and could not find them anywhere. Tommy helped me look for them in the house and in our front yard, but they were nowhere to be found. We gave up the search, but Tommy comforted me, saying, "I know you'll find them, Dad."

We lost Tommy on Wednesday, December 31.

When I woke up on Sunday, January 3, I was more miserable than I have ever been, just engulfed in grief, tragedy, and darkness. I went outside to our front porch in the early morning cold to see if there was food in our bird feeder. There was not. I was about to go fill it up when, out of nowhere, a huge flock of birds landed in our front yard—but all different kinds of birds: robins and orioles and blue jays and the most beautiful cardinals I had ever seen. They came and swooped en masse into our yard.

It was just spectacular and breathtaking. I called for Sarah, and she ran out to the porch.

I had never seen anything like it before, and neither had she.

And despite being a man of science and reason, one not easily given to mystical thinking, I was suddenly seized with the thought, flooded with the feeling, and immersed in the overwhelming physical sensation that *Everything is going to be all right. Tommy is going to take care of us.*

I moved off the porch and into the yard toward the birds, just to get closer to them, just to say hi for a moment, and as suddenly as they arrived, they took flight as if they were one, flapping away into the unknown sky.

For a split second, I was bereft again, but then I looked down at my feet, and there I found my glasses.

ACKNOWLEDGMENTS

I hasten to acknowledge all those who made *Unthinkable* possible, beginning with my most discerning readers and soulful critics: my wife, Sarah; my daughters, Hannah and Tabitha; and my big sister, Erika. I love them boundlessly and have treasured their advice (even when they tried to edit their own descriptions and appearances in the book).

I was blessed to work with Matt Harper of HarperCollins (no relation, they say), a sharp-eyed and imaginative literary editor with a ruthless talent for cutting flab and a dazzling memory for text and the inner structure of the manuscript. Given my day job, this book was a labor of love through the nighttime and weekend hours, and Matt showed endless patience for my insomniac writing, texting, and calling habits. I will miss working with him. He is a gentleman.

My thanks to Lisa Sharkey, director of creative development at HarperCollins, who believed passionately in this book long before I wrote a single word of it. Thanks to the whole first-class team at HarperCollins: Jonathan Burnham, Tina Andreadis, Kate D'Esmond, and Rebecca Holland. I'd also like to thank my great and patient audio team: Louis Milgrom, William Lowe, Brian Jacobs, Beth Ebisch, and Chris Blood.

I had two meticulous and visionary research assistants in my friends, the dynamic duo of Norman and Kimberly Sandridge ("Sandridge Analytica," as we came to call them), the former a professor of classics in the political science department at Howard University and the latter the director of early childhood education at the Evergreen School in Silver Spring and a fanatical grammarian.

Profuse thanks go to Zoie Lafis and the Center for Hellenic Studies in Washington, DC, for giving me a sacred space within which to think and a solid table upon which to write. I am forever grateful.

Special thanks to Riki and Michael Sheehan, Susan and Gary Vogel, Tamara Weiss, Iya Labunka, and John and April Delaney, all of whom provided me at different points in this journey a beautiful refuge and workspace for the writing of my book.

Profound thanks to Lucy Kaplansky, James Montfort, Cheryl Whitman, Lucia Mondesir, Lauren Doney, Donna D'Amico, Nils and Amy Lofgren, Suzanne Berne, Laura Franco, Tom Cove, Whitney Ellenby, Kevin Barr, Julia Sweig, Reed Thompson, and Debra Winger for helping me (in ways big and small that may be known or unknown to them).

And final thanks to the magnificent people of Maryland's Eighth District who have always had my back and always helped me to write the story of our times.

ABOUT THE AUTHOR

JAMIE RASKIN represents Maryland's Eighth Congressional District in the U.S. House of Representatives. In 2021, he was the lead manager in the second impeachment trial of Donald Trump, which resulted in a 57–43 vote to convict in the United States Senate, ten votes short of a two-thirds majority. Raskin is a member of the House Judiciary Committee, the Rules Committee, the Oversight and Reform Committee, and the Administration Committee. He is also a member of the new Select Committee to Investigate the January 6th Attack on the United States Capitol. The former Majority Whip of the Maryland State Senate, where he served for ten years, Raskin led the successful fights there for marriage equality, abolition of the death penalty, passage of the National Popular Vote Interstate Compact, and dozens of other progressive changes in his state.

A graduate of Harvard College and Harvard Law School, where he was an editor of the *Harvard Law Review*, Raskin was a professor of constitutional law at American University's Washington College of Law for twenty-five years. He has written several books, including *We the Students: Supreme Court Cases for and about Students* and *Overruling Democracy: The Supreme Court vs. the American People*. Raskin is the vice-chair of the Democratic Congressional Campaign Committee and the leader of

Democracy Summer, a nationwide program to engage young people in training in political activism and leadership. He and his wife, Sarah, are the parents of three children, Hannah Grace, Thomas Bloom (1995–2020), and Tabitha Claire. Jamie and Sarah live in Takoma Park, Maryland, with their dogs, Toby and Potter.